Mathematical Sciences Research Institute
Publications 3

Mathematical Sciences Research Institute Publications

Volume 1	D. Freed and K. Uhlenbeck: Instantons and Four-Manifolds
Volume 2	S.S. Chern (ed.): Seminar on Nonlinear Partial Differential Equations
Volume 3	J. Lepowsky, S. Mandelstam, and I.M. Singer (eds.): Vertex Operators in Mathematics and Physics
Forthcoming	S.S. Chern and P. Griffiths: Essays on Exterior Differential Systems
	V. Kac (ed.): Infinite Dimensional Lie Groups
	C.C. Moore (ed.): Group Representations, Ergodic Theory, Operator Algebras, and Mathematical Physics: Proceedings of a Conference in Honor of G.W. Mackey

Vertex Operators in Mathematics and Physics

Proceedings of a Conference November 10-17, 1983

Edited by
J. Lepowsky · S. Mandelstam · I.M. Singer

With 37 Illustrations

Springer Science+Business Media, LLC

J. Lepowsky
Department of Mathematics
Rutgers University
New Brunswick, NJ 08903 U.S.A.

S. Mandelstam
Department of Physics
University of California
Berkeley, CA 94720 U.S.A.

I.M. Singer
Department of Mathematics
University of California
Berkeley, CA 94720 U.S.A.

Mathematical Sciences Research Institute
2223 Fulton Street, Room 603
Berkeley, CA 94720 U.S.A.

AMS Classification: 17-02, 20-02, 81-02, 05A19, 17B65, 20D08, 35Q15, 35Q20, 58F07, 81E99, 81G99, 82A15, 83E50

Library of Congress Cataloging in Publication Data
Main entry under title:
Vertex operators in mathematics and physics.
(Mathematical Sciences Research Institute publications ; 3)
Includes proceedings from the Conference on Vertex Operators in Mathematics and Physics, held at the Mathematical Sciences Research Institute, Nov. 10–17, 1983.
Bibliography: p.
1. Nonassociative algebras—Congresses. 2. Groups, Theory of—Congresses. 3. Lie algebras—Congresses. 4. Quantum field theory—Congresses. I. Lepowsky, J. II. Mandelstam, Stanley. III. Singer, I. M. (Isadore Manuel). IV. Conference on Vertex Operators in Mathematics and Physics (1983: Mathematical Sciences Research Institute) V. Mathematical Sciences Research Institute (Berkeley, Calif.) VI. Series.
QA252.V47 1985 512′.24 84-26828

The camera-ready manuscript of this book was received by Springer-Verlag in October, 1984.

The Mathematical Sciences Research Institute wishes to acknowledge support from the National Science Foundation.

ISBN 978-1-4613-9552-2 ISBN 978-1-4613-9550-8 (eBook)
DOI 10.1007/978-1-4613-9550-8

9 8 7 6 5 4 3 2 1

PREFACE

In January, 1972, F. Dyson delivered the J. W. Gibbs Lecture at the Annual Meeting of the American Mathematical Society.[1] Called "Missed Opportunities," it was an inspiring encouragement to mathematicians and physicists to communicate with one another. In the talk, Dyson described several examples of mathematical discoveries which were delayed because mathematicians were not paying sufficient attention when "two disparate or incompatible mathematical concepts were juxtaposed in the description of a single situation." In such a case, he said, mathematicians should try to "create a wider conceptual framework within which the pair of disparate elements would find a harmonious coexistence."

As an example, Dyson mentioned his and I. G. Macdonald's independent work on identities for certain powers of Dedekind's eta-function. He regarded this episode as a personal "missed opportunity" for two reasons -- his failure to notice the connection with finite-dimensional simple Lie algebras "just because the number theorist Dyson and the physicist Dyson were not speaking to each other" and his failure to discuss his eta-function work with Macdonald, because it never occurred to him that he might be studying the same problem as a mathematician. He called this "a trivial episode from my own experience, which illustrates vividly how the habit of specialization can cause us to miss opportunities," asserting that even in his undergraduate days at Cambridge "it was clear ... that number theory in the style of Hardy and Ramanujan was old-fashioned and did not have a great and glorious future ahead of it." He concluded by suggesting that Macdonald had explained the juxtaposition of the "two disparate concepts" of Lie algebras and modular functions and that the subject was all but finished.

---

[1]Bulletin of the American Mathematical Society 78 (1972), 635-652.

Vertex Operators in Mathematics and Physics - Proceedings of a Conference November 10-17, 1983. Publications of the Mathematical Sciences Research Institute #3, Springer-Verlag, 1984.

But as we now know, the subject had hardly started. In fact, the eta-function work of Dyson and Macdonald triggered an explosion of ideas which now blend the "disparate concepts" of infinite-dimensional Lie algebras, modular functions, the "dual-string" theory in particle physics, two-dimensional quantum field theory, the Rogers-Ramanujan identities, soliton theory and Monstrous Moonshine into a rapidly developing theory.

In the last few years, interesting connections have been discovered between the affine Kac-Moody Lie algebras and the dual-string theory, through the use of vertex operators. In an effort to explore these connections, a conference entitled "Vertex Operators in Mathematics and Physics" was held November 10-17, 1983 as part of the 1983-84 program on Kac-Moody algebras at the Mathematical Sciences Research Institute.

The present time is especially appropriate for such a conference, since the great current interest in Kac-Moody algebras coincides with a renewed interest in string models and their relation with supergravity. The interaction between the physical and mathematical theories has already had fruitful applications in mathematics; one may hope for further applications in both mathematics and physics. Some recently discovered symmetries of certain supergravity models appear to bear a striking resemblance to the affine Kac-Moody algebra related to the vertex function of the associated string model. This may indicate the presence of a hitherto undiscovered symmetry or broken symmetry of the string model; such a symmetry could be of crucial importance for the physicist.

Many two-dimensional models in quantum field theory or statistical mechanics are exactly soluble. It has been found that most if not all such models possess Bäcklund transformations and, more recently, it has been shown that at least some of these models have an affine Kac-Moody algebra, possibly without central extension, as a symmetry algebra. Possible extensions to four dimensions have been proposed. This is obviously another area where the interplay between work in mathematics and physics may well be fruitful.

The present volume contains the proceedings of the conference, including some papers by authors who were invited but were unable to attend. We have provided an introduction to help the reader with terminology, notation and historical perspective.

There remain many mysteries to be explained. We hope this volume will introduce the novice to the subject of the conference and stimulate the expert toward deeper investigations.

We thank the Mathematical Sciences Research Institute for inviting us to organize this conference and for providing us with a cheerful and expert staff to run the conference and prepare this volume.

Berkeley, July, 1984

J. Lepowsky
S. Mandelstam
I. M. Singer

CONFERENCE ON
VERTEX OPERATORS IN MATHEMATICS AND PHYSICS
NOVEMBER 10-17, 1983
MATHEMATICAL SCIENCES RESEARCH INSTITUTE
SCHEDULE OF TALKS

Thursday, November 10

Vertex operators, $\mathscr{Z}$-algebras and the Fischer-Griess Monster
James Lepowsky

Introduction to string models and vertex functions
Stanley Mandelstam

Algebras, lattices and strings
Peter Goddard

Sporadic groups and nonassociative algebras
Robert L. Griess, Jr.

Friday, November 11

Structure of the standard $A_1^{(1)}$-modules (principal picture)
Robert Lee Wilson

Structure of the standard $A_1^{(1)}$-modules (homogeneous picture)
Mirko Primc

Algebras, lattices and strings, II
David Olive

A natural module for the Fischer-Griess Monster with the modular function J as character, I
Igor B. Frenkel

A natural module for the Fischer-Griess Monster with the modular function J as character, II
Arne Meurman

Saturday, November 12

Vertex operators and standard modules for some affine Lie algebras
Kailash C. Misra

Solitons and infinite-dimensional Lie algebras
Michio Jimbo

On a duality of branching rules
Tetsuji Miwa

Generalization of the Riemann-Hilbert problem and the KP, TL hierarchy
Kimio Ueno

Monday, November 14

Vacuum vector representations of the Virasoro algebra
Alvany Rocha-Caridi

Another viewpoint for studying instantons
Howard Garland

Local charge algebra of quantum gauge field theory
Itzhak Bars

Matrix coefficients of the wedge representation
Dale H. Peterson

Supergeometry and Kac-Moody algebras
Bernard Julia

Tuesday, November 15

Some applications of vertex operators to Kac-Moody algebras
Alex J. Feingold

Massive Kaluza-Klein theories and bound states in Yang-Mills
Louise Dolan

Towards a catalog of two-dimensional conformal field theories using the representation theory of the Virasoro algebra
Daniel Friedan

Integrability and hidden symmetries
Bernard Julia

Infinite dimensional formal Lie groups - A bridge connecting the linear world and the non-linear world
Motohico Mulase

Wednesday, November 16

Lax pairs, the Riemann-Hilbert transform and Kac-Moody algebras
Yong-Shi Wu

TABLE OF CONTENTS

TABLE OF CONTENTS (CONT'D)

INTRODUCTION

James Lepowsky[†]

The search for symmetry in nature has for a long time provided representation theory with perhaps its chief motivation. According to the standard approach of Lie theory, one looks for infinitesimal symmetry -- Lie algebras of operators or concrete realizations of abstract Lie algebras. A central theme in this volume is the construction of affine Lie algebras using formal differential operators called vertex operators, which originally appeared in the dual-string theory. Since the precise description of vertex operators, in both mathematical and physical settings, requires a fair amount of notation, we do not attempt it in this introduction. Instead we refer the reader to the papers of Mandelstam, Goddard-Olive, Lepowsky-Wilson and Frenkel-Lepowsky-Meurman.

We have tried to maintain consistency of terminology and to some extent notation in the articles herein. To help the reader we shall review some of the terminology. We also thought it might be useful to supplement an earlier fairly detailed exposition of ours [37] with a brief historical account of vertex operators in mathematics and their connection with affine algebras. Since we were involved in the development of the subject, the reader should be advised that what follows reflects our own understanding. For another view, see [29].[1]

---

† Partially supported by the National Science Foundation through the Mathematical Sciences Research Institute and NSF Grant MCS 83-01664.

---

1 We would like to thank Igor Frenkel for his valuable comments on the first draft of this introduction.

Vertex Operators in Mathematics and Physics - Proceedings of a Conference November 10-17, 1983. Publications of the Mathematical Sciences Research Institute #3, Springer-Verlag, 1984.

We first define the term "affine algebra", a term which has been used in a more restricted way in the past. We work over the field $\mathbb{C}$ of complex numbers. Given a Lie algebra $\mathfrak{g}$ and given an invariant symmetric bilinear form $\langle\cdot,\cdot\rangle$ on $\mathfrak{g}$, the *untwisted affine (Lie) algebra of* $\mathfrak{g}$ *(and* $\langle\cdot,\cdot\rangle$*)* is the infinite-dimensional Lie algebra

$$\tilde{\mathfrak{g}} = \mathfrak{g} \otimes \mathbb{C}[t,t^{-1}] \oplus \mathbb{C}c \oplus \mathbb{C}d.$$

Here $\mathbb{C}[t,t^{-1}]$ is the infinite-dimensional associative algebra of Laurent polynomials in an indeterminate t, c is a nonzero central element of $\tilde{\mathfrak{g}}$, and $\tilde{\mathfrak{g}}$ has Lie brackets determined by:

$$[x \otimes t^m, y \otimes t^n] = [x,y] \otimes t^{m+n} + \langle x,y\rangle m\delta_{m+n,0}c,$$

$$[d, x \otimes t^m] = mx \otimes t^m$$

for $x,y \in \mathfrak{g}$ and $m,n \in \mathbb{Z}$. The subalgebra

$$\hat{\mathfrak{g}} = \mathfrak{g} \otimes \mathbb{C}[t,t^{-1}] \oplus \mathbb{C}c$$

of $\tilde{\mathfrak{g}}$ is also referred to as an affine algebra. The quotient of $\hat{\mathfrak{g}}$ by $\mathbb{C}c$ is called the *untwisted affine algebra of* $\mathfrak{g}$ *without central extension* or the *untwisted (polynomial) loop algebra of* $\mathfrak{g}$. Denote it by $\bar{\mathfrak{g}}$. It is essentially a Lie algebra of maps from the circle $S^1 = \{e^{2\pi i\theta} \mid \theta \in \mathbb{R}\}$ to $\mathfrak{g}$ under pointwise bracket, where we identify t with the periodic function $e^{2\pi i\theta}$. Also set $\check{\mathfrak{g}} = \tilde{\mathfrak{g}}/\mathbb{C}c$. Then we have the diagram (see [8])

$$\begin{array}{ccc} \hat{\mathfrak{g}} & \longrightarrow & \bar{\mathfrak{g}} \\ \big\downarrow & & \big\downarrow \\ \tilde{\mathfrak{g}} & \longrightarrow & \check{\mathfrak{g}} \end{array}$$

where the horizontal maps are quotient maps and the vertical maps are embeddings.

Given an automorphism ν of g such that ν preserves $\langle\cdot,\cdot\rangle$ and $\nu^p = 1$ for some positive integer p, let $g_{(n)}$ be the $e^{2\pi in/p}$-eigenspace of ν for $n \in \mathbb{Z}$. Then the Lie subalgebra

$$\tilde{g}(\nu) = (\bigoplus_{n\in\mathbb{Z}} g_{(n)} \otimes t^n) \oplus \mathbb{C}c \oplus \mathbb{C}d$$

of $\tilde{g}$ (or its subalgebra $\hat{g}(\nu)$, without $\mathbb{C}d$) is called the *ν-twisted affine algebra of* g. It may or may not be isomorphic to $\tilde{g}$. The corresponding *ν-twisted loop algebra of* g, $\bar{g}(\nu) = \tilde{g}(\nu)/\mathbb{C}c$, is identified with the subalgebra of $\bar{g}$ containing only the maps f from S^1 to g which are equivariant in the sense that

$$f(e^{2\pi i/p}e^{2\pi i\theta}) = \nu f(e^{2\pi i\theta})$$

for $\theta \in \mathbb{R}$. Defining $\check{g}(\nu) = \tilde{g}(\nu)/\mathbb{C}c$, we have a diagram analogous to the one above.

Affine algebras are essentially the same as current algebras on the circle, introduced by physicists in the 1960's.

According to standard mathematical terminology, given a vector space V, an *affine automorphism* of V is the composition of a linear automorphism with a translation of V. Assume that g is finite-dimensional and let G be a matrix Lie group with Lie algebra g. Let $\bar{G}$ (respectively, $\bar{g}^\infty$) be the loop group (respectively, loop algebra) consisting of the C^∞ maps from S^1 to G (respectively, g), under pointwise multiplication (respectively, pointwise bracket). Besides the obvious action of $\bar{G}$ on $\bar{g}^\infty$ by pointwise conjugation, there is a natural affine action given by the following formula, which has no analogue in the finite-dimensional case:

$$g\cdot f = gfg^{-1} - ag'g^{-1}$$

for $g \in \bar{G}$ and $f \in \bar{g}^\infty$, where $g' = dg/d\theta$. Here $a \in \mathbb{C}$, and the case $a = 0$ gives ordinary conjugation. These affine actions, known to physicists as gauge transformations, motivate the term "affine algebra". They have been identified as coadjoint actions of a central extension of the loop group when g is semisimple ([8],[54]).

A finite-dimensional semisimple Lie algebra $\mathfrak{g}$ of rank ℓ is generated by 3ℓ elements subject to certain standard relations determined by the Cartan matrix of $\mathfrak{g}$ (see [2], [19], [21], [55]). Replacing the Cartan matrix by a more general matrix and defining the analogous Lie algebra, divided by a certain radical, V. G. Kac ([22], [23], [24]), R. V. Moody ([49], [50], [51]) and (from a different viewpoint) I. L. Kantor ([32], [33]) began the detailed study of what were termed "Kac-Moody Lie algebras" in [17].

An untwisted or twisted affine algebra of a finite-dimensional semisimple Lie algebra (a *semisimple-affine* algebra) is an important example of a Kac-Moody algebra (see [23],[24],[25] (cf. [20]),[49],[51], for the loop algebras and [30] for the twisted affine algebras). It turns out that the radical (mentioned above) is zero for a semisimple-affine algebra ([16], [58]) and in fact, more generally, for any Kac-Moody algebra whose "generalized Cartan matrix" is symmetrizable [14]. The terms "affine (or Euclidean) Lie algebra" and "affine Kac-Moody algebra" have often referred only to a *simple-affine* algebra -- the untwisted or twisted affine algebra of a finite-dimensional simple Lie algebra, -- but as several people have suggested, it is more natural to call $\tilde{\mathfrak{g}}$ and $\tilde{\mathfrak{g}}(\nu)$ affine algebras for any Lie algebra $\mathfrak{g}$. The Weyl group of a simple-affine algebra $\hat{\mathfrak{g}}$ can be identified with the classical affine Weyl group of $\mathfrak{g}$, acting as affine automorphisms of a Cartan subalgebra of $\mathfrak{g}$ ([26],[52]; see also [45]).

Kac-Moody algebras attracted increased attention as a result of the link between the simple-affine algebras and Macdonald's identities [45]; see the interesting reviews [5],[57]. The connection was established in [26] and independently noticed in [52]. Adapting the Bernstein-Gelfand-Gelfand proof [1] of Weyl's character formula, Kac [26] generalized Weyl's formula from the finite-dimensional irreducible representations of finite-dimensional semisimple Lie algebras to the standard (integrable highest weight) representations of Kac-Moody algebras with symmetrizable generalized Cartan matrices. A detailed exposition of the general formula was given in [36]. The formula (recalled in the first Jimbo-Miwa article, which reports a duality of branching coefficients), includes Macdonald's identities as the case of the trivial one-dimensional modules for the simple-affine algebras, either untwisted or twisted by a Dynkin diagram automorphism.

By describing the "shape" of a standard module, the character formula has helped in the discovery of operator constructions of simple-affine algebras. In fact, the "principally specialized character" ([38]; see also [35]), a one-variable formal power series which has a natural infinite product expansion ([27], [36]; cf. Misra's article), was used in the first such construction results.

A search for operator realizations of affine Kac-Moody algebras led to an unexpected result [40] -- a construction of the affine algebra $A_1^{(1)} = \mathfrak{sl}(2,\mathbb{C})\hat{}$ as a Lie algebra of differential operators acting on a polynomial algebra in infinitely many variables, identified with one of the standard modules, now called a basic or level 1 standard module. (The "level" is the scalar by which the suitably normalized central element c acts.) The main ingredient was a kind of exponential generating function. Its Fourier components, together with an infinite-dimensional Heisenberg algebra of creation and annihilation operators, spanned a Lie algebra of operators on a Fock space. (The article of Lepowsky-Wilson includes a self-contained exposition of this construction.) H. Garland recognized that the generating function was similar to the vertex operator used in the string model in quantum field theory to describe the interaction of certain kinds of strings at a "vertex". See Mandelstam's paper for a discussion of the original string model, Alvarez's article for an exposition of A. M. Polykov's recent string model, and Curtright's note for references on the related quantum Liouville theory.

The $A_1^{(1)}$-construction was generalized in [30] to realize "most" of the simple-affine algebras via differential operators on their basic modules (cf. Misra's paper). This construction is sometimes referred to as the "principal picture" because the algebras are realized in a twisted form, the twisting automorphisms being "principal" automorphisms, in the sense of [34]. Using the vertex operators of the string model, I. B. Frenkel and Kac [11] and independently G. Segal [54] found a second construction of the affine algebras of types $A_\ell^{(1)}$, $D_\ell^{(1)}$, $E_\ell^{(1)}$ as operators on their basic modules (the "homogeneous picture"; here the automorphism is the identity, and the algebras are untwisted). In particular, the commutators of both kinds of vertex operators (using suitable parameters) "close up" to form affine Lie algebras. In the article of Goddard-Olive, the latter construction is extended to arbitrary integral lattices.

Spin constructions of certain affine algebras as operators on their basic modules were also found ([9],[31]), leading to further links with quantum field theory [10]. The spin approach was extended in [6] to all "classical" affine algebras. An account of this work and of certain tensor product decompositions is given in Feingold's paper.

The structure of standard modules is intimately related to the theory of partitions. For instance, the weight multiplicities of the basic $A_1^{(1)}$-modules were found to be precisely the values of the classical partition function (see [7]), a fact closely related to the homogeneous picture construction ([11], [54]) in the $A_1^{(1)}$-case. In principle, each module construction implies a character and weight multiplicity formula, and if one is lucky, a character formula might suggest a module construction. When one has two essentially different constructions of the same module, or a character formula together with an "unrelated" module construction, then one obtains a combinatorial identity, typically an equality of two formal power series. Moreover, one can sometimes invoke known combinatorial identities to discover new structural information about modules and algebras. In this spirit, a Lie-theoretic interpretation and proof of the classical Rogers-Ramanujan identities was obtained in [41], [43] (see also [38]) by means of concrete constructions of certain non-basic standard $A_1^{(1)}$-modules. This work led to realizations of all the standard $A_1^{(1)}$-modules by means of "$\mathfrak{Z}$-algebras" ([42],[44]), which provide a framework for attacking the general problem of constructing all standard modules for all semisimple-affine algebras. Very roughly, the action of the affine algebra is factored into two commuting parts, consisting of operators resembling vertex operators and "$\mathfrak{Z}$-operators", whose algebraic structure can be analyzed. This approach, continued in [39], [46], [47], is explained and applied to obtain module constructions and combinatorial identities in the papers of Lepowsky-Wilson, Lepowsky-Primc and Misra.

The vertex operator [40] was recognized in [4] as the infinitesimal Bäcklund transformation for generating the soliton solutions of the Korteweg-de Vries equation, in R. Hirota's bilinear formalism. The affine algebra constructions [40] and [30] were among the tools exploited by E. Date, M. Jimbo, M. Kashiwara and T. Miwa [3] to illuminate and extend many results in soliton theory. Related work is surveyed in the second Jimbo-Miwa article.

Another approach to completely integrable systems is based on the Birkhoff decomposition: A suitable matrix-valued function on the unit circle can be expressed as the product of a function holomorphic inside the unit disk with a function holomorphic outside. Finding such a decomposition is known as the Riemann-Hilbert problem. The Birkhoff decomposition is intrinsically related to the Bruhat decomposition of the corresponding loop group (see [53]). Using it, one can define an action of the loop group on the space of functions holomorphic inside the disk, by multiplication followed by projection. This action is often called the Riemann-Hilbert transform. Bäcklund transformations constitute an important case of this transform. This approach and its infinitesimal version underlie the methods used in the papers of Ueno, Chau, Wu-Ge and Dolan to construct solutions of important nonlinear systems. Affine symmetries similar to those obtained in these papers also appear in supergravity theories; see Julia's article. Quantum systems are considered in Bars's paper.

The Virasoro algebra, a central extension of the Lie algebra of polynomial vector fields on the circle, acts naturally on any standard $\hat{\mathfrak{g}}$-module ($\mathfrak{g}$ simple) by "Segal operators" (cf. [10]). Especially important is the case of the basic $A_1^{(1)}$-module, in which a family of irreducible representations of the Virasoro algebra can be described in terms of highest weight spaces for the $\mathfrak{sl}(2,\mathbb{C})$ subalgebra of $A_1^{(1)}$ (see [28], [54]). This provides a connection between vertex operators and the representation theory of the Virasoro algebra. The articles of Thorn, Friedan-Qiu-Shenker, Rocha-Caridi and Wallach deal respectively with the no-ghost theorem for string models, unitarity of representations and critical exponents in two-dimensional systems, characters of representations, and a relationship between characters of the unitary groups and highest weight vectors.

In [15], [16], Garland developed a theory of "complete Chevalley groups" for the untwisted simple-affine algebras using an integral basis of the universal enveloping algebra. As Garland and R. L. Wilson observed, this basis turned out to be closely related to vertex operators, a fact which allowed for its deeper understanding and generalization [48]; see also [56].

Vertex operators are closely related to at least one sporadic finite simple group as well as to Chevalley groups. A natural description of R. L. Griess's commutative nonassociative algebra was given with the help of a new "twisted vertex operator picture" generalizing the principal picture for $A_1^{(1)}$ (see [12]). Griess had used this algebra to prove the existence of the Monster (or Friendly Giant) sporadic group [18]. (See Griess's article for a brief exposition of the finite simple groups.) A natural infinite-dimensional representation of the Monster with the modular function J as character is described in the paper of Frenkel-Lepowsky-Meurman (see also [13]). Vertex operators in both the homogeneous and twisted pictures play fundamental roles in this construction.

The excitement of discovering surprising connections between different disciplines has become a rule in the subject of this volume. We hope and believe that the present round of interaction between mathematicians and physicists has not yet reached its peak.

REFERENCES

[1] I. N. Bernstein, I. M. Gelfand and S. I. Gelfand, Structure of representations generated by highest weight vectors, Funkcional Anal. i Prilozhen. 5 (1971), 1-9. English transl., Functional Anal. Appl. 5 (1971), 1-8.

[2] C. Chevalley, Sur la classification des algèbres de Lie simples et de leurs représentations, C. R. Acad. Sci. Paris 227 (1948), 1136-1138 and 1197.

[3] E. Date, M. Jimbo, M. Kashiwara and T. Miwa, Transformation groups for soliton equations - Euclidean Lie algebras and reduction of the KP hierarchy, Publ. Research Inst. for Math. Sciences, Kyoto University 18 (1982), 1077-1110.

[4] E. Date, M. Kashiwara and T. Miwa, Vertex operators and τ functions - transformation groups for soliton equations II, Proc. Japan Acad. Ser. A Math. Sci. 57 (1981), 387-392.

[5] M. Demazure, Identités de Macdonald, Séminaire Bourbaki, 1975/76, 28e annee, no. 483 (1976).

[6] A. Feingold and I. B. Frenkel, Classical affine algebras, Advances in Math. (1984).

[7] A. Feingold and J. Lepowsky, The Weyl-Kac character formula and power series identities, Advances in Math. 29 (1978), 271-309.

[8] I. B. Frenkel, Orbital theory for affine Lie algebras, Ph.D. thesis, Yale University, 1980 and Invent. Math. (1984).

[9] I. B. Frenkel, Spinor representations of affine Lie algebras, Proc. Nat. Acad. Sci. U.S.A. 77 (1980), 6303-6306.

[10] I. B. Frenkel, Two constructions of affine Lie algebra representations and boson-fermion correspondence in quantum field theory, J. Funct. Analysis 44 (1981), 259-327.

[11] I. B. Frenkel and V. G. Kac, Basic representations of affine Lie algebras and dual resonance models, Invent. Math. 62 (1980), 23-66.

[12] I. B. Frenkel, J. Lepowsky and A. Meurman, An E_8-approach to F_1, Proceedings of the 1982 Montreal Conference on Finite Group Theory, ed. by J. McKay, Springer-Verlag Lecture Notes in Mathematics (1984).

[13] I. B. Frenkel, J. Lepowsky and A. Meurman, A natural representation of the Fischer-Griess Monster with the modular function J as character, Proc. Nat. Acad. Sci. U.S.A. 81 (1984), 3256-3260.

[14] O. Gabber and V. G. Kac, On defining relations of certain infinite-dimensional Lie algebras, Bull. Amer. Math. Soc. (New Series) 5 (1981), 185-189.

[15] H. Garland, The arithmetic theory of loop algebras, J. Algebra 53 (1978), 480-551.

[16] H. Garland, The arithmetic theory of loop groups, Publ. Math. IHES 52 (1980), 5-136.

[17] H. Garland and J. Lepowsky, Lie algebra homology and the Macdonald-Kac formulas, Invent. Math. 34 (1976), 37-76.

[18] R. L. Griess, Jr., The Friendly Giant, Invent. Math. 69 (1982), 1-102.

[19] Harish-Chandra, On some applications of the universal enveloping algebra of a semisimple Lie algebra, Trans. Amer. Math. Soc. 70 (1951), 28-96.

[20] S. Helgason, Differential geometry, Lie groups and symmetric spaces, Academic Press, New York, 1978.

[21] N. Jacobson, Lie algebras, Wiley-Interscience, New York, 1962.

[22] V. G. Kac, Simple graded Lie algebras of finite growth, Funkcional. Anal. i Prilozhen. 1 (1967), 82-83. English transl., Functional Anal. Appl. 1 (1967), 328-329.

[23] V. G. Kac, Graded Lie algebras and symmetric spaces, Funkcional. Anal. i Prilozhen. 2 (1968), 93-94. English transl., Functional Anal. Appl. 2 (1968), 182-183.

[24] V. G. Kac, Simple irreducible graded Lie algebras of finite growth, Izv. Akad. Nauk SSSR 32 (1968), 1323-1367. English transl., Math. USSR Izv. 2 (1968), 1271-1311.

[25] V. G. Kac, Automorphisms of finite order of semi-simple Lie algebras, Funkcional. Anal. i Prilozhen. 3 (1969), 94-96. English transl., Functional Anal. Appl. 3 (1969), 252-254.

[26] V. G. Kac, Infinite-dimensional Lie algebras and Dedekind's η-function, Funkcional. Anal. i Prilozhen. 8 (1974), 77-78. English transl., Functional Anal. Appl. 8 (1974), 68-70.

[27] V. G. Kac, Infinite-dimensional algebras, Dedekind's η-function, classical Möbius function and the very strange formula, Advances in Math. 30 (1978), 85-136.

[28] V. G. Kac, Contravariant form for infinite-dimensional Lie algebras and superalgebras, Springer-Verlag Lecture Notes in Physics 94 (1979), 441-445.

[29] V. G. Kac, Infinite dimensional Lie algebras, Birkhäuser, Boston, 1983.

[30] V. G. Kac, D. A. Kazhdan, J. Lepowsky and R. L. Wilson, Realization of the basic representations of the Euclidean Lie algebras, Advances in Math. 42 (1981), 83-112.

[31] V. G. Kac and D. H. Peterson, Spin and wedge representations of infinite dimensional Lie algebras and groups, Proc. Nat. Acad. Sci. U.S.A. 78 (1981), 3308-3312.

[32] I. L. Kantor, Simple graded infinite-dimensional Lie algebras, Dokl. Akad. Nauk SSSR 179 (1968), 534-537. English transl., Soviet Math. Dokl. 9 (1968), 409-412.

[33] I. L. Kantor, Graded Lie algebras, Trudy Sem. Vect. Tens. Anal., Moscow State University, No. 15 (1970), 227-266 (in Russian).

[34] B. Kostant, On Macdonald's η-function formula, the Laplacian and generalized exponents, Advances in Math. 20 (1976), 179-212.

[35] J. Lepowsky, Macdonald-type identities, Advances in Math. 27 (1978), 230-234.

[36] J. Lepowsky, Lectures on Kac-Moody Lie algebras, Université Paris VI, Spring, 1978.

[37] J. Lepowsky, Affine Lie algebras and combinatorial identities, Proc. 1981 Rutgers Conference on Lie Algebras and Related Topics, Springer-Verlag Lecture Notes in Mathematics 933 (1982), 130-156.

[38] J. Lepowsky and S. Milne, Lie algebraic approaches to classical partition identities, Advances in Math. 29 (1978), 15-59.

[39] J. Lepowsky and M. Primc, Standard modules for type one affine Lie algebras, Number Theory, New York, 1982, Springer-Verlag Lecture Notes in Mathematics 1052 (1984), 194-251.

[40] J. Lepowsky and R. L. Wilson, Construction of the affine Lie algebra $A_1^{(1)}$, Comm. Math. Phys. 62 (1978), 43-53.

[41] J. Lepowsky and R. L. Wilson, The Rogers-Ramanujan identities: Lie theoretic interpretation and proof, Proc. Nat. Acad. Sci. U.S.A. 78 (1981), 699-701.

[42] J. Lepowsky and R. L. Wilson, A new family of algebras underlying the Rogers-Ramanujan identities and generalizations, Proc. Nat. Acad. Sci. U.S.A. 78 (1981), 7254-7258.

[43] J. Lepowsky and R. L. Wilson, A Lie theoretic interpretation and proof of the Rogers-Ramanujan identities, Advances in Math. 45 (1982), 21-72.

[44] J. Lepowsky and R. L. Wilson, The structure of standard modules, I: Universal algebras and the Rogers-Ramanujan identities, Invent. Math. (1984).

[45] I. G. Macdonald, Affine root systems and Dedekind's η-function, Invent. Math. 15 (1972), 91-143.

[46] K. C. Misra, Structure of certain standard modules for $A_n^{(1)}$ and the Rogers-Ramanujan identities, J. Algebra 88 (1984), 196-227.

[47] K. C. Misra, Structure of some standard modules for $C_n^{(1)}$, J. Algebra 90 (1984).

[48] D. Mitzman, Integral bases for affine Lie algebras and their universal enveloping algebras, Ph.D. thesis, Rutgers University, 1983 and Contemporary Math., Amer. Math. Soc., to appear.

[49] R. V. Moody, Lie algebras associated with generalized Cartan matrices, Bull. Amer. Math. Soc. 73 (1967), 217-221.

[50] R. V. Moody, A new class of Lie algebras, J. Algebra 10 (1968), 211-230.

[51] R. V. Moody, Euclidean Lie algebras, Canad. J. Math. 21 (1969), 1432-1454.

[52] R. V. Moody, Macdonald identities and Euclidean Lie algebras, Proc. Amer. Math. Soc. 48 (1975), 43-52.

[53] A. N. Pressley, Decompositions of the space of loops on a Lie group, Topology 19 (1980), 65-79.

[54] G. Segal, Unitary representations of some infinite-dimensional groups, Comm. Math. Phys. 80 (1981), 301-342.

[55] J.-P. Serre, Algèbres de Lie semi-simples complexes, Benjamin, New York, 1966.

[56] J. Tits, Résumé de cours, Annuaire du Collège de France, 1980-81, 75-87.

[57] D.-N. Verma, Review of I. G. Macdonald's paper, "Affine root systems and Dedekind's η-function," Math. Reviews 50 (1975), 1371-1374.

[58] R. L. Wilson, Euclidean Lie algebras are universal central extensions, Proc. 1981 Rutgers Conference on Lie Algebras and Related Topics, Springer-Verlag Lecture Notes in Mathematics 933 (1982), 210-213.

Rutgers University
New Brunswick, NJ 08903

Mathematical Sciences Research Institute
Berkeley, CA 94720

INTRODUCTION TO STRING MODELS AND VERTEX OPERATORS

S. Mandelstam[†]

INTRODUCTION

My aim in this talk is to give a very brief and qualitative outline of string models (or dual-resonance models) and to indicate how vertex operators appear in such models. The basic idea is that the fundamental objects of the theory instead of being point-like, as they are in most other theories, are string-like.[1]

String models are very tightly constrained and possess remarkable properties. They were originally discovered by virtue of these properties, and it was only realized later that they could be obtained from a simple string picture. The original intent was to apply them to hadronic physics, i.e., to sub-nuclear physics, whose domain of applicability involves distances of the order of magnitude of 10^{-13} cm. Physicists working in this regime have now largely abandoned the models, partly because an alternative theory ("quantum chromodynamics") has been found, but also because they possesed certain properties which rendered them inapplicable to hadronic physics. I may mention two such properties:

i) They turned out to be theories of gravity. Gravitational interactions are known to be completely negligible in hadronic physics.

ii) They only worked in unphysical numbers of space-time dimensions; twenty-six or ten. Polyakov[2] has recently proposed a new string model which may overcome these defects, and which may be equivalent to quantum chromodynamics. There will be talks on the Polyakov model later on in the conference; I shall not describe it today.

---

† Partially supported by NSF No. PHY 81-18547.

Vertex Operators in Mathematics and Physics – Proceedings of a Conference November 10–17, 1983. Publications of the Mathematical Sciences Research Institute #3, Springer-Verlag, 1984.

Physicists have recently been studying theories whose distances of applicability are very much smaller than 10^{-13} cm. Quantum gravity, which becomes important at about 10^{-32} cm, is now a subject of investigation. Among theories being studied are those of the Kaluza-Klein type, in which space-time has more than four dimensions, all but four of the dimensions forming a compact space of about 10^{-32} cm in diameter. The features of string models mentioned above are no longer defects; the gravitational properties of the theory are precisely what we require.[3] In fact, string models are the *only* presently known theories of gravity which do not possess short-distance inconsistences (they are the only theories which have a renormalizable perturbation series). Renewed attention is therefore being paid to string models.

There are two basic types of string theories. The first, the generalized Veneziano (G.V.) model, works in twenty-six dimensions, and has no degrees of freedom other than the position of the string. In the second model, the Neveu-Schwarz-Ramond (N.S.R.) model, which works in ten dimensions, there exist "spin" degrees of freedom distributed over the string. The N.S.R. model led to the concept of supersymmetry. Green and Schwarz[4] have recently shown that the model is in fact a physical system with supersymmetry and supergravity properties. They and their collaborators are actively studying the model. It possesses a richer structure than the G.V. model, and has greater possibilities with regard to the elementary-particle problem. Unfortunately, due to limitations of time, I shall not be able to describe the N.S.R. model. My talk will be devoted to the simpler G.V. model.

As is usual when considereing this type of problem, we shall divide it into several stages. We shall first consider the classical (i.e., unquantized) motion of a free, non-interacting string. Next we shall examine the corresponding quantum system which is still, from the physicists point of view, rather trivial, since strings do not interact. Finally we shall consider how we can allow the strings to interact in the simplest possible way, by joining and splitting. The vertex operator appears when we consider such interactions. We shall briefly examine ramifications and generalizations of the vertex operator.

String models have been treated by two methods. The first has the advantage of being manifestly Lorentz invariant. However, it has several disadvantages; one has to use an unphysical Hilbert space with non-positive-definite metric. Furthermore, interactions can only be treated in a rather ad hoc manner; the covariant approach does not lead to a real theory of interacting strings. We shall therefore use the non-covariant approach. Some mathematicians in the audience may have read papers which employ the covariant aproach but, as far as I am aware, the features in which mathematicians are interested are common to both aproaches. A generalization of the vertex operator, which may be of future mathematical interest, occurs only in the non-covariant approach.

CLASSICAL MECHANICS OF A FREE STRING

A string moving in space consists of a two-dimensional sheet moving in d-dimensional space-time (d = 26 for the G.V. model). Though I have emphasized that string theories are theories of gravity, the connection is not quite manifest and will not appear in this exposition. We thus have a curved two-dimensional surface, with index 1, embedded in a flat d-dimensional space with index 1. We parametrize the surface with two arbitrary co-ordinates σ and τ. The d dependent variables x^μ are thus functions of two independent variables σ and τ.

The equations of motion of a point particle moving in space-time without interactions can be found from the variational principle that the length of the path traced out be stationary. The obvious generalization to the motion of a string is that the area be stationary. Thus:

$$(1)\quad \delta\int d\sigma d\tau\left\{\left[\sum_\mu \frac{\partial x^\mu}{\partial\sigma}\cdot\frac{\partial x_\mu}{\partial\tau}\right]^2 - \left[\sum_\mu \frac{\partial x^\mu}{\partial\sigma}\cdot\frac{\partial x_\mu}{\partial\sigma}\right]\left[\sum_\nu \frac{\partial x^\nu}{\partial\tau}\cdot\frac{\partial x_\nu}{\partial\tau}\right]\right\}^{1/2} = 0$$

The equations obtained from (1) are rather unwieldy, and we therefore restrict the choice of co-ordinates. A convenient restriction is that the metric on the surface be pseudo-conformally flat, i.e.,

(2a) $$\sum_\mu \frac{\partial x^\mu}{\partial \sigma} \cdot \frac{\partial x_\mu}{\partial \tau} = 0,$$

(2b) $$\sum_\mu \left\{ \frac{\partial x^\mu}{\partial \sigma} \cdot \frac{\partial x_\mu}{\partial \sigma} + \frac{\partial x^\mu}{\partial \tau} \cdot \frac{\partial x_\mu}{\partial \tau} \right\} = 0 \ .$$

The conditions (2) do not fix the co-ordinates uniquely, since the conditions are invariant under the pseudo-conformal transformations from co-ordinates σ,τ to σ',τ', where

(3a) $$\frac{\partial \sigma'}{\partial \sigma} = \frac{\partial \tau'}{\partial \tau}$$

(3b) $$\frac{\partial \sigma'}{\partial \tau} = \frac{\partial \tau'}{\partial \sigma} \ .$$

In the covariant approach, one does not add further conditions to define the co-ordinates uniquely. In the non-covariant approach which we shall follow, we first replace the d co-ordinates x^0 and x^{d-1} by "light-like" co-ordinates

(3c) $$x^+ = \frac{1}{\sqrt{2}} (x^0 - x^{d-1}) \qquad x^- = \frac{1}{\sqrt{2}}(x^0 - x^{d-1}).$$

We leave the d-2 "transverse" co-ordinates x^i ($1 \leqslant i \leqslant d-2$) unchanged. (Greek indices μ,ν etc. take all values from 0 to d-1, Latin indices i,j etc. all values from 1 to d-2.) We consider x^+ as a "time" co-ordinate, x^- and x^i as "space" co-ordinates. *We then fix the co-ordinate τ to be equal to the "time" x^+.* It is not difficult to show that this can be done consistently with the conditions (2).

In fact, we may use Eqs. (2), together with the relation $\tau = x^+$, to express x^- and $\frac{\partial x^-}{\partial \tau}$ in terms of the x^i's:

(4a) $$x^- = \int d\sigma \sum_i \frac{\partial x^i}{\partial \sigma} \frac{\partial x^i}{\partial \tau} \ ,$$

(4b) $$\frac{\partial x^-}{\partial \tau} = \frac{1}{2} \int d\sigma \sum_i \left\{ \left[\frac{\partial x^i}{\partial \sigma}\right]^2 + \left[\frac{\partial x^i}{\partial \tau}\right]^2 \right\} .$$

Thus, apart from a constant of integration, the variables x^- and $\frac{\partial x^-}{\partial t}$ are functions of the x^i's. The independent variables x^i satisfy the equations of motion, obtained from the variational principle:

$$\frac{\partial^2 x^i}{\partial \tau^2} - \frac{\partial^2 x^i}{\partial \sigma^2} = 0 . \tag{5}$$

(Similar equations hold in the co-variant approach, with i replaced by μ.)

In a system with equations of motion (5), the momentum density π^i, i.e., the momentum per unit "length" σ, is proportional to $\frac{\partial x^i}{\partial \tau}$. It is convenient to choose our units in such a way that the constant of proportionality is $\frac{1}{2\pi}$. Thus

$$\pi^i = \frac{1}{2\pi} \frac{\partial x^i}{\partial \tau} , \tag{6a}$$

$$\pi^- = \frac{1}{2\pi} \frac{\partial x^-}{\partial \tau} = \frac{1}{4\pi} \sum_i \left\{ \left[\frac{\partial x^i}{\partial \sigma} \right]^2 + \left[\frac{\partial x^i}{\partial \tau} \right]^2 \right\} , \tag{6b}$$

$$\pi^+ = \frac{1}{2\pi} \frac{\partial x^+}{\partial \tau} = \frac{1}{2\pi} . \tag{6c}$$

Many members of the audience will recognize the right-hand side of (6b) as the Hamiltonian density. Eq. (6c) tells us that the momentum density associated with the + co-ordinate is $\frac{1}{2\pi}$. In other words, the co-ordinate σ, which we have called the "length," is not a physical length, but is the momentum in the + direction. The difference between the values of σ for two points on a string measures the amount of + momentum between them.

The variational principle (1) is also to be applied to variations in the end-points of the string. This leads to the equations

$$\left. \frac{\partial \sigma}{\partial \tau} \right|_{\text{endpoint}} = 0 . \tag{7}$$

It is convenient to rescale the variables σ, τ in such a way that the σ-co-ordinates of the ends of the string are 0 and π. We thus define

(8a) $$\eta = a\sigma ,$$

(8b) $$\xi = a\tau ,$$

a being determined by the above condition.

For simplicity we shall only examine open strings. One may also consider closed, i.e., endless strings. An interacting theory without closed string is not consistent, and the gravitational properties of the theory involve the closed strings. Limitations of time do not allow us to include them in this summary.

QUANTIZATION OF THE FREE STRING

In the quantum theory the variables x^i become replaced by operators in a Hilbert space. We shall simply give the result of the simple (to physicists) quantization procedure.

It is convenient to expand x^i in a Fourier series in η:

(9) $$x^i(\eta,\xi) = \tilde{q}_0^i(\xi) + \sum_{n>0} \frac{1}{\sqrt{2n}} \tilde{\alpha}_n^i(\xi) \cos n\eta .$$

Application of the equations of motion (5) then leads to the following ξ-dependence of the right-hand side:

(10) $$x^i(\eta,\xi) =$$

$$q_0 + \sqrt{2}\, \alpha_0^i \xi + \sqrt{2}\, i \sum_{n>0} \frac{1}{n} [\alpha_n^i e^{-in\xi} - \alpha_{-n}^i e^{in\xi}] \cos n\eta .$$

The operators satisfy the commutation relations:

(11a) $$[\alpha_n^i, \alpha_m^j] = n\delta_{ij}\delta_{n,-m},$$

(11b) $$[\alpha_n^i, q_0^j] = -\sqrt{2} i \delta_{ij}\delta_{n0} .$$

Such "Heisenberg systems" are familiar to physicists and

mathematicians. The "ideal elements" of the Hilbert space are labelled by a non-negative integer N_n^i for each i and each "mode number" $n \geqslant 1$ and, in addition, by a continuous momentum p^i, which is $1/\sqrt{2}$ the eigenvalue of α_0^i. (Genuine elements are obtained by multiplying by an L_2 function of p^i and integrating.) The N's thus classify the states of the string, i.e., the particles in the model. The element for which all N's are zero is called the ground state, written $|0,p^i\rangle$. It satisfies the equations

(12a) $$\alpha_n^i \; |0,p^i\rangle = \langle 0,p^i|\, \alpha_{-n}^i = 0, \quad (n > 0)$$

(12b) $$\alpha_0^i \; |0,p^i\rangle = \sqrt{2}p^i \; |0,p^i\rangle \; .$$

An important operator in any quantum system is the Hamiltonian, i.e., the generator of the equations of motion according to the relation:

(13) $$i\frac{\partial x^i}{\partial \tau} = [x^i,H] \; .$$

It is easily seen that

(14a) $$H = \frac{1}{4\pi} \int d\sigma \; \sum_i :\left\{ \left[\frac{\partial x^i}{\partial \sigma}\right]^2 + \left[\frac{\partial x^i}{\partial \tau}\right]^2 \right\}: - \frac{1}{a^2 \pi} \; ,$$

(14b) $$= a^{-1} \sum_i \left\{ (\alpha_0^i)^2 + \sum_{n>0} \alpha_{-n}^i \alpha_n^i - 1 \right\}$$

Colons always mean "normal ordering," i.e., the expression is to be written as a function of the α's, and each term $\alpha_{-\ell}\alpha_{\ell}$ is to be reordered $\alpha_{-|\ell|}\alpha_{|\ell|}$. The last, purely numerical term on the right of (14) is, at the moment, arbitrary.

In the string theory it is of interest to consider transformations much more general than (13). We wish to examine infinitesimal pseudo-conformal transformations of the co-ordinates σ and τ, i.e., transformations which maintain the equations of motion (5) and the co-ordinate conditions (2), but relax the condition $x^+ = \tau$. We are only interested in transformations which transform the lines $\eta = 0, \pi$

onto themselves; the general such transformation is

(15a) $$x^i(\eta,\xi) \longrightarrow x^i(\eta',\xi') \ ,$$

(15b) $$(\xi\pm\eta)' = \xi\pm\eta + i\epsilon\sum c_n e^{in(\xi\pm\eta)}, \qquad (c_n = -c^*_{-n} \text{ for reality}),$$

where the c_n are parameters and ϵ is an infinitesimal constant. To generate these transformation we introduce the non-covariant analogue of the *Virasoro operators*

(16) $$L_n = \frac{1}{2}\sum_{n=-\infty}^{\infty}\sum_i : \alpha^i_{-m}\sigma^i_{m+n} \ .$$

Let us write the solution of (5) in the form:

$$x^i(\eta,\xi) = x^i(\xi+\eta) + x^i(\xi-\eta) \ .$$

It is not difficult to verify the commutation relations

(17a) $$\left[L_n, x^i_r(\xi\pm\eta)\right] = -ie^{in(\xi\pm\eta)}\frac{\partial}{\partial(\xi\pm\eta)}\, x^i_r(\xi\pm\eta), \qquad (r=1,2)$$

so that

(17b) $$x^i(\eta',\xi') = x^i(\eta,\xi) - \epsilon\sum_n c_n[L_n, x^i(\eta,\xi)] \ .$$

We notice that the operator L_0 is simply equal to the Hamiltonian, apart from a constant term. The L's satisfy the commutation relations

(18) $$[L_n, L_m] = (n-m)L_{n+m} + \frac{1}{12}(d-2)n(n^2-1)\delta_{n,-m}.$$

The sub-group of the transformations (16) which transform the σ-τ sheet bijectively onto itself is the SU(1,1) sub-group generated by the operators $L_0, L_{\pm 1}$.

In order for the theory to be physically acceptable it must be Lorentz invariant. The classical string theory in our present form is not manifestly co-variant but, since it was derived from the co-variant variational principle (1), it must be Lorentz invariant. However, Lorentz invariance may be lost in the quantization procedure. We cannot really examine the question until we have introduced interactions between strings. A provisional examination shows that Lorentz invariance is likely to be lost for all values of d except d = 26; at this value of d the Lorentz-violating terms cancel. We thus require d to be equal to 26 in a physically acceptable theory. The numerical term on the right of the Hamiltonian (14) is also necessary for Lorentz invariance.*

INTERACTIONS BETWEEN STRINGS.

We have already stated that the fundamental interaction takes place by the splitting or joining of strings. To express such interactions mathematically we require a new Hilbert space, since our previous Hilbert space $\mathcal{H}$ was the space of states of a single string. The new Hilbert space will be:

$$\mathcal{H}' = I \oplus \mathcal{H} \oplus (\mathcal{H} \otimes \mathcal{H})_S \oplus (\mathcal{H} \otimes \mathcal{H} \otimes \mathcal{H})_S \oplus \dots \tag{19}$$

The subscript S indicates that we only allow elements symmetric under interchange of the spaces $\mathcal{H}$. Physicists often use the technique of "second quantization" to study the space $\mathcal{H}'$, but we shall *not* use this technique here.

The dynamics of any system are specified by the Hamiltonian. To the free Hamiltonian (14) (generalized in the obvious way to the Hilbert space $\mathcal{H}'$), we now add an "interaction Hamiltonian" which has "fundamental" non-zero matrix elements between states in the subspaces $\mathcal{H}$ and $(\mathcal{H} \otimes \mathcal{H})_S$ of (19). (More generally, it will have matrix elements between states in any two successive terms of (19); such matrix elements will be the direct product of the fundamental matrix element and the identity).

---

* The negative sign of this term means that the model is physically unacceptable, since the particles would have imaginary mass. The supersymmetric NSR model does not have this defect.

We shall write down the matrix element between strings in definite *positions*, i.e. between states in the Hilbert spaces $\mathcal{H}$ and $(\mathcal{H}\otimes\mathcal{H})_S$ which are eigenstates of the operators $x^i(\eta,\xi)$ (Eq. 10) for a given ξ. (The operators $x^i(\eta,\xi)$ are not well-defined, but the result may be re-expressed in terms of the operators $\alpha^i_{-n}e^{-in\xi} + \alpha^i_{-n}e^{in\xi}$). The matrix element will only be non-zero if the initial string and the final pair of strings coincide along their entire lengths. In other words, it is equal to a constant g' multiplied by a product of delta functions $\delta(\mathbf{x} - \mathbf{x}')$ or $\delta(\mathbf{x}-\mathbf{x}'')$*, one for each element $d\sigma$ of the string in Fig. 1(a)

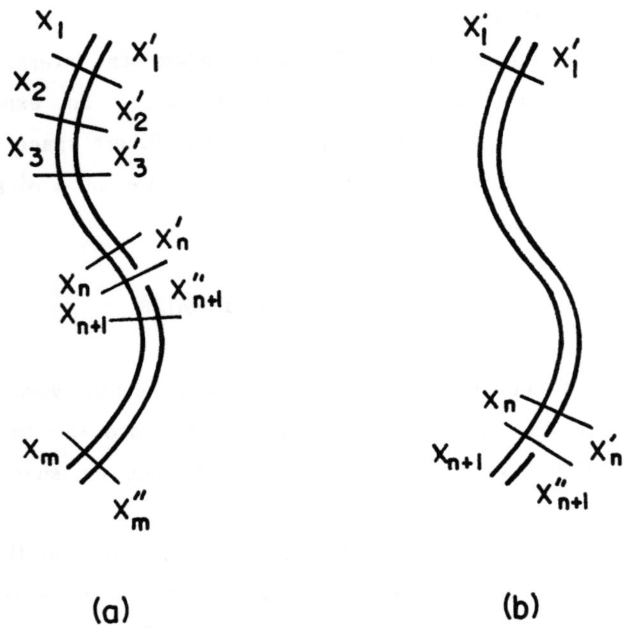

Fig. 1. A splitting string

---

* $\delta(\mathbf{x}-\mathbf{x}') = \prod_{i-1}^{d-2} \delta(x^i - x^{i'})$.

Let us first examine the particular case where the string with the doubly primed co-ordinates is infinitely "short" (i.e., has a very small + momentum). We then have the situation depicted in Fig. 1(b). Furthermore, we shall suppose that the "short" string is in its ground state. It can then be shown that all terms on the right of (10) except the first two may be neglected in the expansion of x'. The "short" string is thus characterized by one pair of (d−2)-dimensional canonical co-ordinates q_0 and α_0 satisfying the commutation relations (11b). The interaction Hamiltonian connects the spaces $\mathcal{H}'\otimes\mathcal{H}''$ but, since the state in the space $\mathcal{H}''$ is given, we may regard $\mathcal{H}$ and $\mathcal{H}'$ as the same Hilbert space; the interaction Hamiltonian will be an operator in this space. The product of δ-functions $\delta(\mathbf{x}'-\mathbf{x}'')$ will be the identity operator, and we are left with the single δ-function

$$g'\delta[\mathbf{x}(0,\xi) - \mathbf{x}''(\xi)] \ . \tag{20}$$

The expression (20) is the interaction Hamiltonian for the case where the "short" string is an eigenstate $|\mathbf{x}''\rangle$ of the co-ordinate with eigenvalue $\mathbf{x}''$ (Thus $\mathbf{x}''$ is a number, $\mathbf{x}(0,\xi)$ an operator in the Hilbert space of the "long" strings.) We are usually interested in the case where the "short" string is an eigenstate $|\mathbf{p}\rangle$ of the momentum operator $\frac{1}{\sqrt{2}}\alpha_0^i$, with eigenvalue p^i. The states $|\mathbf{p}\rangle$ and $|\mathbf{x}''\rangle$ are related by the equation

$$|p\rangle = \int d^{d-2}x^{i\prime\prime}\,\mathrm{Exp}[i\sum_i p^i x^{i\prime\prime}]\,|\mathbf{x}''\rangle \ .$$

It follows that the interaction Hamiltonian for emission of a "short" string with momentum p^i is given by the formula

$$\begin{aligned} &g\int d^{d-2}x^{i\prime\prime}\ \delta[\mathbf{x}(0,\xi) - \mathbf{x}''(\xi)]\mathrm{Exp}[i\sum_i p^i x^{i\prime\prime}] \\ &\qquad = g'\mathrm{Exp}[i\sum_i p^i x^i(0,\xi)]. \end{aligned} \tag{21}$$

In this and all subsequent formulas, note that the x's (or α's and g's) are operators while the p's are numbers.

The expansion (10) may be inserted in (21). The resulting

operator is not well defined, but it can be made well defined by normal ordering, i.e., by replacing the exponential of the last term on the right of (10) by a product of exponentials, the first involving the operators α_{-n} and the second the operators α_n (n>0). The constant g is thereby changed by an infinite factor. Thus:

$$\tilde{V}(p^i,\xi) =$$

(22)
$$g\,\mathrm{Exp}\{-\sqrt{2}\sum_i p^i \sum_{n>0} \frac{1}{n}\alpha^i_{-n} e^{in\xi}\}\,\mathrm{Exp}\{i\sum_i p^i(q^i_0 + \sqrt{2}\alpha^i_0 \xi)\}$$
$$\times\ \mathrm{Exp}\{\sqrt{2}\sum_i p^i \sum_{n>0} \frac{1}{n}\alpha^i_n e^{-in\xi}\}\ .$$

The operator $\tilde{V}$ is the vertex operator.

When performing calculations physicists often find it useful to continue analytically to imaginary time ($\xi' = i\xi$). One may also make a change of variables

(23)
$$z = e^{\xi'}\ .$$

Expressed in terms of the new variables, the vertex operator becomes

$$V(p^i,z) =$$

(24a)
$$\tilde{V}(p^i,\xi) = g\,\mathrm{Exp}\{-\sqrt{2}\sum_i p^i \sum_{n>0} \frac{1}{n}\alpha^i_{-n} z^n\}\,\mathrm{Exp}\{i\sum_i p^i(q^i_0 - \sqrt{2}i\alpha^i_0 \ln z)\}$$
$$\times\ \mathrm{Exp}\{\sqrt{2}\sum_i p^i \sum_{n>0} \frac{1}{n}\alpha^i_n z^{-n}\}\ ,$$

(24b)
$$\equiv gV_1V_2V_3\ .$$

When we continue to imaginary time, the two variables ξ and η may be replaced by a single complex variable $\zeta = \xi' + i\eta$. The pseudo-conformal transformations now become conformal transformations. In terms of the variable $z = e^{\zeta}$, Equ. (17a) take the form:

$$(25) \qquad [L_n, x_1^i] = z^{n+1} \frac{\partial}{\partial z} x_1^i, \qquad [L_n, x_2^i] = -(z^*)^{n+1} \frac{\partial}{\partial z} x_2^i .$$

The subgroup of conformal transformations generated by the operators L_0, $L_{\pm 1}$ becomes the sub-group of projective transformations in z.

S-MATRIX ELEMENTS

An important concept in physics is that of the S-matrix, which enables one to calculate the result of a scattering experiment. Let us consider the process represented by Fig. 2.

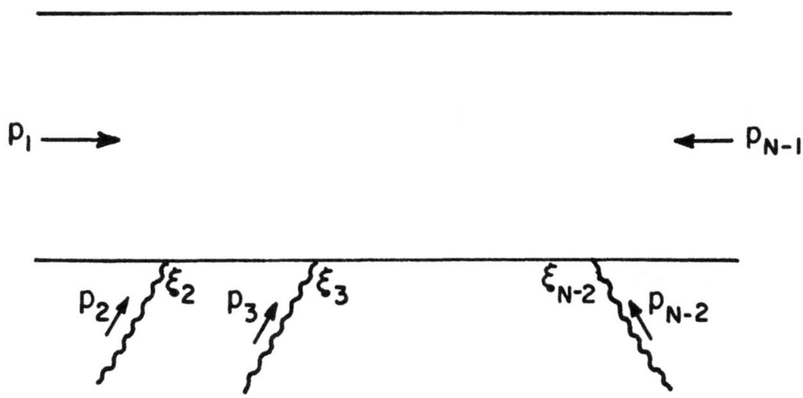

Fig. 2. Scattering process involving one "long" string and several "short" strings.

The strip represents a "long" string, the wavy lines "short" strings. At each vertex $\xi_2, \ldots, \xi_{N-1}$ two strings joing to form one, or one string splits to form two. We shall take the simple case where all strings, long and short, are in their ground states. In the particular process represented by Fig. 2 there are three incoming strings, one "long" and two "short", and four outgoing strings, one "long and three "short".

We shall not attempt to define the S-matrix. It is a function of the N momenta p_N^i (defined negatively for incoming strings) and, to lowest order in g, it can be calculated from the following formula (apart from kinematic factors):

$$A(p_1,\ldots,p_N) = g^{N-2}\int_{-\infty}^{\xi_{N-1}} d\xi_{n-2}\int_{-\infty}^{\xi_{N-2}} d\xi_{n-3}\cdots$$

$$\int_{-\infty}^{\xi_3} d\xi_2 \ \mathrm{Exp}\{-ip_2^- \xi_2 \ldots -ip_{N-1}^- \xi_{N-1}\}$$

(26)

$$\times \langle 0,p_N^i \,|\, \tilde{V}(p_{N-1}^i,\xi_{N-1})\tilde{V}(p_{N-2}^i\,\xi_{N-2})\ldots\tilde{V}(p_2^i,\xi_2)\,|\,0,p_1^i\rangle \ .$$

Notice that we integrate over all ξ's except one; the result will be independent of the value of ξ_{N-1}. For simplicity we shall assume that all the p^-'s are zero; the general case has slight complications of detail but not of principle. If we analytically continue the integrand to imaginary ξ and make the transformation (23), we obtain the formula:

$$A(p_1,\ldots,p_N) = g^{N-2}\int_0^{z_{N-1}} \frac{dz_{N-2}}{z_{N-2}} \int_0^{z_{N-2}} \frac{dz_{N-3}}{z_{N-3}} \cdots \int_0^{z_3} \frac{dz_2}{z_2}$$

(27)

$$\times \langle 0,p_N^i \,\big|\, V(p_{N-1}^i,z_{N-1})\ldots V(p_2^i,z_2) \,\,|\, 0,p_1^i\rangle \ .$$

The expression on the right of (27) appears to treat the variables p_1^i and p_N^i very differently from the other variables. The result should be unchanged under cyclic interchange of the p's; the values of the "lengths" of the string are altered by Lorentz transformations, and the fact that two strings are "long" and the others "short" should make no difference. We shall shortly calculate the integrand of (27) explicitly, but we shall obtain greater insight if we verify the symmetry of (27) in all momenta without explicit calculation.

We first notice that

(28a) $$|0,p_1^i\rangle = \lim_{z_1\to 0} z_1^{-p_1^2} V(p_1^i,z_1)|0\rangle = \lim_{z_1\to 0} z_1^{-1} V(p_1^i,z_1)|0\rangle,$$

(28b) $$\langle 0,p_N^i| = \lim_{z_N\to\infty} z_N^{-p_N^2}\langle 0|V(p_N^i,z_N) = \lim_{z_N\to\infty} z_N\langle 0|V|p_N^i,z_N),$$

where $|0\rangle$ denotes the ground state with $p^i = 0$. In (28a), the last factor of V gives unity when acting on the ground state, the first factor is unity in the limit $z_1\to 0$. It is not difficult to see that the middle factor gives the result quoted. Eq. (14) implies that p^2 for the ground state is 1 if $p^- = 0$. Eq. (28b) is proved in a similar manner. We may therefore write:

(29) $$A(p_1,\ldots p_N) = g^{N-2} \lim_{z_1\to 0} \lim_{z_N\to\infty} \frac{z_N}{z_1} \int_0^{z_{n-1}} \frac{dz_{N-2}}{z_{N-2}} \cdots \int_0^{z_3} \frac{dz_2}{z_2}$$
$$\times\langle 0|V(p_N^i\ ,z_N)\ldots V(p_1^i,z_1)|0\rangle\ .$$

Eq. (29) looks more symmetric in the N momenta, except for the fact that the variables z_1, z_{N-1} and z_N are fixed at 0, an arbitrary value, and ∞ respectively, while the other variables are integrated over. We can remove the remaining asymmetry by projectively transforming the variables. We replace the V's by the factors

(30a) $$UV(p_i,z)U^{-1},$$

where

(30b) $$U = 1-\epsilon(c_{-1}L_{-1}+c_0L_0+c_1L_1)\ ,$$

ϵ being an infinitesimal constant. The expression on the right of (30) is thereby unchanged, since the factors U^{-1} and U between V's cancel one another, and $U^{-1}|0\rangle = |0\rangle$, $\langle 0|U = \langle 0|$. We have already explained that the operator U simply enforces a projective transformation on the variables z ($\equiv e^{\xi'+i\eta}$) in (10). The effect on V is slightly more complicated, since V is not quite a function of x (it is a product of exponentials rather than a single exponential). A straightforward calculation shows that the analogue of (25) is the following:

(31a) $$[L_n, V] = z^{n+1}\frac{\partial V}{\partial z} + np^2 V = z^{n+1}\frac{\partial V}{\partial z} + nV.$$

Therefore

(31b) $$UV(p_i, z)U^{-1} = [1-\epsilon(-c_{-1}z^{-1} + c_1 z)]V(p_i, z'),$$

where

(31) $$z' = z - \epsilon(c_{-1}+c_0 z+c_1 z^2).$$

We can use the invariance of the integrand under the transformations (30) to make an infinitesimal projective transformation of the z's. The factor in the braces of (31), as well as the changes in the z-factors of (29), are found to compensate the Jacobian. It follows that the amplitude (29) is invariant under an infinitesimal projective transformation of the z's. The variables z_1, z_{N-1} and z_N may therefore be fixed at any values in cyclic order. This being the case it can easily be shown that the amplitude remains unchanged if a different triple of successive z's is kept fixed; z_1 is regarded as succeeding z_N. The amplitude (29) is therefore unchanged under cyclic reordering of the p's.

Finally let us calculate the integrand of (29) explicitly. By applying the Campbell-Hausdorff formula to the factor V_2 of (24), we can rewrite V in the form

(32) $$V(p^i, z) = gzV_1'V_2' \qquad (p^2 = 1),$$

where

(33a) $$V_1' = V_1 \mathrm{Exp}\{i\sum_i p^i q_0^i\},$$

(33b) $$V_2' = \mathrm{Exp}\{\sqrt{2}\sum_i p^i \alpha_0^i \ln z\} V_2.$$

We then commute all factors V_2 to the right and all factors V_1 to the left. The effect of interchanging the order of factors is easily found from the Campbell-Hausdorff formula:

(34) $$V_2'(p_r^i, z_r) V_1'(p_s^i, z_s) = (z_r - z_s)^{2p_r \cdot p_s} V_1'(p_s^i, z_s) V_2'(p_r^i, z_r).$$

After interchanging the factors this way the V_1's and V_2's may be dropped, since

(35) $$V_2' |0\rangle = |0\rangle \qquad \langle 0| V_1' = \langle 0| \mathrm{Exp}\{i\sum_i p^i q_0^{i'}\},$$

and the sum of all terms in braces is zero due to momentum conservation. We are left with the factors $(z_r - z_s)^{2p_r \cdot p_x}$ from (34) and the z factors in (29) and (32). Therefore

(36) $$A(p_1,\ldots,p_N) = g^{N-2} \int_0^{z_{N-1}} dz_{N-2} \cdots \int_0^{z_3} dz_2 \prod_{r>s} (z_r - z_s)^{2p_r \cdot p_s}.$$

The subscripts r and s in (36) range from 1 to N-1, with $z_1 = 0$. All factors of involving z_N cancel as a result of the equations $\sum_{r=1}^{N} p_r = 0$, $p_N^2 = 1$.

It is not difficult to check directly that we can make a projective transformation of the variables to rewrite (36) in the form, symmetric in all p's

$$A(p_1,\ldots,p_N) = g^{N-2}(z_{N-1}-z_1)(z_N-z_{N-1})(z_1-z_N)\int_0^{z_{N-1}} dz_{N-2}\cdots$$

(37)

$$\int_0^{z_3} dz_2 \prod_{r>s} (z_r-z_s)^{2fr\cdot fs} .$$

The subscripts r and s range from 1 to N, with z_1, z_{N-1} and z_N fixed at arbitrary values. Eq. (37) is the Koba-Nielsen formula.

MORE GENERAL AMPLITUDES

Let us now consider the process represented in Fig. 3. Horizontal lines denote the ends of strings, which are moving in time τ. The transverse dimensions are perpendicular to

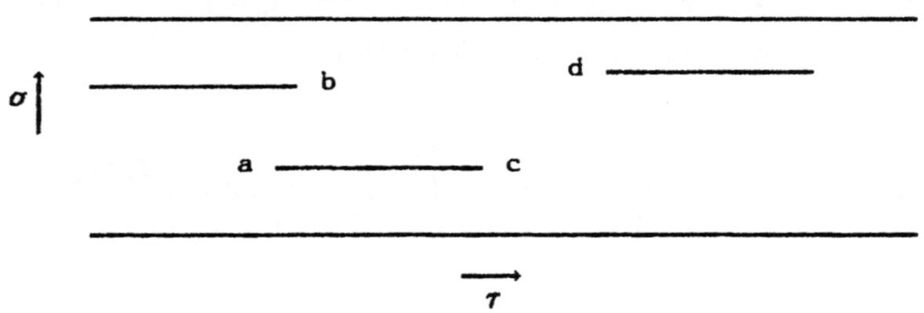

Fig. 3 General Scattering process .

the plane of the paper, and they are discontinuous across the solid lines. Thus two strings enter the diagram, one splits at a, one of the strings so formed joins the other at b, and so on. The process depicted in Fig. 3 is more general than that depicted in Fig. 2 in two respects.

i) all the strings have finite "length".

ii) A similar scattering process could occur without the interactions at a and c. In other words, the process is not of lowest order in g.

The contribution to the S-matrix represented by Fig. 3 is obtained as an integral over the position of the points a,b,c,d in the σ-τ diagram. The easiest method of calculating the integrand is probably to perform a Feynman functional integral (Feynman-Kac formula). One integrates over all possible motions of the string with the given boundary conditions, each configuration being weighted by a factor

$$\text{(38)} \qquad \mathrm{Exp}\left\{-\frac{1}{4\pi}\int d\sigma d\tau' \sum_{i=1}^{d-2} \left\{ \left[\frac{\partial x^i}{\partial\sigma}\right]^2 + \left[\frac{\partial x^i}{\partial\tau'}\right]^2 \right\}\right\} .$$

The variable τ' is $i\tau$, i.e., we integrate over the imaginary τ-axis. The functional integration can be performed, since the integrand is a Gaussian. Besides a factor representing the determinant of the Laplacian, for Fig. 3, raised to the power $-\frac{1}{2}$, there are factors obtained from the boundary conditions which depend on the momenta and mode numbers of the incoming and outgoing strings. The latter factors are exponents of a sum of terms depending on the Neumann function for Fig. 3.

The determinant of the Laplacian was originally evaluated by various indirect methods. Later we discovered that mathematicians had been looking at the problem at about the same time.[5]

One can also evaluate the S-matrix-element corresponding to Fig. 3 by taking traces of products of vertex operators. A comparison of the two methods may possibly lead to information concerning such traces.

The vertices occuring in Fig. 3 correspond to the process of Fig 1(a) rather than the limiting case of Fig. 1(b). The generalization of the formula (25) to such vertex operators has been found.[6] One can either evaluate the product of δ-functions corresponding to Fig. 1 in terms of the operators α_n, or one can use the Neumann-function technique, referred to above, applied to the process shown in Fig. 4. To our knowledge this general vertex function has not

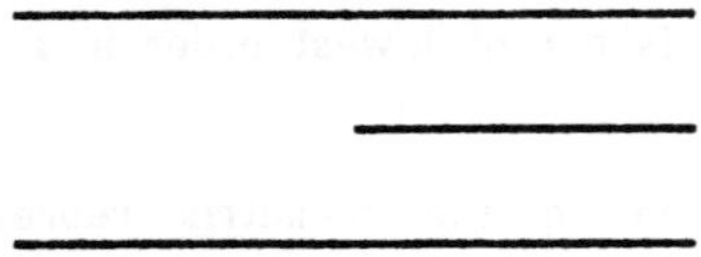

Fig. 4. String diagram for the general vertex operator.

yet found applications in the theory of Kac-Moody algebras; I leave it to the present audience to evaluate the possibilities of future applications.

REFERENCES

[1] V. Alessandrini, D. Amati, M. LeBellac and D.I. Olive, Phys. Reports 1C, 170(1971).
P. Frampton, Dual Resonance Models (Benjamin, 1974).
S. Mandelstam, Phys. Reports 13C, 259(1974).
S. Mandelstam, in "Structural Analysis of Collision Amplitudes" (Les Houches, 1975, June Institute) p. 593.
C. Rebbi, Phys. Reports 12C, 1(1974).
J. Scherk, Rev. Mod. Phys. 47, 123(1975).
J.H. Schwarz, Phys. Reports 8C, 269(1973).
G. Veneziano, Phys. Repots 9C, 199(1974).

The articles in Physics Reports have been collected in a reprint volume "Dual Models" (North Holland, 1974). The review articles (1) contain further references.

[2] A.M. Polyakov, Phys. Lett. 103B, 207, 211(1981).

[3] J. Scherk and J.H. Schwarz, Nucl. Phys. B36, 155(1972).

[4] See the review article by J.H. Schwarz, Phys. Repots 89, 223(1982).

[5] H.P. McKean and I.M. Singer, J. Diff. Geom. 1, 43(1967)
D.B. Ray and I.M. Singer, Adv. Math. 7, 145 (1971); Ann Math 98, 154(1973).
H.P. McKean, Comm. Pure Appl. Math 25, 225 (1972).

[6] M. Ademollo, E. Del Giudice, P. DiVecchia and S. Fubini, Nuovo Cimento 19A, 181(1974).
S. Mandelstam, Nucl. Phys. B 64, 205(1973).
J. Goldstone, Private Communication.
E. Cremmer and J.-L Gervais, Nucl. Phys. B76, 209(1974).

The first of these papers is different in spirit from the approach followed here.

Department of Physics
University of California
Berkeley, California 94720

AN INTRODUCTION TO POLYAKOV'S STRING MODEL

Orlando Alvarez[+]

This talk is an introduction to Polyakov's string model. The objective is to demonstrate how the Liouville field theory enters into the string model. Thom Curtright will discuss the quantum Liouville theory in the next talk.

The major ideas in this talk are due to Polyakov [1]. The global viewpoint is due to Friedan [2]. The Polyakov model in an arbitrary topology has been analyzed in [3]. Complications introduced by the inclusion of boundaries may be found in [3] and [4]. A good review article which discusses the different string models is the one by Fradkin and Tseytlin [5].

AN INTRODUCTION TO POLYAKOV'S STRING MODEL

Several years ago Polyakov introduced a new quantization of the string model. It was hoped that some of the difficulties of the Nambu string model could be cured by the new quantization scheme. Polyakov's price for a manifestly Lorentz covariant quantization is the introduction of a new longitudinal degree of freedom. This new degree of freedom is described by the quantum Liouville field theory. The equation of motion for this theory is the equation for a constant curvature metric on a two dimensional surface.

The temporal evolution of a string like object is a two dimensional surface. Nambu and Goto proposed that the equations of motion should be derived from a natural variational principle. They chose the minimization of the area of the two dimensional surface generated by the time development of the string.

[+] Partially supported by NSF Grant PHY 81-18547 and an A.P. Sloan Research Fellowship.

Vertex Operators in Mathematics and Physics - Proceedings of a Conference November 10-17, 1983. Publications of the Mathematical Sciences Research Institute #3, Springer-Verlag, 1984.

Throught this paper we will work in the Euclidean space formulation of the theory. Let M be a two dimensional orientable manifold and let $x: M \longrightarrow \mathbb{R}^d$ be an embedding of M in $\mathbb{R}^d$. The Einstein summation convention will be used with the Greek indices μ, ν, ρ, σ taking values 1,2,...,d and Latin indices a,b,c,d taking values 1 and 2.

The Nambu and Goto ansatz for the action of a string theory is

$$(1) \qquad J = \int_M d^2z\ (\det \gamma_{ab})^{1/2},$$

where z denotes coordinates on M, and

$$(2) \qquad \gamma_{ab} = \frac{\partial x^\mu}{\partial z^a} \frac{\partial z^\mu}{\partial z^b}$$

$$= \partial_a x^\mu\ \partial_b x^\mu,$$

is the induced metric on M due to the embedding. The equations of motion deduced from (1) are minimum area equations:

$$(3) \qquad \partial_a [(\det \gamma)^{1/2}\ \gamma^{ab}\ \partial_b x^\mu] = 0.$$

The classic paper on the quantization of the nambu-Goto model is by Goddard, Goldstone, Rebbi and Thorn [6].

Polyakov proposes that one should quantize a different theory which leads to the same classical equations of motion. He considers a theory where in addition to the degrees of freedom associated with the position of the surface $x^\mu(z)$, there is an additional dynamical field given by an "intrinsic" metric on the manifold M. The variational principle that Polyakov proposes is

$$(4) \qquad I[x,g] = \frac{1}{2} \int_M d^2z\ g^{1/2}\ g^{ab}\ \partial_a x^\mu\ \partial_b x^\mu,$$

where g_{ab} is the intrinsic metric on the surface. The above choice for the action is invariat under the diffeomorphism group of M. It is also invariant under conformal transformations of the metric:

$$(5) \qquad g_{ab}(z) \longrightarrow g_{ab}(z)\ \exp[2c(z)].$$

The classical equations of motion that one obtains by varying (4) with respect to g and x are equivalent to the minimum area equations (3).

Path integral methods are used to quantize (4). An important quantity is the partition function Z defined by

$$(6) \qquad Z = \int [dx]\ [dg]\ \exp(-I[x,g]),$$

where the integration is over all positions of the surface and all possible metrics on M. Polyakov demonstrated that equation (6) is equivalent to the quantum Liouville field theory. Let us see how this comes about.

The first step is to understand equation (6) in more detail. The "deformations" dx and dg are turned into Hilbert spaces by defining metrics

$$(7) \qquad \|dx\|^2 = \int d^2z\ g^{1/2}\ dx^{\mu}(z)\ dx^{\mu}(z),$$

and

$$(8) \qquad \|dg\|^2 = \int d^2z\ g^{1/2}\ (G^{abcd} + ug^{ab}g^{cd})\ dg_{ab}\ dg_{cd},$$

where $u > 0$ and G^{abcd} is the identity operator on the space of symmetric traceless tensors:

$$(9) \qquad G_{ab}{}^{cd} = \tfrac{1}{2}(\delta_a{}^c\delta_b{}^d + \delta_a{}^d\delta_b{}^c - g_{ab}g^{cd}).$$

In equation (6) one should view [dx] and [dg] as the volume elements induced by (7) and (8). The measures [dx] and [dg] are invariant under the action of the diffeomorphism group Diff M. The measures [dx] and [dg] are not invariant under conformal transformations.

Two diffeomorphic metrics are physically equivalent. A physical state is an equivalence class (x,g) with respect to the action of Diff

M. One would like to isolate the role of Diff M in equation (6). Qualitatively one can say that in equation (6) one is overcounting by a factor of the volume of Diff M. There is a procedure called the Faddeev-Popov ansatz which makes this statement more precise. We now turn to a discussion of the Faddeev-Popov ansatz applied to the Polyakov string model.

A classic theorem on Riemann surfaces states that every metric is conformal to a constant curvature metric. We will use this fact to implement the Faddeev-Popov procedure. The action of Diff M on the metrics is implemented by the Lie derivative. If V is the infinitesimal generator of the diffeomorphism then the charge in the metric is given by

(10) $$\delta g_{ab} = \nabla_a V_b + \nabla_b V_a .$$

Naively, the metric depends on three functions, and the diffeomorphisms are parametrized by the two functions that specify the vector field V. We would like to trade two of the variables in g for the vector field V. The remaining variable in g will be taken to be a conformal factor. Instead of parametrizing the space of metrics by g_{11}, g_{12}, g_{22}, we would like to specify it by V^1, V^2, and σ. The way to do this is as follows: pick a fiducial constant curvature metric $\hat{g}$. The physical metrics of interest will be of the form

(11) $$g = e^{2\sigma}\hat{g}.$$

These metrics depend on one function $\sigma(z)$. The volume element $[dg]$ will be written as

(12) $$[dg] = [dV]\,[d\sigma]\,J$$

where J is the Jacobian of the transformation that relates g to V and σ. Let us see how this works in more detail.

It is convenient to decompose dg into

(13) $$dg_{ab} = 2g_{ab}(dc) + dh_{ab}$$

where dh_{ab} is traceless, $g^{ab}dh_{ab} = 0$. The symmetric traceless deformations dh belong to a two dimensional space. These are the objects that will be traded for diffeomorphisms. Notice that (13) defines an orthogonal decomposition therefore

$$[dg] = [dh]\,[dc]. \tag{14}$$

The change in the metric induced by a diffeomorphism is given by equation (10), and the change induced by a conformal transformation is a simple consequence of equation (11):

$$dg_{ab} = 2(d\sigma)\, g_{ab}. \tag{15}$$

Using orthogonal decomposition (13) and putting together (10) and (15) one obtains

$$2(dc) = 2(d\sigma) + g^{ab}\nabla_a V_b, \tag{16}$$

$$\begin{aligned} dh_{ab} &= 2\, G_{ab}{}^{cd}\, \nabla_c V_d \\ &= (PV)_{ab}. \end{aligned} \tag{17}$$

The operator P maps vector fields into symmetric traceless tensor fields. Changing variables from τ and h to σ and V one finds

$$[d\tau]\,[dh] = [d\sigma]\,[dV] \left| \frac{\partial(\tau, h)}{\partial(\sigma, V)} \right|. \tag{18}$$

The "Jacobian" is easily computed

$$\begin{aligned} \left| \frac{\partial(\tau, h)}{\partial(\sigma, V)} \right| &= \left| \det \begin{pmatrix} 1 & \text{stuff} \\ 0 & P \end{pmatrix} \right| \\ &= |\det P| \\ &= (\det P^{+}P)^{1/2} \end{aligned} \tag{19}$$

where $P^\dagger$ is the adjoint of P. The determinant is infinite and it has to be regularized. These manipulations allow one to write the partition function in the form:

$$(20) \qquad Z = \int [d\sigma]\ [dV]\ [dx]\ )\ (\det P^\dagger P)^{1/2} \exp(-I[x,g]).$$

One can express what was just done with terminology commonly used in the study of Riemann surfaces. Choose conformal (isothermal) coordinates then $g = \Omega^2(dzd\bar{z} + d\bar{z}dz)$. One can show that quadratic differentials of the type

$$(21) \qquad Q = Q_{zz}dzdz + Q_{\bar{z}\bar{z}}d\bar{z}d\bar{z}$$

correspond to the complexification of the space of symmetric traceless tensors. An object of the form

$$(22) \qquad R = R_{z\bar{z}}\, dz\, d\bar{z}$$

corresponds to the trace of a symmetric tensor.

The derivation of equation (20) has missed several important technical points. Assume there is a V such that PV = 0. Such a V is called a conformal Killing vector. An alternative statement is that PV = 0 are the Cauchy-Riemann equations. A conformal Killing vector corresponds to a global analytic vector field. Equations (16) and (17) tell us that a diffeomorphism generated by a conformal Killing vector is equivalent to a change in the conformal factor. Since we want to count each deformation of the metric once, such a diffeomorphism should be omitted. The zero eigenvalues of $P^\dagger P$ should be omitted in equation (20). The correct determinant in (10) should be $\det' P^\dagger P$ where the prime denotes the omission of the zero eigenvalues. This determinant is called the Faddeev-Popov determinant. A consequence of the exclusion of the zero eigenvalues is that the integrand in (20) has a residual invariance under diffeomorphisms generated by conformal Killing vectors.

A second error is that we implicitly assumed that all deformations dh_{ab} were expressible as infinitesimal diffeomorphisms. This is not true in general. The following figure illustrates the situation.

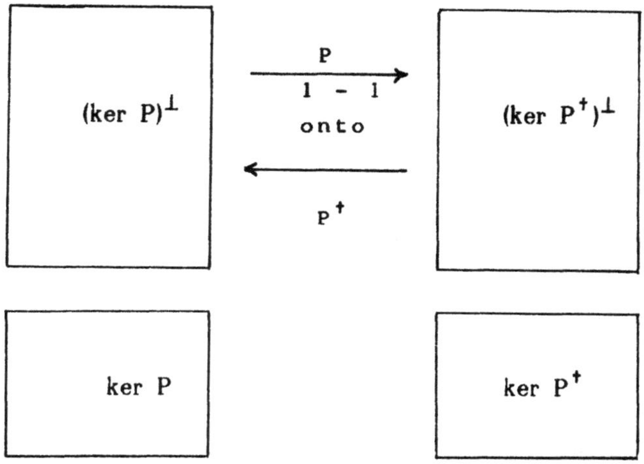

Diffeomorphisms Metrical Deformations

Only those metrical deformations in $(\ker P^{\dagger})^{\perp}$ are expressible as infinitesimal diffeomorphism. The elements of $\ker P^{\dagger}$ are analytic tensors of type $Q_{zz}\, dz\, dz$ or anti-analytic tensors of type $A_{\bar{z}\bar{z}}\, d\bar{z}\, d\bar{z}$. Such tensor fields correspond to a tangent vector of the Teichmuller space of the manifold. The Riemann-Roch theorem states that

$$\dim_{\mathbb{R}} \ker P - \dim_{\mathbb{R}} \ker P^{\dagger} = 3\,\chi(M) \tag{23}$$

where $\chi(M)$ is the Euler characteristic of the manifold. If $M = S^2$ then dim ker $P = 6$, and dim ker $P^{\dagger} = 0$. If $m = T^2$ then dim ker P $= 2$, and dim ker $P = 2$. If $\chi(M) < 0$ then dim ker $P = 0$, and dim ker $p^{\dagger} = -3\chi(M)$. In complicated topologies one has to worry about the Teichmuller deformations. The dimension of ker $P^{\dagger}$ corresponds to the number of distinct constant curvature metrics on the manifold M.

A third technical worry is that the volume element [dV] depends on the conformal factor σ. This may be seen by looking at the metric on the space of infinitesimal diffeomorphisms.

$$\|dV\|^2 = \int d^2z\, g^{1/2}\, g_{ab} (dV)^a\, (dV)^b. \tag{24}$$

This is not a conformally invariant expression. The fact that [dV] is not conformally invariant leads to a major problem. The whole idea of the Faddeev-Popov procedure is that by isolating the diffeomorphism group one would like to write (20) as

$$(25) \qquad Z = (\text{vol Diff } M)\int [d\sigma][d\ \text{Teich}][dx](\det{}'P^{+}P)^{1/2}\ \exp(-I[x,g]).$$

In the above [d Teich] denotes integration over the Teichmuller directions. In this way the new integral counts each fiber once. The (vol Diff M) term is an overall divergent factor that expresses the invariance of the theory under diffeomorphisms. We have seen that in principle one cannot do this in the Polyakov model. Neither would one be able to do a similar thing in any theory of gravitation. Note that Yang-Mills theory is not burdened by this problem. It is possible to determine the behavior of [dV] as a function of σ except that there is no unambiguous regularization. One has to reach a decision about what to do about [dV].

Since the rules for quantization are not known, one reaches a decision. Let's extract [dV] or let's throw out [dV]. This is a purely ad-hoc decision. The partition function of interest is then given by

$$(26) \quad Z = \int [d\sigma]\ [d\ \text{Teich}]\ [dx]\ (\det{}'P^{+}P)^{1/2}\ \exp(-I[x,g])$$

The x integration is gaussian and leads to

$$(27) \quad Z = \int [d\sigma]\ [d\ \text{Teich}]\ (d\xi^{\mu})(\det{}'P^{+}P)^{1/2}(\det{}'\Delta)^{-d/2}.$$

In the above ξ^{μ} are "collective coordinates" that one has to introduce because of the non-zero kernel of Δ. For simplicity I will drop all issues associated with kernels, Teichmuller spaces, etc. and work at the simplest possible level.

$$(28) \qquad Z = \int [d\sigma]\ (\det P^{\dagger}P)^{1/2}\ (\det \Delta)^{-d/2}.$$

It is possible to evaluate the dependence of the determinants on σ. This is done by writing down a differential equation for σ and afterwards integrating the equation. The outcome of this procedure is

$$(29) \quad \ln \det P^{\dagger}P = -\frac{1}{2\pi\epsilon}\int_M d^2z\ \hat{g}^{1/2}\exp(2\sigma) + \frac{4}{3}\chi(M)\ln\epsilon$$

$$-\frac{13}{3\pi}\Big[\frac{1}{2}\int_M d^2z\ \hat{g}^{1/2}\hat{g}^{ab}\partial_a\sigma\partial_b\sigma + \frac{1}{2}\int_M d^2z\ \hat{g}^{1/2}\hat{R}\ \sigma\Big] + \cdots$$

$$(30) \quad \ln \det \Delta = -\frac{1}{4\pi\epsilon}\int_M d^2z\ \hat{g}^{1/2}\exp(2\sigma) + \frac{1}{6}\chi(M)\ln\epsilon$$

$$-\frac{1}{6\pi}\Big[\frac{1}{2}\int_M d^2z\ \hat{g}^{1/2}\hat{g}^{ab}\partial_a\sigma\partial_b\sigma + \frac{1}{2}\int_M d^2z\ \hat{g}^{1/2}\hat{R}\ \sigma\Big] + \cdots$$

In the above ϵ is a short distance cutoff required to regulate the determinant. The ellipsis denotes terms that involve information about the kernels and some $\hat{g}$ dependence.

By putting all the information together one learns that the partition function may be written as

$$(31) \qquad Z = \int [d\sigma]\ \exp(\ -\ S_{eff}[\sigma]\ )$$

where

$$(32) \quad S_{eff}[\sigma] = \mu_0^2\int_M d^2z\ \hat{g}_t^{1/2}\exp(2\sigma)$$

$$+\frac{26-d}{12\pi}\Big[\frac{1}{2}\int_M d^2z\ \hat{g}^{1/2}\hat{g}^{ab}\partial_a\sigma\partial_b\sigma + \frac{1}{2}\int_M d^2z\ \hat{g}^{1/2}\hat{R}\sigma\Big] + \cdots$$

In the above μ_0^2 is a "divergent bare mass".

The effective action of equation (32) defines a field theory known as the Liouville field theory. Note the important factor 26-d. The action is positive semidefinite if and only if $d \leqslant 26$. The main result of Polyakov is that his model naturally incorporates a new degree of freedom which could be very useful in eliminating the problems which faced the old string model. This requires an understanding of the quantum Liouville theory.

The equations of motion for the classical Liouville theory are very interesting:

$$R = \text{neg. const.} \tag{33}$$

where R is the scalar curvature in the metric $g = \hat{g} \exp(2\sigma)$. R is related to $\hat{R}$ by

$$R = e^{-2\sigma}(\hat{R} + 2\hat{\Delta}\sigma), \tag{34}$$

where ^ refers to quantities calculated with respect to $\hat{g}$. The classical Liouville model is a theory of surfaces of constant negative curvature. The quantum Liouville model is a theory of surfaces that fluctuate about a constant curvature surface.

Let me finish this talk with the following figure.

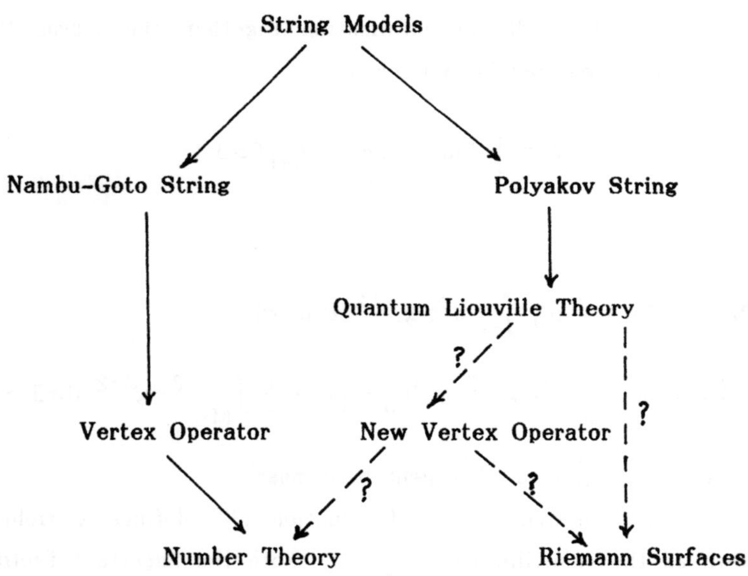

REFERENCES

[1] A.M. Polyakov, Phys. Lett. 103B (1981) 207.

[2] D. Friedan, Talk at Copenhagen Workshop on String Models, October 1981, unpublished.

[3] O. Alvarez, Nucl. Phys. B216 (1983) 125.

[4] B. Durhuus, P. Olesen, J.L. Petersen, Nucl. Phys. B198 (1982) 157.

[5] E.S. Fradkin, A.A. Tseytlin, Ann. Phys. 143 (1982) 413.

[6] P. Goddard, J. Goldstone, C. Rebbi, C. Thorn, Nucl. Phys. B56 (1973) 109.

Department of Physics
University of California
Berkeley, California 94720

CONFORMALLY INVARIANT FIELD THEORIES IN TWO DIMENSIONS

Thomas Curtright[†]

The talk of Dr. Curtright dealt primarily with properties of the Liouville quantum field theory in two-dimensional spacetime. The material in the talk may be found in a series of papers by Dr. Curtright and his collaborators. Those papers are:

"Conformally Covariant Quantization of the Liouville Theory," T. Curtright and C. Thorn, Phys. Rev. Lett. 48 (1982) 1309.

"Quantum Backlund Transformation for the Liouville Theory," E. Braaten, T. Curtright, and C. Thorn, Phys. Lett. 118B (1982) 115.

"An Exact Operator Solution of the Quantum Liouville Field Theory," E. Braaten, T. Curtright, and C. Thorn, Ann. Phys. (NY) 147 (1983) 365.

"A Class of Conformally Invariant Quantum Field Theories," E. Braaten, T. Curtright, G. Ghandour, and C. Thorn, Phys. Lett. 125B (1983) 301.

"Nonperturbative Weak-coupling Analysis of the Liouville Quantum Field Theory," E. Braaten, T. Curtright, G. Ghandour, and C. Thorn, Phys. Rev. Lett. 51 (1983) 19.

"Nonperturbative Weak-coupling Analysis of the Quantum Liouville Field Theory," E. Braaten, T. Curtright, G. Ghandour, and C. Thorn, Ann. Phys. (NY) to appear (1984).

[†] Partially supported by the National Science Foundation through the Mathematical Sciences Research Institute

Vertex Operators in Mathematics and Physics - Proceedings of a Conference November 10-17, 1983. Publications of the Mathematical Sciences Research Institute #3, Springer-Verlag, 1984.

"Weak-coupling Analysis of the Supersymmetric Liouville Theory," T. Curtright and G. Ghandour, Phys. Lett. B, to appear December 1983.

These papers contain adequate references to the work of others on the subject.

In addition to the above, further information on conformally invariant field theories in two-dimensional spacetime may be found in the lectures given by Dr. Curtright in the Proceedings of the Tata Institute Winter School at Mahabaleshwar, India, 4--19 January 1984.

Department of Physics
University of Florida
Gainesville, FL 32611

ALGEBRAS, LATTICES AND STRINGS

P. Goddard[†] and D. Olive[†]

Abstract

A unified construction is given of various types of algebras, including finite dimensional Lie algebras, affine Kac-Moody algebras, Lorentzian algebras and extensions of these by Clifford algebras. This is done by considering integral lattices (i.e. ones such that the scalar product between any two points is an integer) and associating to the points of them the square of whose length is 1 or 2, the contour integral of the dual model vertex operator for emitting a "tachyon". If the scalar product is positive definite, the algebra of these quantities associated with the points of length 2 closes, when the momenta are included, to form a finite dimensional Lie algebra. If the scalar product is positive semi definite, this algebra closes to an affine Kac-Moody algebra when the vertex operators for emitting "photons" are added. If the scalar product is Lorentzian, the algebra closes if the vertex operators for all the emitted states in the dual model are added. Special lattices in 10, 18, and 26 dimensional Lorentzian space are discussed and implications of the dual model no ghost theorem for these algebras are mentioned. This framework links many physical ideas, including concepts in magnetic monopole theory and the fermion-boson equivalence as well as the dual model. (Knowledge of dual models is not assumed but familiarity with aspects of the theory of Lie algebras is presumed in the latter part of this paper.)

[†] Partially supported by the National Science Foundation through the Mathematical Sciences Research Institute.

Vertex Operators in Mathematics and Physics - Proceedings of a Conference November 10-17, 1983. Publications of the Mathematical Sciences Research Institute #3, Springer-Verlag, 1984.

1. INTRODUCTION

Many of the ideas occurring in the quantum theory of the relativistic string (or dual model), such as supersymmetry or dimensional reduction, have found applications outside its immediate context. (For reviews of the subject, see the collection of ref. [1], or ref [2], for example.) Although exciting and important developments are still being made [3], interest in the theory was most intense in the period 1968 to 1974. The objectives of those working on it progressively broadened from the phenomenological description of high energy scattering (resonances and Regge behavior in particular) to the construction of a completely consistent theory of the strong, and possibly other, interactions. The demand for consistency led to the realization that each dual model or string theory should be considered in a particular space-time dimension, 26 for the original model of Veneziano and others and 10 for the theory of Neveu, Schwarz and Ramond, which includes fermions and introduced supersymmetry. Although formulated in spaces of nonphysical dimensions, these theories possess a high level of consistency and contain very rich algebraic structures; for instance, Yang-Mills gauge theory and supergravity appear as "subtheories" by taking suitable limits. One explanation for why these theories moved from the center of interest is that the technical difficulties presented by handling amplitudes with many fermions proved insuperable on the time scale that theoretical physicists usually expect to solve their problems. Nor could help be found in the mathematical literature because mathematicians were only just discovering the sort of structures which have to be exploited in order to gain an economical understanding of string theories. Subsequently the trade has so far been mainly the other way, with mathematicians taking advantage of the constructions made by physicists.

One of the ideas to have proved useful in mathematics is that of vertex operators $U(r,z)$, which are analytic operators functions of momentum r and a complex variable, z. They occur naturally in dual models and they, or rather their moments

$$(1.1) \qquad A_n(r) = \frac{1}{2\pi i}\oint\frac{dz}{z}z^n\, V(r,z)\ ,$$

are important in constructing representations [4] of affine Kac-Moody Algebras [5]. Here the "momenta" r are Euclidean vectors and correspond to the roots of a simply-laced Lie algebra. In dual theory, the $n = 0$ operators are particularly interesting because they create the physical states of the model by application to some basic state. This is because they commute with the operators, L_n, of the Virasoro algebra [6], which define gauge conditions in the model. They were crucial in the work which led to the proof of the absence of ghosts (i.e. physically unacceptable negative norm states) from the model [7-10].

In this paper we shall show how these physical state creation operators can be made to associate to any integral lattice a Lie algebra. [This describes an approach we developed last winter; after completing this work we learned of unpublished work of I.B. Frenkel which adopted a very similar point of view and who obtained some further results [11]. Our approach is expressed in a formalism which might be familiar to physicists.] If the construction is applied to a Euclidean Lattice Λ, we obtain a finite dimensional compact Lie algebra, g_Λ, which will be semi-simple if and only if the points of squared length 2 in Λ, Λ_2, span a space of the same dimension as Λ. In this case, if we extend the lattice to $\Lambda' = \Lambda\oplus\mathbb{Z}$ where the new direction is taken to be null, the construction applied to Λ' yields the affine Lie algebra $\tilde{g}_\Lambda$ associated with g_Λ.

For a lattice Λ in a real vector space V with an inner product (i.e. symmetric bilinear form), which is not necessarily positive definite, or even non-singular, the construction produces a Lie algebra of rank equal to the dimension of V, with roots corresponding to points v, with $v^2 \leqslant 2$, on the even lattice Λ_R generated by the set Λ_2 of point r of Λ with $r^2 = 2$. The root spaces corresponding to $r \in \Lambda_2$ have dimension one and, if the inner product is non-singular, those corresponding to non-zero $v \in \Lambda_R$ with $v^2 = 0$ have dimension at most dim Λ_R-2.

A particularly interesting lattice to consider is the twenty-six dimensional even Lorentzian lattice $II^{25,1}$. Conway and Sloane [12] have shown that one can take as a set of simple roots for this lattice a set of points, ℓ, isometric to the Leech Lattice. (A set of simple roots is a minimal set of vectors having the property that the reflections in the hyperplanes perpendicular to them generate the Weyl group of the lattice, that is the group of all reflections which are automorphisms of the lattice.) It has been suggested [13] that these simple roots can be used to construct an infinite rank Lie algebra L_∞, which might be related to the Fischer-Griess Monster group. Application of the construction to $II^{25,1}$ yields a representation for L_∞ which has rank 26. However it seems that a more elaborate setting [14] is necessary, which is in fact related to the vertices which describe the interactions of dual models with electromagnetic currents [15], in order to accommodate the action of the group.

In this case the algebra obtained has a direct interpretation in terms of the relativistic string theory. The lattice $II^{25,1}$ can be interpreted as the momentum space for a (particular) toroidal version of 26-dimensional Minkowski space. This is especially intriguing because the dimension 26 is a critical dimension for the relativistic string model. In higher dimensions the spectrum of physical states has unacceptable states with negative norms, which give rise to nonphysical negative probabilities. (In lower dimensions, it seems, at least in any interacting version of the theory, that the spectrum has other features which are at best unnattractive.) Here the construction gives rise to physical creation operators for the relativistic string. Thus they can be represented in the Fock space of physical states. This space has a non-negative inner-product. It is tempting to suppose that the root spaces can be identified with the physical state creation operators for the corresponding momentum. In fact Frenkel [11] has shown that one can set up an isomorphism between those root spaces and those physical states generated by the physical state creation operators from the single particle states. (These are not necesarily all the physical states for number theoretic reasons.) This puts bounds on the dimensions of the root spaces.

Our paper is organized as follows. In Section 2 we review the background information we shall need about lattices and we establish some notation. In Section 3 we give our construction of a Lie algebra associated with an integral lattice. In Section 4 we discuss the rank and the roots of this algebra. Then, in Section 5, we consider the relationship of this construction to the Virasoro algebra and the way that previous representations of affine Lie algebras are subsumed in it. The following section applies the construction to even Lorentzian lattices and, in particular, a Lie algebra L_∞ to which Conway, Queen and Sloane draw attention. In Section 7 we extend the construction of Section 3 to include anticommuting operators and in Section 8 we apply the idea of associating an algebra with a lattice to enable us to obtain information about self-dual lattices.

2. LATTICES AND THEIR PROPERTIES

To set the scene for our subsequent discussions we shall survey briefly the main facts about lattices that will be relevant for us (see e.g. [16]). Suppose that V is a real N-dimensional vector space with an inner-product, that is a symmetric bilinear form, not necessarily non-singular, denoted $x \cdot y$ for $x,y \in V$. We shall be principally interested in the cases where V is Euclidean space, $\mathbb{R}^N$, or Minkowski space, $\mathbb{R}^{N-1,1}$. We can define a *lattice* in V as a set of points of the form

$$(2.1) \qquad \Lambda = \{ \sum_{i=1}^{N} n_i \underline{e}_i : n_i \in \mathbb{Z} \}$$

where $\underline{e}_i$, $1 \leqslant i \leqslant N$, forms a basis for V; it is also called a basis for Λ. The lattice Λ will be *Euclidean* or *Lorentzian* in the cases that V is Euclidean or Minkowski space, respectively. It is said to be *unimodular* if $|\det(\underline{e}_i \cdot \underline{e}_j)| = 1$. This is equivalent to the condition that Λ should contain one point of V per unit volume. We can also consider the situation where Λ spans a subspace rather than the whole of V.

For any lattice $\Lambda \subset V$, we define the *dual* of Λ, denoted Λ^*, to be the set of points $y \in v$ for which $x \cdot y$ is integral for all $x \in \Lambda$. If Λ spans V, and the inner product is non-singular, Λ^* is a lattice, the *dual lattice* to Λ. In that case we can form a basis for Λ^* by taking the basis $\underline{e}_i^*$, $1 \leqslant i \leqslant N$, for V dual to $\underline{e}_i$, $1 \leqslant i \leqslant N$, so that $\underline{e}_i \cdot \underline{e}_j^* = \delta_{ij}$. The lattice Λ is *integral* if $x \cdot y$ is an integer for every $x, y \in \Lambda$. This is equivalent to the condition $\Lambda \subset \Lambda^*$. Since Λ and Λ^* have reciprocally related numbers of points per unit volume in V, the condition that Λ be both integral and unimodular is equivalent to the condition that it be self-dual, i.e. $\Lambda = \Lambda^*$.

A vector $x \in \Lambda$ is said to be *primitive* if x is not a multiple of any other point of the lattice, that $x \neq ny$ for any $y \in \Lambda$, $n \in \mathbb{Z}$ with $|n| \geqslant 2$. Having a basis it is easy to see that x is primitive if and only if $x \cdot x' = 1$ for some x' in the dual lattice Λ^*.

A lattice is the direct sum of lattices $\Lambda_i \subset \Lambda$, $1 \leqslant i \leqslant r$ if Λ_i is perpendicular to Λ_j, for $i \neq j$, and any $x \in \Lambda$ can be written in the form

$$x = \sum_{i=1}^{r} x_i, \quad x_i \in \Lambda_i \; . \tag{2.2}$$

A lattice is *indecomposable* if it can not be expressed as the direct sum of two non zero sublattices. A Euclidean lattice has a unique decomposition into a direct sum of indecomposable sublattices but, in spaces of indefinite signature, such decompositions are not unique, in general.

The simplest examples of self-dual lattices are the cubic lattices $\mathbb{Z}^{m,n}$, consisting of those points in $\mathbb{R}^{m,n}$ with integral coordinates (with respect to some orthonormal basis). Any integral Euclidean lattice is isomorphic to one of the form $\mathbb{Z}^m \oplus \Lambda$, for some m, where Λ contains no points of length one. So, in classifying integral Euclidean lattices, it is only necessary to consider those whose nonzero points have squared length at least 2. Amongst these are the <u>even</u> lattices, that is the integral lattices with $x^2 \in 2\mathbb{Z}$ for all x. Even lattices are sometimes called type II, whilst the remainder are called <u>odd</u> or type I. This terminology also applies to non-Euclidean lattices.

If we consider self-dual Euclidean lattices, the choice in low dimensions is limited (see, e.g. [17]). The first non-trivial example is in dimension 8, the root lattice of the group E_8, which is even.

The next is a weight lattice associated with D_{12}, the algebra of SO(24), and this is odd. Even self-dual Euclidean lattices only occur in dimensions which are multiples of 8. There are two in dimension 16 and 24 in dimension 24 where, for the first time there is a lattice whose minimum nonzero squared length is 4. This is called the Leech lattice.

The possibilities for self-dual Lorentzian lattices are even more limited. If Λ is an odd self-dual lattice in a space of indefinite signature, $\mathbb{R}^{m,n}$, it can be shown that it is isomorphic to $\mathbb{Z}^{m,n}$. Thus, in particular, there is just one odd self-dual Lorentzian lattice in each dimension. Even Lorentizan lattices exist only in dimensions of the form $N = 8n+2$, n an integer, and again such lattices are determined by their dimension. These lattices we denoted by $II^{N-1,1}$, and they can be defined as consisting of those x for which

$$\text{(2.3a)} \qquad \text{(i) either } x \in \mathbb{Z}^{N-1,1} \text{ or } x - \ell \in \mathbb{Z}^{N-1,1}$$

and

$$\text{(2.3b)} \qquad \text{(ii) } x \cdot \ell \in \mathbb{Z}$$

where $\ell = (\frac{1}{2},\frac{1}{2},\ldots,\frac{1}{2};\frac{1}{2})$.

We can understand the relationship between the constraints on the dimensions of even self-dual Euclidean and Lorentzian lattices by considering a light-like vector k in a Lorentzian lattice in Minkowski space V; so $k^2 = 0$ but $k \neq 0$. If V_k denotes the subspace orthogonal to k, consisting of $x \in V$ with $x \cdot k = 0$, V_k contains K_k the one-dimension of subspace of vectors parallel to k. The restriction to V_k of the inner product on V is singular but non-negative, with null space K_k. We can form a Euclidean space V_k/K_k of dimension two less than V, by identifying vectors of the form $x + \lambda k$ for different $\lambda \in \mathbb{R}$ and fixed $x \in V_k$. The length of such a vector $(x+\lambda k)^2$ equals x^2, independent of λ. We might as well take k to be primitive. Then $V_k \cap \Lambda$ defines a lattice, which, which becomes a Euclidean lattice Λ_k in V_k/K_k when we identify all points of the form $x + mk$

for different $m \in \mathbb{Z}$. If Λ is even and self-dual, then so is Λ_k, showing that dim $\Lambda = 8n+2$, given the result on the dimensionality of even self-dual Euclidean lattices.

Two vectors in a lattice λ are *equivalent* if they are related by an automorphism of the lattice, that is a (pseudo)-orthogonal transformation R of V such that $R(\Lambda) = \Lambda$. Clearly equivalent light-like vectors in a self-dual Lorentzian lattice Λ (i.e. $II^{8n+1,1}$ for some n) yield isomorphic self-dual Euclidean lattices Λ_k. Conversely, it is not sufficient to check that if isomorphic Euclidean lattices are obtained from different light vectors in Λ, the isomorphism can be extended to an automorphism of the whole lattice Λ, under which the light-vectors are equivalent. Further, any self-dual Euclidean lattice can be obtained in this way since, if Λ' is such a lattice, $\Lambda = \Lambda' \oplus II^{1,1}$ is a self-dual Lorentzian lattice and, if k $= (0,\ell)$ where $\ell = (\frac{1}{2};\frac{1}{2}) \in II^{1,1}$, $\Lambda_k = \Lambda'$. Hence the problem of classifying self-dual Euclidean lattices is equivalent to that of classifying the inequivalent primitive light-like vectors in $II^{8n+1,1}$. Hence, it follows from the results we quoted on self-dual Euclidean lattices, that there is just one such vector for n = 1, two for n = 2 but 24 for n = 3. Conway and Sloane have shown that, taking k to be the particular light-like vector

$$w = (0,1,2,\dots,24;70) \tag{2.4}$$

produces a Λ_k isomorphic to the Leech lattice.

3. THE LIE ALGEBRA ASSOCIATED WITH AN INTEGRAL LATTICE

We give a construction which associates to any integral lattice Λ, a Lie algebra g_Λ, of rank $N = \dim \Lambda$, which is finite dimensional if Λ is Euclidean. This construction also works in the more general circumstance that $\Lambda \subset \Lambda_R^*$ where Λ_R is the sublattice of Λ generated by the points Λ_2 of squared length 2 in Λ. If the inner product, on the vector space V containing Λ, is singular, or if Λ_R

does not span all of V, the requirement that $\Lambda \subset \Lambda_R^*$ places no restriction on the components, of points of Λ, orthogonal to points of Λ_R.

We define the algebra g_Λ by means of a Fock space representation. To this end we choose a basis for V with respect to which the metric tensor is $g^{\mu\nu}$. We introduce an infinite set of annihilation and creation operators α_m^μ, $m \in \mathbb{Z}$, $1 \leqslant r \leqslant N$, satisfying the commutation relations

(3.1) $$[\alpha_m^\mu, \alpha_n^\nu] = m g^{\mu\nu} \delta_{m,-n}$$

and the hermiticity condition

(3.2) $$\alpha_m^{\mu\dagger} = \alpha_{-m}^\mu$$

We interpret $p^\mu = \alpha_0^\mu$ as a momentum operator and introduce orthonormal vectors Ψ_γ, $\gamma \in \Lambda$, of momentum γ,

(3.3) $$p^\mu \Psi_\gamma = \gamma^\mu \Psi_\gamma,$$

(3.4) $$\langle \Psi_\gamma, \Psi_{\gamma'} \rangle = \delta_{\gamma\gamma'},$$

(3.5) $$\alpha_n^\mu \Psi_\gamma = 0, \qquad n > 0$$

The Fock space, $\mathcal{H}$, in which we work is that generated from the vectors Ψ_γ by the operators α_m^μ $(m < 0)$. It is the direct sum of the momentum eigenspaces $\mathcal{H}_\gamma = \{\psi \in \mathcal{H}: p\psi = \gamma\psi\}$.

We can introduce operators $e^{i\gamma\cdot q}$, $\gamma \in \Lambda$, which generate momentum

(3.6) $$e^{i\gamma\cdot q}\Psi_{\gamma'} = \Psi_{\gamma+\gamma'}.$$

Finally, we have the commutation relations

(3.7) $$[q^\mu, p^\nu] = i g^{\mu\nu}.$$

The whole of $\mathcal{H}$ is generated from the vacuum vector Ψ_0 by α^μ_m, $m<0$, and $e^{i\gamma\cdot q}$, $\gamma \in \Lambda$.

Now introduce

$$(3.8) \qquad Q^\mu(z) = q^\mu - ip^\mu \log z + i \sum_{n\neq 0} \frac{1}{n} \alpha^\mu_n z^{-n}$$

so that

$$(3.9) \qquad p^\mu(z) \equiv iz\frac{dQ^\mu}{dz} = \sum_n \alpha^\mu_n z^{-n}$$

Let

$$(3.10a) \qquad Q^\mu_>(z) = i \sum_{n>0} \frac{1}{n} \alpha^\mu_n z^{-n}, \qquad Q^\mu_<(z) = i \sum_{n<0} \frac{1}{n} \alpha^\mu_n z^{-n}$$

and

$$(3.10b) \qquad Q^\mu_0 = q^\mu - ip^\mu \log z.$$

Then the vertex operator of "momentum" r is defined by the normal ordered expression

$$U(r,z) = z^{r^2/2} : \exp\{ir\cdot Q(z)\}:$$

$$(3.11) \qquad = z^{r^2/2} \exp\{ir\cdot Q_<(z)\} e^{ir\cdot q} z^{r\cdot p} \exp\{ir\cdot Q_>(z)\}$$

$$= \exp\{ir\cdot Q_<(z)\}\exp\{ir\cdot Q_0(z)\}\exp\{ir\cdot Q_>(z)\}$$

Note in the normal ordered expression $Q_>$ is moved to the right of $Q_<$ and p to the right of q. Defined in this way $U(r,z)$, $r^2 = 2$ has a well defined action on $\mathcal{H}$ as $r\cdot\gamma \in \mathbb{Z}$ for $r \in \Lambda_2$, $\gamma \in \Lambda$ if $\Lambda \subset \Lambda^*_k$ and so there is no branch cut at $z = 0$. Note the hermiticity property

$$U(r,z)^\dagger = U(-r,1/z^*)$$

The physical state creation operator corresponding to the

vertex U is defined by

$$A_r = \frac{1}{2\pi i} \oint \frac{dz}{z} U(r,z) \tag{3.12}$$

with the integration contour positively encircling $z = 0$. It follows from the hermiticity property that

$$A_r^\dagger = A_{-r} \tag{3.13}$$

To investigate the commutation relations of the A_r, $r \in \Lambda_2$ we need to normal order $U(r,z)U(s,\zeta)$. This is a standard dual model calculation followed in evaluating particle scattering amplitudes for example. We find

$$:\exp[ir.Q(z)]::\exp[is.Q(\zeta)]: = \tag{3.14}$$

$$:\exp[ir.Q(z)+is\cdot Q(\zeta)]:(z-\zeta)^{r-s} \text{ for } |\zeta| < |z|$$

because

$$\exp[ir\cdot Q_>(z)]\exp[is\cdot Q_<(\zeta)] = \tag{3.15}$$

$$(1-\zeta/z)^{r\cdot s} \exp[is\cdot Q_<(\zeta)]\exp[ir\cdot Q_>(z)] \text{ if } |\zeta| < |z|$$

and

$$z^{r\cdot p}e^{is\cdot q} = e^{is\cdot q}z^{r\cdot p}\, z^{r\cdot s}$$

Note that the right hand side of equation (3.14) is defined by analytic continuation for all z,ζ $(z \neq 0,\ \zeta \neq 0,\ z \neq \zeta)$ and that simultaneous interchange of r with s and z with ζ changes it by a factor $(-1)^{r\cdot s}$. Hence

$$(3.16) \qquad A_r A_s - (-1)^{r\cdot s} A_s A_r =$$

$$= \frac{1}{(2\pi i)^2}\left[\oint_{|z|>|\zeta|}\frac{dz}{z}\oint\frac{d\zeta}{\zeta} - \oint_{|\zeta|>|z|}\frac{dz}{z}\oint\frac{d\zeta}{\zeta}\right] z^{r^2/2}\zeta^{s^2/2}$$

$$:\exp[ir\cdot Q(z)+is\cdot Q(\zeta)]:(z-\zeta)^{r\cdot s}$$

$$= \frac{1}{(2\pi i)^2}\oint_0\frac{d\zeta}{\zeta}\oint_\zeta\frac{dz}{z}\,(z-\zeta)^{r\cdot s}z^{r^2/2}\zeta^{s^2/2}:\exp[ir\cdot Q(z)+is\cdot Q(\zeta)]:$$

where the z integral is taken on a contour positively encircling ζ, excluding $z = 0$ and the ζ contour is then taken positively about $\zeta = 0$. The integrand is non singular at $z = \zeta$ if $r\cdot s \geqslant 0$, it has a simple pole if $r\cdot s = -1$, in which case $r+s \in \Lambda_2$ and a double pole if $r\cdot s = -2$ which happens in particular if $r = -s$. Thus

$$(3.17) \qquad A_r A_s - (-1)^{r\cdot s} A_s A_r = 0 \quad \text{if } r\cdot s \geqslant 0$$

$$(3.18) \qquad = A_{r+s} \quad \text{if } r\cdot s = -1$$

$$(3.19) \qquad = \frac{1}{2\pi i}\oint\frac{d\zeta}{\zeta}r\cdot P(\zeta) = r\cdot p \quad \text{if } r\cdot s = -2 \text{ and if } r+s = 0$$

In addition we have

$$(3.20) \qquad [p^\mu, A_r] = r^\mu A_r$$

Now we wish to modify equations (3.17) – (3.19) so they become commutators. This is done by introducing quantities C_r, $r \in \Lambda_R$, commuting with the previously mentioned oscillators and satisfying

$$\left.\begin{aligned} C_u C_v &= (-1)^{u\cdot v} C_v C_u \\ C_u C_v &= \epsilon(u,v) C_{u+v} \end{aligned}\right\} \quad u, v \in \Lambda_R$$

where $\epsilon(u,v)$ takes values ± 1. Such quantities can be constructed for any even integral lattice. Generalizations to odd integral lattices and more discussion will be found in the appendix. The C_u can be taken to be functions of momentum, $C_u\Psi_\gamma = \epsilon(u,\gamma)\Psi_\gamma$, so that we are not increasing the size of the representation space, or to be some generalisation of Dirac γ matrices.

If we set

$$(3.21) \qquad e(r) = A_r C_r \qquad r \in \Lambda_2$$

we find

$$(3.22a) \qquad [e(r),e(s)] = 0 \qquad \text{if } r\cdot s \geq 0$$

$$(3.22b) \qquad = \epsilon(r,s)e(r+s) \qquad \text{if } r\cdot s = -1$$

$$(3.22c) \qquad = r\cdot p \qquad \text{if } r = -s$$

Further it follows from eqn. (3.20) that

$$(3.23) \qquad [p^\mu,e(r)] = r^\mu e(r)$$

The algebra g_Λ associated with the lattice Λ is defined to be the Lie algebra generated by $e(r)$, $r \in \Lambda_2$ p^μ, $1 \leq \mu \leq N$. In the given representation, the generators have the hermiticity property

$$(3.24) \qquad e(r)^\dagger = e(-r) \qquad p^{\mu\dagger} = p^\mu$$

If Λ_R is Euclidean, we always have $(r\pm s)^2 \geq 0$ with zero only if $r\pm s = 0$. Hence the cases covered by equation (3.22) exhaust all the possibilities. In this case $e(r)$, $r \in \Lambda_2$, p^μ, $1\leq\mu\leq N$ close under commutation and form a basis for g_Λ. It is semisimple if and only if $\dim \Lambda_R = \dim \Lambda$ and simple if in addition Λ_R is indecomposable. The semisimple part of g_Λ is always simply laced (i.e. has roots of equal length).

From equations (3.6) and (3.11) we find

(3.25) $$A_r\Psi_\lambda = \oint\frac{dz}{2\pi i z}\, z^{r^2/2 + r\cdot\lambda}\, \exp[ir\cdot Q_<(z)]\, \Psi_{\lambda+r}$$

As $Q_<(z)$ contains only positive powers of z we see, given $r^2 = 2$, that this expression vanishes if and only if

(3.26) $$r\cdot\lambda \geqslant 0$$

When g_Λ is a semisimple finite dimensional Lie algebra all its finite dimensional irreducible representations possess unique "highest weight states" annihilated by the step operators for positive roots. The corresponding weight, which is dominant, can be used to label the irreducible representation. We deduce in this case from equation (3.25) and (3.26) that our construction (3.11) furnishes a reducible representation which includes in its decomposition all irreducible representations of g_Λ whose highest weights occur in Λ.

More generally if Λ_R is Euclidean, g_Λ is the Lie algebra of a compact Lie group G. A specific global structure is obtained for G when the antihermitian linear combinations of the generators are exponentiated. This global form is the one for which Λ is the lattice of weights.

Finally let us complete the calculation of $A_rA_s - (-1)^{r\cdot s}A_sA_r$ when $(r+s)^2$ vanishes without r+s necessarily vanishing. Of course this can only happen if Λ is not Euclidean. Because $r\cdot s = -2$ we have

$$A_rA_s - A_sA_r = \oint\frac{d\zeta}{2\pi i}\frac{d}{dz}: \exp[ir\cdot Q(z) + is\cdot Q(\zeta)]:\Big|_{z=\zeta}$$

$$= \oint\frac{d\zeta}{2\pi i\zeta} : r\cdot P(\zeta)\, \exp[i(r+s)Q(\zeta)]:$$

by (3.9). This does not appear to exhibit the required antisymmetry in r and s until we realize that

$$\frac{1}{\zeta}\left(\frac{r+s}{2}\right): P(\zeta)\exp[i(r+s)Q(\zeta)]:$$

is a total derivative and therefore integrates to zero. Hence

$$(3.27) \quad A_rA_s - A_sA_r = \oint \frac{dz}{2\pi iz} \left(\frac{r-s}{2}\right) \cdot P(z) \exp\{i(r+s)Q(z)\}$$

Note that normal ordering has been omitted. It is unnecessary as

$$(3.28) \qquad (r+s)^2 = 0, \qquad (r-s)\cdot(r+s) = 0$$

so that all quantities in (3.27) mutually commute. The operator on the right side of (3.27) is known as the DDF operator (7) and played an important role in dual theory (8,9,10). It is the integral of the "photon emission vertex." The polarization of the "photon" is $\frac{r-s}{2}$ and is automatically transverse to the photon momentum $(r+s)$ by (3.28) as it must be in a physical situation.

4. THE RANK AND ROOTS OF g_Λ

We have seen that, in the Euclidean case, the rank of g_Λ is $N = \dim \Lambda$ and its roots are just the points of Λ_2. We shall now establish that, in the non-Euclidean case, its rank is still N and its roots are points $v \in \Lambda_R$ with $v^2 \leqslant 2$.

The algebra g_Λ is spanned by elements of the form

$$(4.1) \qquad e' = [e(r_1), [e(r_2),\ldots[e(r_{m-1}),e(r_m)]\ldots]]$$

where each $r_i \in \Lambda_2$ (together with the p^μ, which are necessary if $\dim \Lambda_R < \dim \Lambda$, which we shall call the non-semi-simple case). Then e' belongs to the simultaneous eigenspace of the p labelled by

$$(4.2) \qquad v = \sum_{i=1}^{m} r_i.$$

We first show that $e' = 0$ if $v^2 > 2$. To this end we calculate

$$U_{r_1,r_2,\ldots r_m}(z_1,z_2,\ldots z_m) = U(r_1,z_1)U(r_2,z_2)\ldots U(r_m,z_m)$$

$$(4.3) \qquad = \exp\{i\sum_{j=1}^{m} r_jQ_<(z_j)\}e^{ir\cdot q}\prod_{j=1}^{m} z_j^{r_j\cdot p+1} \exp\{i\prod_{j=1}^{m} r_jQ_>(z_j)\}$$

$$\prod_{i<j}(z_i-z_j)^{r_i\cdot r_j}$$

$$\text{if } |z_1| > |z_2| > \ldots > |z_m|$$

Arguments similar to those used in eq. (3.14), show that the multiple commutator (4.1) equals, up to a sign,

$$(4.4)\qquad \frac{1}{(2\pi i)^m}\oint_0 \frac{dz_m}{z_m}\oint_{z_n}\frac{dz_{m-1}}{z_{m-1}}\ldots\oint_{z_2}\frac{dz_1}{z_1}\,U_{r_1,r_2,\ldots,r_m}(z_1,z_2,\ldots,z_m)$$

Here the integration contours are such that z_j encircles z_{j+1} positively excluding z_i for $j+1 < i \leq m+1$, where $z_{m+1} = 0$. Now since A_r takes a finite occupation number state (i.e. one of the form $\alpha_{-m_1}^{\mu_1}\alpha_{-m_2}^{\mu_2}\ldots\alpha_{-m_M}^{\mu_M}\Psi_\gamma$) into a finite linear combination of such states, to establish that e' vanishes, we need only prove that all the matrix elements of (4.4) between such states vanish. These matrix elements have the form

$$(4.5)\qquad \frac{1}{(2\pi i)^m}\oint_0\frac{dz_m}{z_m}\oint_{z_m}\frac{dz_{m-1}}{z_{m-1}}\ldots\oint_{z_1}\frac{dz_1}{z_1}\prod_{i<j}(z_i-z_j)^{r_i\cdot r_j}$$

$$f(z_1,z_2,\ldots,z_m)$$

where f is analytic provided that all the z_i are nonzero and finite. It is easy to see that the integral over z_{m-1} (or one of the previous integrals over z_i, $1 \leq i \leq m-2$) will vanish unless

$$(4.6)\qquad \sum_{i<j} r_i\cdot r_j \leq -(m-1),$$

that is

$$(4.7)\qquad v^2 = 2m + 2\sum_{i<j} r_i\cdot r_j \leq 2,$$

our desired result.

Any element of g_Λ commuting with p^μ, $1 \leq \mu \leq N$, would have to be a linear combination of them together with expressions of the form (4.1) with $v = 0$, or, equivalently of expressions of the form $[e(-v),e']$ where $v^2 = 2$. In this case the integral of eq. (4.5) is

(4.8) $$\frac{\lambda}{2\pi i} \oint \frac{dz}{z} f(z,z,\ldots,z)$$

where λ is a constant (possibly zero) depending only on $r_1,r_2,\ldots,r_n$. We see this as follows. In evaluating the integral we take residues at the poles $(z_i-z_j)^{r_ir_j}$. If we have a multiple pole we either differentiate f or the remaining terms in the product $\Pi(z_i-z_j)^{r_i \cdot r_j}$. If we have differentiated f M times the integral over z_{m-1} has a factor $(z_{m-1}-z_m)^L$ where

(4.9) $$\begin{aligned} L &= \sum_{i<j} r_i \cdot r_j + (m-2) + M \\ &= \tfrac{1}{2}v^2 + M-2 = M-1 \end{aligned}$$

so to have any pole left we must have $M = 0$; all derivatives must be applied to the product $\Pi(z_i - z_j)^{r_i \cdot r_j}$ giving the result (4.8). Thus, in this case, e' equals $\lambda e(v)$ up to a sign. From this we see that any element of g_Λ commuting with the p^μ is a linear combination of them. In consequence, the rank of g_Λ is dim Λ and the roots of g_Λ are vectors of Λ_R with $v^2 \leq 2$. If g_Λ^v is the corresponding root space, we have established that

(4.10) $$\dim g_\Lambda^v = 1 \quad \text{if } v^2 = 2$$

If we consider those e' with $v^2 = 0$, arguments of the sort we have just used show that e' is of the form

(4.11) $$e' = \xi \cdot a(v)C_v$$

where

$$(4.12)\qquad a^{\mu}(k) = \frac{1}{2\pi i}\oint \frac{dz}{z} : P^{\mu}(z)\exp\{ik\cdot Q(z)\}:$$

for some vector $\xi \in V$, or rather that subspace spanned by Λ_R. We shall show in the next section that

$$(4.13)\qquad \xi\cdot v = 0$$

(as in equation (3.27) and (3.28). Further, since $v\cdot a(v) = 0$,

$$(4.14)\qquad \dim g_{\Lambda}^{v} \leqslant \dim \Lambda_R - 2 \quad \text{if } v^2 = 0$$

and, if Λ_R is irreducible, we would expect equality. (In terms of dual theory, $a^{\mu}(k)$ are physical creation operators for "photons" and eq. (4.13) the transversality condition on the "photon polarization" vector ξ.)

5. THE VIRASORO ALGEBRA AND AFFINE LIE ALGEBRAS

The Virasoro Algebra

If the inner product on V is non-singular, we can define a representation of the Virasoro algebra in $\mathcal{H}$ in the familiar way, by introducing the operators

$$(5.1)\qquad L_n = \tfrac{1}{2}: \Sigma\, \alpha_m\cdot\alpha_{n-m}:$$

It is the algebra with basis L_n, $n \in \mathbf{Z}$, together with the central element 1; they satisfy the commutation relations

$$(5.2)\qquad [L_m,L_n] = (m-n)L_{m+n} + \frac{N}{12} m(m^2-1)\delta_{m,-n}\ .$$

The vertex operators U(r,z) were constructed specifically so that they have the property

$$(5.3)\qquad [L_n,U(r,z)] = \{z^{n+1}\frac{d}{dz} + n\frac{r^2}{2}z^n\}U(r,z)$$

and $r^2 = 2$ has been chosen because then

$$(5.4) \qquad [L_n, e(r)] = \frac{1}{2\pi i} \oint \frac{d}{dz} \{z^n U(r,z)\} C_r = 0$$

Also L_n clearly commutes with p^μ so that the subspace $\mathcal{H}_\lambda$ of $\mathcal{H}$ defined by the equations

$$(5.5) \qquad L_0 \psi = \lambda \psi, \qquad L_n \psi = 0, \qquad n > 0 ,$$

is an invariant subspace for g_Λ. In particular cases $\mathcal{H}_\lambda$ may give nontrivial irreducible representations.

We can use the fact that e', as defined by eq. (4.11) commutes with L_n to establish eq. (4.13). Calculations [7] shows that

$$(5.6) \qquad [L_n, a^\mu(k)] = \frac{n^2}{2} k^\mu \oint \frac{dz}{z} z^n \exp\{ikQ(z)\},$$

where normal ordering is actually unnecessary because $k^2 = 0$, so that $\xi \cdot a$ commutes with L_n if and only if eq. (4.13) holds.

<u>Affine Lie algebras</u>

In the Euclidean case, we can extend g_Λ to an affine Lie algebra $\tilde{g}_\Lambda$ by using the Frenkel-Kac construction [4], which amounts to taking other moments of the vertex operator U(k,z) and which we shall describe next. However such algebras can also be obtained directly from our construction by enlarging the lattice Λ to a lattice Λ' and we shall describe this afterwards.

We can construct $\tilde{g}_\Lambda$ by defining[3],

$$(5.7) \qquad e_n(r) = \frac{1}{2\pi i} \oint \frac{dz}{z} z^n U(r,z) C_r$$

for $r \in \Lambda_2$. It has a basis consisting of $e_n(r)$, $r \in \Lambda_2$, $n \in \mathbb{Z}$; $p^\mu_n = \alpha^\mu_n$ $1 \leqslant \mu \leqslant N$, $n \in \mathbb{Z}$ and the central element 1. We then find

$$(5.8) \qquad [p^\mu_n, p^\nu_n] = m g^{\mu\nu} \delta_{m,-n}$$

(5.9) $$[p^{\mu}_m, e_n(r)] = r^r e_{m+n}(r)$$

(5.10a) $$[e_m(r), e_n(s)] = 0 \qquad \text{if } r\cdot s \geq 0$$

(5.10b) $$= \epsilon(r,s)\, e_{m+n}(r+s) \qquad \text{if } r\cdot s = -1$$

(5.10c) $$= r\cdot p_{m+n} + m\delta_{m,-n} \qquad \text{if } r = -s$$

Since we have assumed that Λ_k is Euclidean, these exhaust the possibilities for $r\cdot s$ and the algebra closes.

We also have

(5.11) $$[L_m, e_n(r)] = -ne_{m+n}(r)$$

and

(5.12) $$e_n(r)\Psi_{\gamma} = 0 \text{ if and only if } n+\gamma\cdot r \geq 0$$

by extending the argument leading to (3.26). Making a particular choice of positive roots for g_{Λ}, the positive root spaces of $\tilde{g}_{\Lambda}$ are spanned by $e_0(r)$, $r > 0$; α^{μ}_n, $e_n(s)$, $n > 0$. We see from eq. (5.12) that the states in $\mathcal{H}$ with momentum of the form $\gamma+s$ will give a representation space for $\tilde{g}_{\Lambda}$ with highest weight vector Ψ_{γ} if and only if γ is a minimal fundamental weight [18] (or zero).

Eqs. (5.10) still hold even if Λ_k is not Euclidean, and then the $e_n(r)$, p^{μ} generate an algebra $\tilde{g}_{\Lambda} \supset g_{\Lambda}$.

To obtain $\tilde{g}_{\Lambda}$ from the construction of Section 3 as an algebra $g_{\Lambda'}$ take Λ' to be a lattice consisting of the points $x+nk$, $x \in \Lambda$, $n \in \mathbb{Z}$, in a vector space $V' = V\oplus K_k$ where K_k denotes a one-dimensional space spanned by a null vector k. Since $(x+nk)^2 = x^2$, Λ_2' consists of all the points $r + nk$, $r \in \Lambda_2$, $n \in \mathbb{Z}$. The Lie algebra $g_{\Lambda'}$ is generated by elements $e(r+nk)$, $r \in \Lambda_2$, $n \in \mathbb{Z}$, and p^{μ}, $q \leq \mu \leq N+1$. Writing

(5.13) $$\hat{e}_n(r) = e(r+nk)$$

and

$$\hat{p}^{\mu}_n = \frac{1}{2\pi i}\oint P^{\mu}(z)\ \exp[ink\cdot Q(z)], \tag{5.14}$$

so that $k\cdot\hat{p}_n = k\cdot p\,\delta_{n0}$, we have the algebra

$$[\hat{p}^{\mu}_m,\hat{p}^{\nu}_n] = mg^{\mu\nu}\delta_{m,-n}k\cdot p \tag{5.15}$$

$$[\hat{p}^{\mu}_m,\hat{e}_n(r)] = r^{\mu}\hat{e}_{m+n}(r) \tag{5.16}$$

$$[\hat{e}_m(r),\hat{e}_n(s)] = 0 \qquad \text{if } r\cdot s \geq 0 \tag{5.17}$$

$$= \epsilon(r,s)\hat{e}_{m+n}(r+s) \qquad \text{if } r\cdot s = -1 \tag{5.18}$$

$$= r\cdot\hat{p}_{m+n} + mk\cdot p\,\delta_{m-n} \qquad \text{if } r = -s \tag{5.19}$$

(Note that $g^{\mu\nu}$ is now singular and $k_{\mu}g^{\mu\nu} = 0$.) Thus we see that we have regained the affine Lie algebra of eqs. (5.8–10) with $e_n(r)$ replaced by $\hat{e}_n(r)$, p_m by $\hat{p}_m$ and the central element 1 by $k\cdot p$.

In the expressions for $\hat{e}_n(r)$, $\hat{p}^{\mu}_n$, $k\cdot\alpha_n$ commutes with everything. Therefore we can take the quantities $k\cdot\alpha_n$ to be constant without altering the commutation relation (5.15–5.19). The substitution

$$k\cdot p \longrightarrow 1 \quad k\cdot\alpha_n \longrightarrow 0 \quad \text{if } n \neq 0 \tag{5.20}$$

explicitly yields the Kac–Frenkel construction (5.7).

Note that the derivation d does not occur in our construction unless we extend the lattice to a Lorentzian one containing a vector $\bar{k}$ satisfying $\bar{k}\cdot r = 0$, $\bar{k}\cdot k = 1$ and taking

$$d = \bar{k}\cdot p$$

Again we have equality between the rank of the algebra and the dimension of the lattice.

6. A LORENTZIAN ALGEBRA

If we take $\Lambda = II^{25,1}$, g_Λ is a representation of the algebra L_∞ introduced by Conway, Queen and Sloane [13], which they conjectured might be a natural setting for the Fischer-Griess monster group [19]. They defined L_∞ by means of a set of relations. For this purpose, let ℓ denote the points of Λ satisfying $r \cdot w = -1$ and $r^2 = 2$ where w is the light-like vector of eq. (2.4). These point are called Leech roots and the set ℓ is isometric to the Leech lattice. Then L_∞ has three generators e_r, e_{-r} and h_r for each $r \in \ell$, satisfying the relations

(6.1) $$[h_r, e_s] = r \cdot s e_s \qquad r \in \ell,\ \pm s \in \ell;$$

(6.2) $$[e_r, e_{-s}] = h_r \delta_{rs} \qquad r,s \in \ell;$$

(6.3) $$[h_r, h_s] = 0 \qquad r,s \in \ell;$$

(6.4) $$(\mathrm{ad}\, e_r)^{1-r \cdot s}(e_s) = 0$$

for $r \neq s$ and either both $r,s \in \ell$ or both $-r, -s \in \ell$. (Here $\mathrm{ad}\, e_r(x) = [e_r, x]$.)

Taking $e_r = e(r)$ as defined by eq. (3.14) and $h_r = r \cdot p$ clearly provides a representation of eqs. (6.1-3). Further, if $r \neq s$,

(6.5) $$(s + r(1 - r \cdot s))^2 = (r-s)^2 \geqslant 4$$

for $r,s \in \ell$, because the minimum squared distance between two points of the Leech Lattice is 4. Thus since the roots v of g_Λ have $v^2 \leqslant 2$ we see that eq. (6.5) also holds. Because the e_r, $\pm r \in \ell$, generate g_Λ, it provides a representation of L_∞, of rank 26. This reduction in rank corresponds to process, advocated by Conway <u>et. al.</u>, of setting $\sum n_i h_{r_i} = 0$ whenever $\sum n_i r_i = 0$.

Taking k to be any of the 23 other inequivalent light-like vectors in Λ, g_{Λ_k} provides us with a simply-laced finite dimensional subalgebra of g_Λ associated with the 24-dimensional Euclidean lattice Λ_k. Further taking the lattice $\Lambda_k' = \{x \in \Lambda: x\cdot k = 0\}$ gives us the corresponding affine Lie algebra $g_{\Lambda_k'} = \tilde{g}_{\Lambda_k}$ in Section 5 with $g_{\Lambda_k'} \subset g_{\Lambda_k} \subset g_\Lambda$.

More generally, consider g_Λ where Λ is the even Lorentzian lattice $II^{N-1,1}$. The representation in $\mathcal{H}$ involves negative norm states. If we wish to avoid these dual theory suggests we consider the invariant subspaces $\mathcal{H}$ defined by eqs. (5.5). In fact we have the following result [9,10]

(6.6) $\langle\psi,\psi\rangle \geqslant 0$ for all $\psi \in \mathcal{H}_\lambda$ if and only if $N \leqslant 26$ and $\lambda = 1$

So we can provide a non-negative representation of g_Λ for $\Lambda = II^{25,1}$ by using $\mathcal{H}_1$. (Thorn has given an alternative proof of the "no ghost" theorem (6.6) based on a formula of Kac [20] of which a proof is given in ref [21].)

Frenkel [11] has shown that the adjoint representation of g_Λ is isomorphic to its action on the subspace of $\mathcal{H}$, generated from the states Ψ_γ with $\gamma^2 = 2$ so that the root spaces of g_Λ with root vector v, E_v, have

$$\dim E_v \leqslant d(-\tfrac{1}{2}v^2) \qquad (6.7)$$

where

$$q^{-1}\prod_n(1-q^n)^{-24} = \sum_{m=-1}^{\infty} d(m)q^m. \qquad (6.8)$$

7. OPERATORS ASSIGNED TO LATTICE POINTS OF UNIT LENGTH.

So far we have assigned operators to the points Λ_2 of the lattice Λ (i.e. those with $r^2 = 2$). Now we shall see that this can also be done for the points of $\Lambda_1 = \{r \in \Lambda: r^2 = 1\}$, by virtually the same construction. The difference is that the new operators will tend to anticommute rather than commute and so resemble a generalized Clifford algebra. For ordinary orthogonal Lie groups the generators can be represented as bilinear in gamma matrices (This is how Dirac established the Lorentz covariance of his relativistic electron equation). The analogue of this holds for certain affine Kac-Moody algebras [22] and the basic identities so obtained provide the basis for Skyrme's fermion-boson equivalence theorem [23], obtained by him and others [24] in the context of the Sine-Gordon-massive Thirring model. A difference is that in the Kac-Moody case there are two sorts of Clifford field, recognized by physicists as the "Neveu-Schwarz" [25] and "Ramond" [26] fields of the spinor dual string model. This difference can be understood in terms of the structure of the Dynkin diagrams for the even dimensional orthogonal group and the affine Kac-Moody algebra based on it.

In fact all our previous equations (3.11) - (3.20) all hold good whether r^2 and s^2 equal 1 or 2, provided the action of A_r on $\mathcal{H}$ is well defined. The dangerous factor in U(r,z) which has to be single valued is

$$z^{r^2/2 + r\cdot p} \tag{7.1}$$

When $r^2 = 2$ this requires $p \in \Lambda_R^*$, as we said in section 3. When $r^2 = 1$ the discussion is more complicated. First let us suppose Λ is Euclidean. Then, if e and f $\in \Lambda_1$, $|e.f| \leq 1$ so that Λ_1 must consist of a set $\pm e_i$ where $e_i \cdot e_j = \delta_{ij}$. This is suggestive of the weights of the 2N dimensional representation of D_N and we shall suppose this is g_Λ (For D_4 we could also have the weights of one of the spinor representations). The weight lattice of D_N splits into four cosets with respect to Λ_R, namely

$$\Lambda_R, \quad (\lambda_v + \Lambda_R), \quad (\lambda_s + \Lambda_R), \quad (\lambda_{\bar{s}} + \Lambda_R)$$

where λ_v, λ_s, $\lambda_{\bar{s}}$ are the fundamental (minimal) weights defining the vector (2N dimensional), and the two inequivalent spinor representations respectively. They correspond to the points of the Dynkin diagram of D_N as indicated.

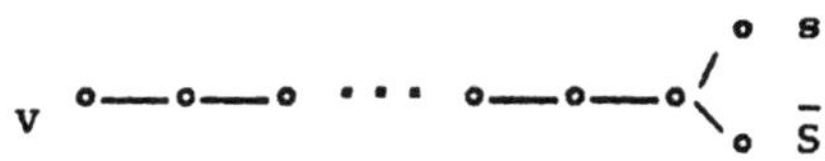

The requirement that (7.1) and hence U(r,z) be single value when $r \in \Lambda_1$ is that $r \cdot p \in \mathbb{Z} + \frac{1}{2}$, i.e.

$$p \in (\lambda_s + \Lambda_R) \cup (\lambda_{\bar{s}} + \Lambda_R) \tag{7.2}$$

This is invariant with respect to addition by $\Lambda_R \cup (\lambda_v + \Lambda_R)$, that is, for D_N, the quotient group of Λ_w by $\Lambda_R \cup (\lambda_v + \Lambda_R$ is the $\mathbb{Z}_2$ group with generators

$$(\Lambda_R + \lambda_s) \cup (\Lambda_R + \lambda_{\bar{s}})$$

This guarantees that A_e $(e^2 = 1)$ is well defined on any state constructed from a product of A_r's, $(r^2 = 1$ or $2)$ operating on Ψ_γ with γ satisfying (7.2).

Now let us adopt the convention $e,f \in \Lambda_1$; $r,s, \in \Lambda_2$. Then we see from (3.17) and (3.18) that

$$(7.3) \quad \begin{cases} A_e^2 = 0 & \\ A_e A_f - A_f A_e = 0 & e \cdot f = 0 \\ A_e A_{-e} + A_{-e} A_e = A_0 = 1 & \end{cases}$$

and

$$(7.4) \quad \begin{cases} A_e A_r - (-1)^{e \cdot r} A_r A_e = 0 & r \cdot e = 0 \text{ or } 1 \\ A_e A_r + A_r A_e = A_{r+e} & r \cdot e = -1 \end{cases}$$

This covers all possibilities in Euclidean space. As explained in the appendix the left hand side of equations (7.3) can be converted into the form of anticommutators (and the left hand side of equations 7.4 into the form of commutators) by a similar construction to the previous one. Then (7.3) becomes the algebra of N fermionic oscillators and can be rewritten as a D_N Clifford algebra by defining

$$(7.5)\quad \gamma_i = A_{e_i} + A_{-e_i}, \quad \gamma_{i+N} = \frac{A_{e_i} - A_{-e_i}}{i} \quad : \ i = 1 \ldots N$$

Now let us discuss the singular lattice obtained by adding to the weight lattice of D_N a single, orthogonal, light-like vector k. Our construction (3.11) applied to Λ_2 yields the Frenkel-Kac constructions for the D_N Kac-Moody algebra on substituting (equation 5.20)

$$(7.6)\qquad k \cdot p \longrightarrow 1 \qquad k \cdot \alpha_n \longrightarrow 0 \quad n \neq 0$$

Λ_1 consists of the points $\pm e_i + nk$, $n \in \mathbf{Z}$, so that making the same substitution (7.6) in our construction (3.11) applied to Λ_1 yields

$$(7.7)\qquad A_{e+nk} \longrightarrow B^n_e = \oint \frac{dz}{2\pi i z} z^{n+\frac{1}{2}} : \exp\{ie \cdot Q(z)\}:$$

of course

$$(B^n_e)^\dagger = B^{-n}_{-e}$$

By equations (3.17) and (3.18) we obtain

$$(7.8)\qquad \left.\begin{array}{ll} B^n_e B^m_e + B^m_e B^n_e = 0 & \\ B^n_e B^m_f - B^m_f B^n_e = 0 & e \cdot f = 0 \\ B^n_e B^{-m}_{-e} + B^{-m}_{-e} B^n_e = \delta_{nm} & \end{array}\right\}$$

Let us discuss the single valuedness of the integrand in (7.7). One possibility generalizes the previous one (7.2)

$$(7.9)\qquad n \in \mathbf{Z}, \quad e_i \cdot p \in \mathbf{Z}+\tfrac{1}{2} \quad \text{so} \quad p \in (\Lambda_R+\lambda_s) \cup (\Lambda_R+\lambda_{\bar{s}})$$

Another is to alter what we said above and let n be a half integer:

$$(7.10) \qquad n \in \mathbb{Z}+\tfrac{1}{2}, \quad e_i p \in \mathbb{Z} \quad \text{so} \quad p \in \Lambda_R \cup (\Lambda_R + \lambda_v)$$

Equations (7.8) apply in both cases and when we convert them into anticommutators and make the substitutions generalizing (7.7) we obtain quantities known to physicists as respectivly the Ramond (7.9) and Neveu-Schwarz (7.10) oscillators, [25,26].

These occur in the spinor dual string model in the context of fermionic and bosonic excitations of the string respectively. More precisely we define

$$(7.11) \qquad B_e(z) = \sum_{r \in \mathbb{Z}+1/2} z^{-r} B_e^r = \sqrt{z} :e^{ieQ(z)}: .$$

and the Neveu-Schwarz field is $H_i(z)$, $i = 1 \ldots 2N$ where

$$H_i(z) = \begin{cases} B_{e_i}(z) + B_{-e_j}(z) & i = 1 \ldots N \\ (B_{e_i}(z) - B_{-e_i}(z))/i & i = N+1, \ldots 2N \end{cases}$$

This field and the correspondingly defined Ramond fields are related by an "intertwining operator" called the "fermion emission vertex" which is much studied in the literature, [27].

We have seen that the generators of the D_N Kac-Moody algebra in the vertex operator representation can only act in highest weight representations whose highest D_N weight is either 0 or one of the D_N minimal weights λ_v, λ_s or $\lambda_{\bar{s}}$. These weights correspond to one of the four tips of the extended Dynkin diagram:

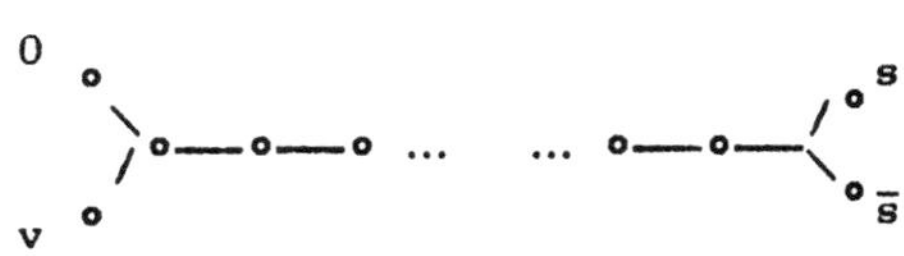

The Neveu-Schwarz oscillators act in the representations defined by 0 and λ_v while the Ramond oscillators act in the representations defined by λ_s and $\lambda_{\bar{s}}$. The fact that the affine Kac-Moody algebra has two "Clifford algebra" representations as compared to the single one for the D_N algebra is evidently due to the enhanced symmetry of the extended Dynkin diagram relative to the ordinary one, drawn above (7.2).

Let us concentrate on the Neveu-Schwarz oscillators. By (3.25) we find

$$B_e^r \Psi_0 = 0 \qquad r \geqslant 1/2\,, \quad \text{all } e^2 = 1.$$

Thus we can think of $B_e^r (r \geqslant 1/2)$ as destruction operators and B_e^r $(r \leqslant -1/2)$ as creation operators and hence define a new normal ordering operation denoted with open dots. From equation (7.8) we find

$$(7.12)\quad B_e(z)B_f(\zeta) = {}^\circ_\circ B_e(z)B_f(\zeta){}^\circ_\circ + \frac{\sqrt{z\zeta}}{z-\zeta}\,\delta_{e+f,0} \qquad |\zeta|<|z|$$

Hence using (7.8)

$$(7.13)\quad \begin{aligned} B_e(z)B_e(z) &= {}^\circ_\circ B_e(z)B_e(z){}^\circ_\circ = 0 \\ B_e(z)B_f(z) &= {}^\circ_\circ B_e(z)B_f(z){}^\circ_\circ = :\exp\{i(e+f)\cdot Q(z)\}: \quad \text{if } e\cdot f = 0 \end{aligned}$$

while if e = -f we have by (7.11), (7.12) and (3.14)

$${}^\circ_\circ B_e(z)B_{-e}(\zeta){}^\circ_\circ = \sqrt{z\zeta}\,(:\exp\{ie(Q(z) - Q(\zeta))\}: - 1)/(z-\zeta)$$

Taking $z \to \zeta$ using l'Hôpital's rule we find

$$(7.14)\qquad {}^\circ_\circ B_e(z)B_{-e}(z){}^\circ_\circ = iz\frac{d}{dz}e\cdot Q(z) = e\cdot P(z)$$

by (3.9). Equations (7.13) and (7.14) show us how the D_N Kac-Moody algebra generators can be expressed as bilinear in the Neveu-Schwarz fields, at least in the representation with highest weights 0 and λ_v. Of course the Ramond fields must be used in the other two representations [22].

Identities (7.11) and (7.14) are also remarkable in relating anticommuting fields B to commuting fields Q. In fact they are the basis for the boson-fermion equivalence originally discovered by Skyrme [23]. In two dimensional field theories there are two identities of the above form one for each light like component.

We think it remarkable that the Kac-Moody theoretical framework enables us to unify such apparently disparate physical ideas.

8. SELF-DUAL LATTICES

Interactions between elementary particles are believed to be controlled by a gauge group, whis is some compact Lie group. The fields describing particles transform according to some representation of the gauge group and the corresponding particles carry quantum numbers which are the weights of the group in that representation. Thus physicists are led to consider weight lattices and root lattices. It has been shown that the possible magnetic monopoles, that can occur in gauge theoreies, carry magnetic quantum numbers that correspond to points of a lattice dual to the weight lattice of the gauge group [28]. It has been conjectured that there may exist a gauge field theory of such monopoles, whose weight lattice is this dual lattice [29]. An interesting symmetry arises if the weight lattice and its dual are the same. The physical ideas are highly speculative but they provide some physical motivation for the study of self-dual lattices.

Euclidean integral lattices

Consider an integral Euclidean lattice Λ and the even lattice Λ_R spanned by the point of Λ with squared length 2, i.e. Λ_2. We shall say Λ is saturated if $\dim \Lambda_R = \dim \Lambda$. According to the construction of section 3, there is a simply laced Lie algebra g_Λ whose root lattice is Λ_R and whose root system is Λ_2. If Λ is saturated, g_Λ is semisimple and

$$\Lambda_R \subset \Lambda \subset \Lambda^* \subset \Lambda_R^* = \Lambda_W \tag{8.1}$$

where Λ_W denotes the weight lattice of the simply-connected Lie group with Lie algebra g_Λ.

Semisimple Lie algebras g are classified; each one is the algebra of a finite number of compact Lie groups. The weight lattices Λ of such groups satisfy (8.1) where Λ_R is the root lattice of the Lie algebra. Thus saturated Euclidean integral lattices can be enumerated by Lie algebra theory. Now

$$(8.2) \qquad \Lambda_W/\Lambda_R = Z(g),$$

the center of the simply connected group G whose Lie algebra is g. Further there must exist subgroups K, K^v of Z(g), with $K^v \subset K$, such that

$$(8.3) \qquad \Lambda = \Lambda(G/K), \qquad \Lambda^* = \Lambda(G/K^v),$$

where $\Lambda(H)$ is the weight lattice of the group H, and

$$(8.4) \qquad \Lambda(G)/\Lambda(G/K^v) = \Lambda_R^*/\Lambda^* \cong K^v,$$

$$(8.5) \qquad \Lambda(G/K)/\Lambda(G/Z) = \Lambda/\Lambda_R \cong Z/K$$

Since these are equal,

$$(8.6) \qquad Z/K \cong K^v$$

These results are of significance in the theory of magnetic monopoles [29]. In particular, if Λ is self-dual, $K \cong K^v$ and so

$$(8.7) \qquad |Z| = |K|^2$$

where $|Z|$ denotes the order of Z.

<u>Self-dual Euclidean lattices whose associated Lie algebras are simple</u>

It is instructive to use the previous results to construct some Euclidean self-dual saturated lattices. The easiest ones for us to find are those whose associated Lie algebras are simple. As it is also simply laced it must be one of E_6, E_7, E_8, D_n, A_n. We saw that in order for a lattice to be self-dual the center of the simply connected group with the associated Lie algebra must have an order which is a perfect square. This means we only need to consider E_8, D_n, A_{n^2-1}, whose centres have orders 1, 4 and n^2, respectively. We consider these in turn:

E_8. As the center is trivial $\Lambda_R(E_8) = \Lambda_W(E_8)$ and hence it is self-dual and even. Clearly it has dimension 8.

A_{n^2-1}. As the centre is $\mathbb{Z}_{n^2}$, the candidate subgroup for K is $\mathbb{Z}_n$ leading to the possibly self-dual lattices $\Lambda(SU(n^2)/\mathbb{Z}_n)$. This lattice is generated by $\lambda_n + \Lambda_R$ where λ_n is the nth fundamental weight and comprises n cosets with respect to Λ_R:

$$m\lambda_n + \Lambda_R, \qquad m = 1,2,\ldots,n \tag{8.8}$$

Now if $x \in m\lambda_n + \Lambda_R$ and $x' \in m'\lambda_n + \Lambda_R$, $x \cdot x'$ differs from $mm'\lambda_n^2$ by an integer. Using an algorithm from magnetic monopole theory [30]

$$2\lambda^2/\alpha^2 = |Z(G_\alpha)|/|Z(G)| \tag{8.9}$$

where λ is a fundamental weight, α the corresponding simple root of G and G_α the simply connected group whose Dynkin diagram is obtained by deleting the point corresponding to α from the Dynkin diagram for G. As $\alpha^2 = 2$ in the present context, we have

$$\begin{aligned}\lambda_n^2 &= |Z(SU(n)\times SU(n^2-n))|/|Z(SU(n^2))| \\ &= n(n^2-n)/n^2 = n-1.\end{aligned} \tag{8.10}$$

Thus $\Lambda(SU(n^2)/\mathbb{Z}_n)$ is integral for all n and even or odd as n is odd or even. As $\mathbb{Z}_{n^2}/\mathbb{Z}_n \cong \mathbb{Z}_n$, it is self-dual. If the lattice is even, with n = 2s+1 say, then its dimension $n^2 - 1 = 4s(s+1)$ which is a multiple of 8. If the lattice is odd, with n = 2s say, it has dimension $4s^2 - 1$, s= 1,2,....

In the cases n = 2 and 3 we have $\lambda_2^2 = 1$ and $\lambda_3^2 = 2$ whilst otherwise $\lambda_n^2 \geqslant 3$. We now discuss the special cases of n = 2 and 3. If n = 2 the points of the lattice with $\lambda^2 = 1$ are the weights of the 6-dimensional (real) representation of SU(4) which is isomorphic to the covering group of SO(6); thus the lattice so obtained is

$$\Lambda(SU(4)/\mathbb{Z}_2) = \mathbb{Z}^3. \tag{8.11}$$

If n = 3, the lattice $\Lambda(SU(9)/\mathbb{Z}_3)$ contains extra points of length squared 2 which when added to the roots of SU(9) yield the root system of a bigger algebra, E_8 in fact. Thus we obtain again the lattice already mentioned,

$$\Lambda(SU(9)/\mathbb{Z}_3) = \Lambda_R(E_8) \tag{8.12}$$

D_n. The centre is $\mathbb{Z}_2 \times \mathbb{Z}_2$ if n is even and $\mathbb{Z}_4$ if n is odd. A spinor weight, λ_{sp} generates a subgroup $\mathbb{Z}_2$ if n is even, but $\mathbb{Z}_4$ if n is odd. The subgroup G_α of the algorithm (8.9) is SU(n) so that

$$\lambda_{sp}^2 = n/4 \tag{8.13}$$

and the corresponding lattices Λ are even and self-dual only if n is a multiple of 8; if n differs by 4 from a multiple of 8, Λ is odd and self-dual. If n is not a multiple of 4, the lattice is not integral. In the special cases of n = 4 and 8, the spinor weight λ_{sp} yields $\mathbb{Z}^4$ and $\Lambda_R(E_8)$ again respectively.

The vector fundamental weight, λ_v, generates a subgroup $\mathbb{Z}_2$ for every n. The corresponding lattices are $\mathbb{Z}^n$, as the algorithm (8.9) show that $\lambda_v^2 = 1$ for each n.

Note that all self-dual even Euclidean lattices found above

have dimensions which are multiples of 8, in agreement with the well-known result [16]. The same methods can be applied to semisimple Lie algebras g, but this is more tedious. It turns out that, at least for even self-dual lattices with dimensions less than 24, the various simple components of the semisimple algebra must have the same Coxeter number. We shall understand this result another way below.

__Relation between Euclidean and Lorentzian even self-dual lattices__

If Λ is an Euclidean even self-dual lattice, then as mentioned in section 2,

$$\Lambda \oplus II^{1,1} = II^{8n+1,1} , \tag{8.14}$$

where dim Λ = 8n, by the uniqueness theorem for even self-dual Lorentzian lattices.

If k denotes one of the two primitive light-like vectors in $II^{1,1}$, we can use it to define $\tilde{g}_\Lambda$, the affine Kac-Moody algebra associated with g_Λ. Its simple roots are those of g_Λ together with $k - \psi_i$, where ψ_i is the highest root of g_i, the i-th simple component of

$$g_\Lambda = \bigoplus_i g_i . \tag{8.15}$$

Thus

$$\Lambda_2 = \Phi(g_\Lambda) \subset \Phi(\tilde{g}_\Lambda) \subset II_2^{8n+1,1} , \tag{8.16}$$

the set of points of squared length 2 on $II^{8n+1,1}$, where $\Phi(g)$ denotes the roots of the Lie algebra g.

Our aim is to deduce more about such lattices Λ and about $II^{8n+1,1}$, from such interrelationships. Our analysis will be incomplete in that we are going to make some plausible assumptions for which we do not yet have general proofs. These assumptions lead to results which we can check, thereby adding credence to the assumptions. We shall assume:

(a) that there exists $\Delta \in II^{8n+1,1}$ such that the simple roots α of $II^{8n+1,1}$ satisfy $\alpha \cdot \Delta = 1$, at least for n = 1,2,3;

b) that the simple roots of the affine Kac-Moody algebra $\tilde{g}_\Lambda$ can also be taken as simple roots of $II^{8n+1,1}$;

(c) $II^{8n+1,1}$ has a simple root orthogonal to all the simple roots of the Lie algebra g_Λ, and so in $II^{1,1}$.

Conway and Sloane [12] have proved (a) for n = 1,2,3,. It follows that if $\Delta^2 < 0$ it lies in the positive Weyl chamber defined on the "mass shell" hyperboloid, $x^2 = -1$, or, if $\Delta^2 = 0$, the light cone $x^2 = 0$.

Let $\bar{k}$ be the unique light-like lattice vector in $II^{1,1}$ satisfying

(8.17) $$\bar{k}^2 = 0, \; k \cdot \bar{k} = 1.$$

then we write

(8.18) $$\Delta = \sum_i \delta_i + nk + m\bar{k}$$

where δ_i is half the sum of the positive roots of g_i, since if α is a simple root of g_Λ it is also a simple root of $\tilde{g}_\Lambda$ and hence, by (b) and (a), $\alpha \cdot \Delta = 1$ which forces Δ to have the stated form. Now

(8.19) $$\sum_i \delta_i = \frac{1}{2} \sum \alpha$$

where the sum is over all the positive roots of g_Λ. The remaining simple roots of the Kac-Moody algebra $\tilde{g}_\Lambda$ are the $k-\psi_i$. So, using assumptions (a) and (b) again,

(8.20) $$1 = \Delta \cdot (k-\psi_i) = 1-h_i + m$$

where h_i is the Coxeter number of g_i. Hence each $h_i = m$, and the Coxeter numbers of each of the components of g have a common value which henceforth will simply be called h; this result was mentioned earlier. It remains to determine the integer n in eq. (8.18), and to do this we use assumption (c). The only roots of $II^{1,1}$ are $\pm(k+\bar{k})$ and the sign must be chosen so that the scalar products with the other simple roots are non positive. In fact

$$-(k+\bar{k})\cdot(k-\psi_i) = -1 \tag{8.21}$$

so the simple root in question must be

$$\alpha_{-1} = -(k+\bar{k}) \tag{8.22}$$

Now $\alpha_{-1}\cdot\Delta = -h-n$ and so, by (c), $n = -h-1$ giving

$$\Delta = \sum_i \delta_i + h\bar{k} - (h+1)k. \tag{8.23}$$

This shows that $\sum \delta_i \in \Lambda$, not just Λ_W. $II^{8n+1,1}$ may well have more simple roots, but the information obtained so far may be used to help construct its Dynkin diagram. Before discussing this we shall derive a "strange formula" for Δ^2. This will be the main result of this section. Clearly,

$$\Delta^2 = \sum_i \delta_i^2 - 2h(h+1)\ . \tag{8.24}$$

By the Frendenthal-de Vries "strange formula",

$$\delta_i^2 = \frac{1}{12}h_i \dim g_i = \frac{1}{12}h_i(h_i+1)\dim g_i. \tag{8.25}$$

But, since $h_i = h$ for each component,

$$\sum \delta_i^2 = \frac{1}{12}h(h+1)\sum_i \operatorname{rank} g_i = \frac{1}{12}h(h+1)\dim \Lambda \tag{8.26}$$

since Λ is assumed saturated. Hence the result

$$\Delta^2 = \frac{1}{12}h(h+1)(\dim \Lambda - 24) \tag{8.27}$$

indicating that 24 is a critical dimension for Euclidean even self-dual lattices.

Notice that Δ is intrinsic to the Lorentzian lattice whilst h is an attribute of the Euclidean lattice Λ imbedded transversely in it. It follows that for $II^{9,1}$ and $II^{17,1}$, i.e. for dim Λ = 8 and 16, h must be the same for all transverse Euclidean self-dual even Lattices of the same dimension; we shall verify this. Formally our formula works for $II^{1,1}$ as there h = 0 so that $\Delta = -k$, though in that case the Δ satisfying (a) is not unique.

Let us discuss $II^{9,1}$. We found the right-dimensional self-dual Euclidean lattice

(8.28) $$\Lambda = \Lambda_R(E_8)$$

with Dynkin diagram

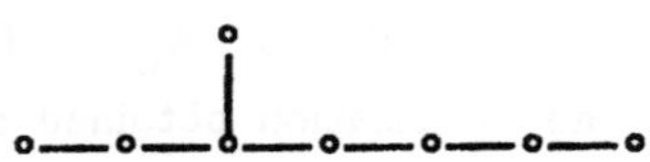

Adding to this the points corresponding to $\alpha_0 = k-\psi$ and α_{-1} yields

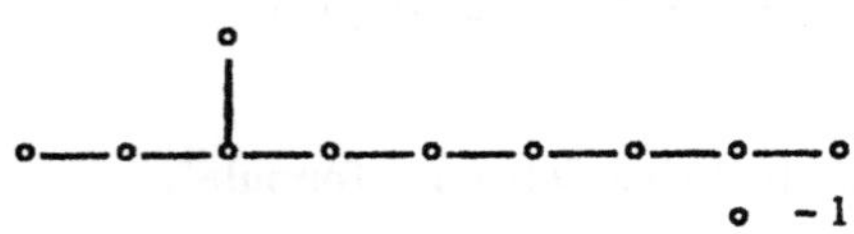

This is indeed the Dynkin diagram of $II^{9,1}$; it can be thought of as E_{10}, being a natural extension of the E series. In view of the recent speculation about Grand Unified Theories being based on N = 4 supersymmetric E_8 gauge theory, it is intriguing that E_8 is related to a lattice in a 10-dimensional Lorentzian space as that is a natural space for N = 4 supersymmetry. In the argument singling out E_8 as a possible gauge group, it was essential that its root lattice was self-dual [31].

Now let us turn to $II^{17,1}$. From our previous work, there are at least two possible 16-dimensional even saturated self-dual Euclidean lattices $\Lambda_R(E_8)\oplus\Lambda_R(E_8)$ and $\Lambda(\widetilde{SO}(32)/\mathbb{Z}_2)$, where $\widetilde{SO}(32)$ is the simply connected group with the Lie algebra of SO(32). Note that D_{10} and E_8 both have the same Coxeter number 30 as "predicted."

The Dynkin diagram of $E_8\oplus E_8$ with the extended roots $\alpha_{0i} = k-\psi_i$, i = 1,2, added is

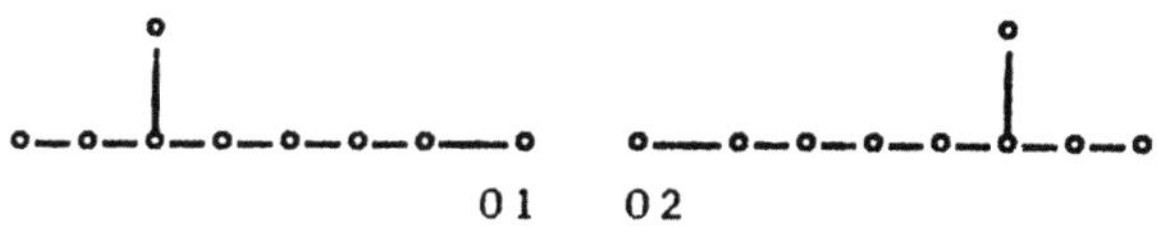

The over extended root α_{-1} is orthogonal to all the simple roots except α_{01} and α_{02} with which it has inner product -1. Hence we deduce that, granted our assumptions (a)-(c), the diagram

describes at least some of the simple roots of $II^{17,1}$. It is in fact the correct diagram.

Now consider starting from the Dynkin diagram for D_{16} instead. Adding the extended and overextended roots yields

i.e. the previous diagram with one point missing. In this case, unlike the previous cases, there exist points of Λ not on $\Lambda_R(D_{16})$. These are defined by a special minimal weight Λ_{sp} of D_{16} as discussed previously. Here λ_{sp} is the highest weight of one of the two spinor representations of D_{16}. Let us now try to construct a root of $II^{17,1}$ by adding to $-\lambda_{sp}$ an element of $II^{1,1}$,

(8.29) $$\beta = -\lambda_{sp} + nk + m\bar{k}$$

Then for the simple roots α_i of D_{16}, $\beta \cdot \alpha_i = 0$ unless α_i is the root corresponding to λ_{sp}, for which it equals -1; so if β is a simple root it is joined to the point of the Dynkin diagram corresponding to λ_{sp}. Now

(8.30) $$\beta^2 = \lambda_{sp}^2 + 2nm = 4 + 2nm,$$

by our previous calculation of λ_{sp}^2 for D_{16}. So if $\beta^2 = 2$, $nm = -1$. Now

$$\beta \cdot \alpha_0 = 1+m$$

as λ_{sp} is minimal (i.e. $\lambda_{sp} \cdot \psi = 1$). Thus to get $\beta \cdot \alpha_0 \leqslant 0$, consistently with $\beta^2 = 2$, we need $m = -1$, $n = 1$. Then β is automatically orthogonal to α_{-1} and so it furnishes the missing point.

Note that E_8 and D_{16} both have Coxeter number 30 as predicted above and this gives $\Delta^2 = -1240$ for $II^{9,1}$ and $\Delta^2 = -620$ for $II^{17,1}$, in agreement with Conway's calculation [17].

Finally we turn to $II^{25,1}$. The procedure described above can be repeated but does not terminate, indicating an infinite Dynkin diagram, as is indeed the case. Note that as $\Delta^2 = 0$ we can not deduce that the different 24-dimensional Euclidean self-dual even lattices have the same Coxeter number. It does follow however that Δ furnishes a light-like vector orthogonal to all the roots of $II^{25,1}$. Thus the lattice transverse to Δ has no roots (and so can not be saturated). It is the Leech lattice. The associated Lie group is $U(1)^{24}$ with $h = 0$, proving that eq. (8.23) for Δ again holds.

We conclude that the formulae (8.23) and (8.24) for Δ and Δ^2 hold in 28 different cases, lending support to assumptions (a), (b) and (c) for dimensions 26 or less. In higher dimensions they must break down as it is easy to find Euclidean even self-dual saturated lattices of the same dimension with different Coxeter numbers.

9. CONCLUSION

Our main point has been to relate integral lattices directly to algebras of various kinds by associating to the points of length squared equal to 1 or 2 (the sets Λ_1 and Λ_2 respectively) the contour integral of the vertex operator for emitting the "tachyon" state of the dual string model. The advantage of this point of view is that we can understand a wide range of results in a unified way, as well as viewing the structure of the dual string model in a new light.

If the scalar product, defining the notion of integrality on the lattice, is positive definite, the Λ_2 operators generate a finite dimensional Lie algebra, requiring only the addition of the momenta p^μ for closure. If the scalar product is positive semi-definite with a unique null direction (a vector in this direction being included in the lattice), the algebra generated is an affine Kac-Moody algebra and requires the addition of "photon" emission vertices for closure. These are derived as commutators of "tachyon" operators, a circumstance which would not generally obtain in the physical theory. If the lattice is Lorentzian (i.e. the scalar product corresponds to a nonsingular matrix with a single negative eigenvalue), the algebra generated is yet larger, with additional roots of length squared equal to $-2, -4, -6, \dots$, whose corresponding step operators are contour integrals of vertices for emitting the massive states of the dual string model. The same construction for the Λ_1 points extends the Λ_2 algebra by a Clifford algebra. Corresponding to the possibilities of nesting lattices in each other, there exist natural subalgebras. The point of view we have advocated here is not entirely new, being similar to that of Frenkel [11], but we hope that our account will at least make more intelligible to physicists. The basic technique is an exercise in normal ordering of free quantum fields which was developed and exploited in the work on the dual model [1,2,3].

We think the picture developed should be of interest to physicists since it correlates in a more uniform way diverse results they already know. We have already mentioned the role of contour integrals of vertex operators for emitting physical states of the dual string model. In section 7 we explained how the Ramond and Neveu-Schwarz fields arose naturally in connection with the Λ_1

points of a positive semidefinite lattice, constituting a generalised Clifford algebra for the D_N Kac-Moody algebra. These "fermi" fields were constructed out of the "bose" fields, a point of view not envisaged by the dual string literature. Nevertheless this relationship is precisely the fermion-boson quantum equivalence established by Skyrme [23] ten years earlier, in relating the quantum Sine-Gordon and massive Thirring models, and subsequently much exploited in other two dimensional field theories. This same fermion-boson equivalence relation (and its generalizations) has been used to construct general solutions of an enormous class of soliton equations, whose classification has recently been related to that of the affine Kac-Moody algebras in what is possibly their most remarkable physical applications so far [32].

Particle physicists are more interested in spontaneously broken gauge theories in four space-time dimensions and possible soliton-like behavior occurring there in the guise of magnetic monopoles. Conjectures have been made concerning alternative electric and magnetic formulations of the theory based on the construction of a "dual group" from the lattice reciprocal to the weight lattice of the original gauge group [29]. The construction of section 3 achieves this in a more concrete way. A key question concerns the analogue of the fermion-boson equivalence theorem which would facilitate the construction of monopole quantum field operators, just as the known construction yields the Sine-Gordon soliton field operators. Subsequent work [33] indicates that if this can be done it is more likely to be in the N=4 supersymmetric theory which naturally lives in a ten dimensional Lorentzian space [34], as it is a limiting case of the spinor string theory which exists there.

In this context, Lie algebraic methods have been used when studying higher gauge groups and many of the concepts developed as useful in the study of magnetic monopoles, e.g. the importance of minimal weights [35,36] and the importance of the quantity δ, equal to one half the sum of positive roots, in grading the roots [17], find echoes in the work of section 8.

These comments show intriguing links between the study of gauge theories and affine Kac-Moody algebras, a point of view put forward by other speakers at this meeting from different standpoints. Possibly some larger algebra, perhaps Lorentzian, which might involve supersymmetry also, plays a role as well, but it is yet to be identified.

Particularly interesting are the Lorentzian algebras $II^{9,1}$, $II^{17,1}$, $II^{25,1}$, named after the lattices to which they correspond. The latter two are not hyperbolic in the sense that the deletion of any point of their Dynkin diagrams leaves an ordinary or extended Dynkin diagram. Nevertheless the discussions of section 6 and 8 indicate that these algebras possess simplifying features. Firstly, the no ghost theorem establishes the existence of an invariant subspace for the algebra, which is positive definite, despite the Lorentzian metric associated with the lattice. Secondly, the existence of the vector Δ, which has unit scalar product with all the simple roots, seems to be a very special feature which these particular Lorentzian algebras share with all finite dimensional semisimple algebras. Thus these algebras may conceivably be simpler in structure than the other Lorentzian algebras and the present tenous connection may be the hint of a more substantial connection.

These comments suggest that both mathematicians and physicists are in the happy state of seeing the glimmerings of a vast new structure, of considerable importance in their respective subjects, which it will be their task to illuminate in the future.

ACKNOWLEDGEMENT

We are grateful to I. Frenkel, V.G. Kac and J. McKay for helpful and stimulating conversations. Most of this work was done while visiting the Mathematics and Physics departments of the University of Virginia, and whilst one of us (DO) was a member of the Institute for Advanced study there. We are very grateful for their hospitality.

APPENDIX

The operator construction determines equations with quantities of the following form on the left hand side:

$$A_r A_s - (-1)^{r \cdot s} A_s A_r \tag{A1}$$

where r^2 and s^2 could equal 1 or 2. We should like to modify the quantities A_r such that instead of (A1) we have an anticommutator if both r^2 and s^2 equal 1 and a commutator otherwise. We now explain how to do this.

Suppose $\{e_i\}$ constitute a basis for the integral lattice in question, containing Λ_2 (and if relevant Λ_1). Suppose further that we can construct corresponding quantities γ_i satisfying

$$\gamma_i \gamma_j = (-1)^{[(e_i \cdot e_j)^2 - e_i^2 e_j^2]} \gamma_j \gamma_i \tag{A2}$$

and commuting with all the A_r's. Then for any point of the lattice

$$u = \Sigma\, n_i e_i$$

define

$$\gamma_u = (\gamma_1)^{n_1} (\gamma_2)^{n_2} \ldots$$

It is easy to check that (A2) is generalized:

$$\gamma_u \gamma_v = (-1)^{[(u \cdot v)^2 - u^2 v^2]} \gamma_v \gamma_u$$

Note that the square bracket in the exponent of (-1) is the square of the area of the parallelogram with sides u and v. Further

$$\gamma_u \gamma_v = \epsilon(u,v) \gamma_{u+v}$$

where $\epsilon(u,v)$ equals ± 1. Then we define

$$E_r = \gamma_r A_r$$

so that

$$E_r E_s - (-1)^{r^2 s^2} E_s E_r = \epsilon(r,s)\gamma_{r+s}(A_r A_s - (-1)^{r\cdot s} A_s A_r)$$

If for example $r\cdot s \geq -1$ we have by the equations of the text

$$= \begin{cases} 0 & r\cdot s \geq 0 \\ \epsilon(r,s)E_{r+s} & r\cdot s = -1 \end{cases}$$

irrespective of the metric. It is understood $E_0 = A_0 = 1$ effectively. This is the desired form of a $\mathbb{Z}_2$ graded algebra with elements odd or even as $r^2 = 1$ or 2. It may be necessary to redefine the γ_i's by a sign to get the conventional Lie algebra signs on the right hand side.

Thus it remains to show that solutions to (A2) exist. For example the lattice $\mathbb{Z}^n$ in Euclidean or Lorentzian space has a basis e_i satisfying $e_i e_j = \pm\delta_{ij}$. A solution to (A2) is given by the Clifford algebra (Dirac gamma matrices).

Let Λ be any Euclidean or Lorentzian self-dual lattice. Then $\Lambda\oplus\mathbb{Z}$ is odd self dual and Lorentzian if the added $\mathbb{Z}$ is respectively time like or space like. By the uniqueness theorem for odd self dual Lorentzian lattices

$$\Lambda\oplus\mathbb{Z} = I_{\dim \Lambda,1} \,.$$

Since γ matrices (A2) exist for $I_{\dim \Lambda,1}$ they do also for Λ by this equation, that is for any self dual Euclidean or Lorentzian lattice. We suspect solutions to (A2) exist for any integral lattice but have not shown it.

Instead of increasing the dimension of the representation space by introducing γ matrices, we can achieve the same effect by considering a function of momentum u, C_u, defined by $C_u\Psi_\gamma = \epsilon(u,\gamma)\Psi_\gamma$ for $\gamma \in \Lambda_R$. Such a function can be extended to $\gamma \in \Lambda$

by choosing a γ_0 in each coset of Λ by Λ_R and setting $\epsilon(u,\gamma) = \epsilon(u,\gamma-\gamma_0)$ for γ in that coset.

REFERENCES

[1] M. Jacob (ed.) Dual Theory (North Holland, Amsterdam, 1975).

[2] J. Scherk, Rev. Mod. Phys. 47, 123 (1975).

[3] J.H. Schwarz, Phys. Report 89, 223 (1982); M.B. Green, Surveys in High Energy Physics 3, 127 (1983)

[4] I.B. Frenkel and V.G. Kac, Inv. Math. 62, 23 (1980).

[5] V.G. Kac, Math USSR-Izv. 2, 1271 (1968); R.V. Moody, J. Algebra 10, 211 (1968)

[6] M.A. Virasoro, Phys. Rev. D1, 2933 (1970).

[7] E. Del Giudice, P. Di. Vecchia and S. Fubini, Ann. Phys. (N.Y.) 70, 378 (1972).

[8] R.C. Brower and P. Goddard, Nuc. Phys. B40, 437 (1972).

[9] R.C. Brower, Phys. Rev. D6, 1655 (1972).

[10] P. Goddard and C.B. Thorn, Phy. Letters 40B, 235 (1972).

[11] I.B. Frenkel, Representations of Kac-Moody algebras and dual resonance models, preprint.

[12] J.H. Conway and N.J.A. Sloane, Lorentzian forms for the Leech lattice, preprint

[13] J.H. Conway, L. Queen and N.J.A. Sloane, A monster Lie algebra?, preprint.

[14] I.B. Frenkel, J. Lepowsky and A. Meurman, An E_8 approach to F_1, preprint.

[15] E. Corrigan and D.B. Fairlie, Nuc. Phys. B91, 527 (1975); M. Green Nucl. Phys. B103, 313 (1976); M. Green and J. Shapiro Phys. Lett. 64B, 454 (1976); R. Horsley, Nuc. Phys. B138, 474 (1978).

[16] J.-P. Serre, A course in arithmetic (Springer Verlag, New York, 1973).

[17] J.H. Conway, The automorphism group of the 26-dimensional even unimodular Lorentizian Lattice, preprint.

[18] J.E. Humphreys, Introduction to Lie Algebras and Representation Theory, (Springer-Verlag, 1972).

[19] R.L. Griess, Inv. Math. 69, 1 (1982).

[20] V.G. Kac in: W. Berginböck, A Böhm and E. Takasugi (eds.) Group Theoretical Methods in Physics, Lecture Notes in Physics, Vol. 94 (Springer-Verlag, New York, 1979).

[21] B.L. Feigin and D.B. Fuks, Funct. Anal. 16, 2, 47 (1982).

[22] I.B. Frenkel, Journal of Funtional Analysis, 44, 259 (1981).

[23] T.H.R. Skyrme, Proc. R. Soc. A247, 260(1958), A252, 236(1959), A260, 127(1961), A262, 237(1961).

[24] R.F. Streater and I.F. Wilde, Nucl. Phys. B24, 561 (1970); S. Coleman, Phys. rev. D11, 2088(1975); S. Mandelstam, Phys. Rev D11, 3026(1975).

[25] A. Neveu and J. Schwarz, Nucl. Phys. B31, 86(1971).

[26] P. Ramond, Phys. Rev. D3, 2415(1971).

[27] C.B. Thorn, Phys. Rev. D4, 1112(1971); E. Corrigan and D. Olive, Nuovo Dim. 11A, 749(1972).

[28] F. Englert and P. Windey, Phys. Rev. D14, 2728 (1976).

[29] P. Goddard, J. Nuyts and D. Olive, Nucl. Phys. B125, 1 (1977).

[30] P. Goddard and D. Olive, Nucl. Phys. B191, 511(1981).

[31] G. Chapline and R. Slansky, Nucl. Phys. B209, 461 (1982);
D. Olive and P. West, Nucl. Phys B217, 1 (1983).

[32] M. Jimbo: talk at the meeting.

[33] D. Olive, Magnetic monopoles and electromagnetic conjectures in Monopoles in Quantum Field theory (e.d N. Craigie et al., world Scientific, Singapore (1982).

[34] F. Gliozzi, D. Olive and J. Scherk, Nucl. Phys. B122, 253(1977).

[35] R. Brandt and F. Neri, Nucl. Phys. B161, 253 (1979);
S. Coleman in Proceedings of the 1981 School of Subnuclear Physics "Ettore Majorana".

[36] P, Goddard and D. Olive, Nucl. Phys. B191, 528 (1981).

[37] N Ganoulis, P. Goddard and D. Olive, Nucl. Phys. B205 [FS5], 601(1982).

P. Goddard
Department of Applied Mathematics and Theoretical Physics
University of Cambridge, England

D. Olive
Blackett Laboratory
Imperial College, London, England

$\mathcal{Z}$-ALGEBRAS AND THE ROGERS-RAMANUJAN IDENTITIES

James Lepowsky[†] and Robert Lee Wilson[†]

1. INTRODUCTION.

Using the vertex operator introduced in [12], we have given two Lie-theoretic interpretations and proofs of the classical Rogers-Ramanujan identities, which state

$$(1.1)\qquad \prod_{n\geq 1}(1-q^{5n-4})^{-1}(1-q^{5n-1})^{-1} = \sum_{n\geq 0} q^{n^2}/(1-q)(1-q^2)\dots(1-q^n),$$

$$(1.2)\qquad \prod_{n\geq 1}(1-q^{5n-3})^{-1}(1-q^{5n-2})^{-1} = \sum_{n\geq 0} q^{n(n+1)}/(1-q)(1-q^2)\dots(1-q^n).$$

Both of these approaches are based on a pair of level 3 standard modules for the simplest affine Kac-Moody Lie algebra $A_1^{(1)}=\mathfrak{sl}(2,\mathbb{C})^\wedge$ and on the principal Heisenberg subalgebra, say $\hat{\mathfrak{h}}$, of $A_1^{(1)}$. Moreover, both approaches interpret the product sides of (1.1) and (1.2) as the principal characters of the vacuum spaces for $\hat{\mathfrak{h}}$ in the standard modules. Both approaches also interpret the sum sides of (1.1) and (1.2) by means of the $\hat{\mathfrak{h}}$-filtration of these vacuum spaces. The second approach [15], [16] has the advantage over the first [13], [14] that the argument which interprets the nth summand on the right-hand side of (1.1) or (1.2) generalizes naturally.

The second approach is based on the new algebras ("$\mathcal{Z}$-algebras") introduced in the announcement [15]. The details of this work are given in [16] for the general theory and for the standard $A_1^{(1)}$-modules of levels 1, 2 and 3, and in [17] for the higher level standard $A_1^{(1)}$-modules in the principal picture. In the

[†] Partially supported by the National Science Foundation through the Mathematical Sciences Research Institute and NSF grant MCS 83-01664.

Vertex Operators in Mathematics and Physics - Proceedings of a Conference November 10-17, 1983. Publications of the Mathematical Sciences Research Institute #3, Springer-Verlag, 1984.

same sense that the level 3 modules are related to the Rogers-Ramanujan identities, the level 2 modules are related to a pair of Euler identities and the higher level modules are related to combinatorial identities of Gordon, Andrews and Bressoud. For all the standard $A_1^{(1)}$-modules, we correspondingly have constructions of $A_1^{(1)}$ generalizing the first vertex operator construction [12] (the case of the basic, or level 1 standard, modules), although for the higher levels, we do not now have an independent proof of the corresponding combinatorial identities, which are interpreted by and used in the module construction.

The present paper is a virtually self-contained exposition of the most important cases of this work - the level 1, 2 and 3 cases - and an introduction to the papers [15]-[17]. The notation and arguments used here usually suggest the general notation and arguments. The reader should also find the earlier expositions [9], [10] useful. For further background and bibliography, we refer the reader to the five papers just mentioned.

Vertex operators and $\mathfrak{Z}$-algebras have turned out to be closely related to work in several different directions, discussed in a number of papers in this volume. Many times during the development of these ideas, new insight has arisen from interaction between apparently distinct theories. One such case is described in Remark 8.11. We hope that the present elementary exposition will facilitate communication between people with different perspectives.

2. THE AFFINE LIE ALGEBRA $A_1^{(1)}$ IN THE PRINCIPAL PICTURE.

Let $\mathfrak{g}$ be the 3-dimensional simple Lie algebra $\mathfrak{sl}(2,\mathbb{C})$, consisting of all 2×2 complex matrices of trace 0. Take the following basis:

$$\beta = \begin{bmatrix} 1 & 0 \\ 0 & -1 \end{bmatrix}, \quad x_\beta = \begin{bmatrix} 0 & 1 \\ 0 & 0 \end{bmatrix}, \quad x_{-\beta} = \begin{bmatrix} 0 & 0 \\ -1 & 0 \end{bmatrix}.$$

The brackets are

$$[\beta, x_{\pm\beta}] = \pm 2x_{\pm\beta}, \quad [x_\beta, x_{-\beta}] = -\beta.$$

The nonsingular symmetric $\mathfrak{g}$-invariant bilinear form

$$\langle x,y\rangle = \mathrm{tr}(xy), \quad x,y \in \mathfrak{g}$$

has the values

(2.1) $\langle\beta,\beta\rangle = 2, \ \langle\beta,x_{\pm\beta}\rangle = 0, \ \langle x_\beta,x_{-\beta}\rangle = -1, \ \langle x_{\pm\beta},x_{\pm\beta}\rangle = 0.$

Consider the Lie algebra involution

$$\nu x = -{}^t x, \quad x \in \mathfrak{g},$$

so that

$$\nu\beta = -\beta, \quad \nu x_{\pm\beta} = x_{\mp\beta}.$$

Denote by $\mathfrak{g}_{(0)}$ and $\mathfrak{g}_{(1)}$ the ± 1-eigenspaces of ν in $\mathfrak{g}$, respectively. We have

$$\mathfrak{g}_{(0)} = \mathbb{C}(x_\beta+x_{-\beta}), \ \mathfrak{g}_{(1)} = \mathbb{C}\beta\oplus\mathbb{C}(x_\beta-x_{-\beta}),$$

$$\mathfrak{g} = \mathfrak{g}_{(0)}\oplus\mathfrak{g}_{(1)}.$$

For $x\in\mathfrak{g}$, write

(2.2) $$x = x_{(0)} + x_{(1)}$$

for the ν-eigenspace decomposition of x, so that

$$x_{(0)} = \frac{1}{2}(x+\nu x), \quad x_{(1)} = \frac{1}{2}(x-\nu x).$$

Then

(2.3) $$\beta_{(0)} = 0, \qquad \beta_{(1)} = \beta.$$

We shall also write

$$\mathfrak{g}_{(2n)} = \mathfrak{g}_{(0)}, \ \mathfrak{g}_{(2n+1)} = \mathfrak{g}_{(1)},$$

(2.4)

$$x_{(2n)} = x_{(0)}, \ \ x_{(2n+1)} = x_{(1)}$$

for $n \in \mathbb{Z}$, $x \in \mathfrak{g}$.

Let t be an indeterminate. Define the Lie algebra

(2.5) $$\hat{\mathfrak{g}} = \mathfrak{g}_{(0)} \otimes \mathbb{C}[t^2,t^{-2}] \oplus \mathfrak{g}_{(1)} \otimes t\mathbb{C}[t^2,t^{-2}] \oplus \mathbb{C}c,$$

spanned by the elements

$$x \otimes t^n, \qquad c \ (\neq 0)$$

where $n \in \mathbb{Z}$, $x \in \mathfrak{g}_{(n)}$, with the brackets determined as follows:

$$[c,\hat{\mathfrak{g}}] = 0 \ ,$$

$$[x \otimes t^m, y \otimes t^n] = [x,y] \otimes t^{m+n} + \frac{1}{2}\langle x,y\rangle m\,\delta_{m+n,0}c$$

for $m,n \in \mathbb{Z}$ and $x \in \mathfrak{g}_{(m)}$, $y \in \mathfrak{g}_{(n)}$.

For $x \in \mathfrak{g}$, $n \in \mathbb{Z}$, we set

(2.6) $$x(n) = x_{(n)} \otimes t^n \in \hat{\mathfrak{g}}$$

(see (2.4)). Then for example $\beta(n)=0$ if $n \notin 2\mathbb{Z}+1$ (see (2.3)). In this notation, the brackets become:

$$[c,\hat{g}] = 0$$

(2.7)

$$[x(m),y(n)] = [x,y](m+n) + \frac{1}{2}\langle x,y\rangle m\,\delta_{m+n,0}c$$

for m,n,x,y as above.

Let d be the derivation of $\hat{g}$ such that

$$d(c) = 0,$$

$$d(x(n)) = n(x(n))$$

for all $n\in\mathbb{Z}$, $x\in g_{(n)}$, and form the semidirect product Lie algebra

(2.8)
$$\tilde{g} = \hat{g}\rtimes\mathbb{C}d \; .$$

Then $\tilde{g}$, or sometimes $\hat{g}$, is called the *affine Kac-Moody Lie algebra* $A_1^{(1)}$. The $\mathbb{Z}$-gradation of $\tilde{g}$

(2.9)
$$\tilde{g} = \coprod_{n\in\mathbb{Z}}\tilde{g}_n,$$

where $\tilde{g}_n$ is the n-eigenspace for d in $\tilde{g}$, is called the *principal gradation.*

Let h be the one-dimensional space

(2.10)
$$h = \mathbb{C}\beta \subset g_{(1)},$$

a ν-stable Cartan subalgebra of g for which $x_{\pm\beta}$ are root vectors. The restriction of the form $\langle\cdot,\cdot\rangle$ to h being nonsingular since $\langle\beta,\beta\rangle = 2$ (2.1), we may correspondingly identify h with its dual h^*. Then $\beta\in h = h^*$ is the root associated with x_β.

Consider the Lie subalgebra

(2.11)
$$\hat{h} = h\otimes t\mathbb{C}[t^2,t^{-2}] \oplus \mathbb{C}c$$

of $\hat{g}$, defined by strict analogy with (2.5), (2.7). This algebra has basis

$$\{c, \beta(n) \mid n \in 2\mathbb{Z}+1\}$$

and brackets given by:

$$[c, \hat{\mathfrak{h}}] = 0,$$

(2.12)

$$[\beta(m), \beta(n)] = m\delta_{m+n,0}c$$

for $m,n \in 2\mathbb{Z}+1$. In particular, $\hat{\mathfrak{h}}$ is an infinite-dimensional Heisenberg subalgebra of $\hat{\mathfrak{g}}$, called the *principal Heisenberg subalgebra*.

The subalgebra $\hat{\mathfrak{h}}$ is principally graded in the sense that

$$\hat{\mathfrak{h}} = \coprod_{n \in \mathbb{Z}} (\hat{\mathfrak{h}} \cap \tilde{\mathfrak{g}}_n)$$

(see (2.9)), or equivalently,

$$[d, \hat{\mathfrak{h}}] \subset \hat{\mathfrak{h}}.$$

We define the subalgebra

(2.13)
$$\tilde{\mathfrak{h}} = \hat{\mathfrak{h}} \rtimes \mathbb{C}d$$

of $\tilde{\mathfrak{g}}$, by analogy with (2.8).

We have the following basis of $\tilde{\mathfrak{g}}$:

$$\{c, d, \beta(m), x_{\beta}(n) \mid m \in 2\mathbb{Z}+1,\ n \in \mathbb{Z}\}.$$

3. THE STRUCTURE OF $\tilde{\mathfrak{g}}$ IN TERMS OF FORMAL VARIABLES.

It will be useful to express the bracket relations (2.7) in terms of formal variables.

Let $\zeta, \zeta_1, \zeta_2, \ldots$ be a (possibly finite) sequence of symbols, viewed as commuting formal indeterminates, and let V be a vector space. Denote by $V\{\zeta_1,\zeta_2,\ldots\}$ the vector space of formal Laurent series in $\zeta_1,\zeta_2,\ldots$ with coefficients in V, i.e., the space of all (possibly infinite) formal sums

$$\Sigma_{n_1,n_2\ldots\in\mathbb{Z}}\, v_{n_1n_2\ldots}\zeta_1^{n_1}\zeta_2^{n_2}\ldots$$

with each $v_{n_1n_2}\ldots\in V$. For a vector space V' and a linear map

$$\pi: V \longrightarrow V',$$

also denote by

$$\pi: V\{\zeta_1,\zeta_2,\ldots\} \longrightarrow V'\{\zeta_1,\zeta_2,\ldots\} \tag{3.1}$$

the linear map which takes the typical element above to

$$\Sigma\ \pi(v_{n_1n_2}\ldots)\zeta_1^{n_1}\zeta_2^{n_2}\ldots\ . \tag{3.2}$$

Consider the subspace $V[\zeta_1,\zeta_1^{-1},\zeta_2,\zeta_2^{-1},\ldots]$ of Laurent polynomials (i.e., finite formal Laurent series) in $\zeta_1,\zeta_2,\ldots$ with coefficients in V. The algebra of differential operators generated by the partial derivatives $\partial/\partial\zeta_1$, $\partial/\partial\zeta_2,\ldots$ and the multiplication operators $\zeta_1,\zeta_1^{-1},\zeta_2,\zeta_2^{-1},\ldots$ acts in a natural way on both this space and $V\{\zeta_1,\zeta_2,\ldots\}$.

If $f\in V\{\zeta\}$, we shall sometimes write $f(\zeta)$ for f, and f', $f'(\zeta)$ or $(d/d\zeta)f$ for $\partial f/\partial\zeta$. We shall take other liberties with notation. For instance, if $a\in\mathbb{C}^*$ and

$$f = \Sigma_{n\in\mathbb{Z}}v_n\zeta^n\in V\{\zeta\}$$

$(v_n\in V)$, we let

$$f(a\zeta) = \Sigma_{n\in\mathbb{Z}}a^nv_n\zeta^n\in V\{\zeta\}$$

and

$$f(\zeta_1/\zeta_2) = \Sigma_{n\in\mathbb{Z}} v_n \zeta_1^n \zeta_2^{-n} \in V\{\zeta_1, \zeta_2\}.$$

Permissible "specializations" such as these will be used freely.

Define the endomorphism D of $V\{\zeta\}$ by the condition

$$D = \zeta(d/d\zeta),$$

so that for $v_n \in V$,

$$D\left[\Sigma v_n \zeta^n\right] = \Sigma n v_n \zeta^n \in V\{\zeta\}.$$

Note that for $a\in\mathbb{C}^*$ and $f\in V\{\zeta\}$,

$$D(f(a\zeta)) = (Df)(a\zeta).$$

Define

$$\delta(\zeta) = \Sigma_{n\in\mathbb{Z}} \zeta^n \in \mathbb{C}\{\zeta\}.$$

This is formally the Fourier expansion of the δ-function at $\zeta=1$, and $\delta(\zeta)$ behaves accordingly:

Let $f(\zeta)\in V[\zeta,\zeta^{-1}]$. Then with respect to the obvious pairing

$$V[\zeta,\zeta^{-1}]\times\mathbb{C}\{\zeta\} \longrightarrow V\{\zeta\},$$

denoted by juxtaposition, we have

$$f(\zeta)\delta(\zeta) = f(1)\delta(\zeta), \tag{3.3}$$

$$f(\zeta)\delta(\zeta^2) = \frac{1}{2}\Sigma_{p=0,1} f((-1)^p)\delta((-1)^p\zeta). \tag{3.4}$$

Indeed, by linearity, it is sufficient to prove the formulas for $f(\zeta) = \zeta^n$ $(n\in\mathbb{Z})$, and both are clear.

We shall reformulate the bracket structure of $\tilde{g}$ in terms of the following "periodic elements" $x(\zeta)$ of $\tilde{g}\{\zeta\}$ (see (2.6)):

(3.5) $$x(\zeta) = \Sigma_{n\in\mathbb{Z}} x(n)\zeta^n$$

for $x\in\mathfrak{g}$. Note that (recalling (2.2), (2.4))

(3.6) $$x(-\zeta) = (\nu x)(\zeta).$$

<u>**Proposition**</u> **3.1.** For all $x,y\in\mathfrak{g}$,

(3.7) $$[c,x(\zeta)] = 0, \quad [d,x(\zeta)] = Dx(\zeta), \quad [c,d] = 0,$$

(3.8) $$[x(\zeta_1),y(\zeta_2)] = \frac{1}{2}\Sigma_{p=0,1}[\nu^p x,y](\zeta_2)(\delta((-1)^p\zeta_1/\zeta_2)$$
$$+\frac{1}{4}c\Sigma_{p=0,1}\langle\nu^p x,y\rangle(D\delta)((-1)^p\zeta_1/\zeta_2).$$

<u>**Proof.**</u> First note that the assertions take place in $\tilde{\mathfrak{g}}\{\zeta\}$ and $\tilde{\mathfrak{g}}\{\zeta_1,\zeta_2\}$, and that they are to be interpreted in the obvious way: by equating the coefficients of ζ^n or $\zeta_1^m\zeta_2^n$ $(m,n\in\mathbb{Z})$ on the two sides. The formulas (3.7) are clear, and for (3.8), we have

$$[x(\zeta_1),y(\zeta_2)] = F+G,$$

with

$$G = \frac{1}{2}c\Sigma_{n\in\mathbb{Z}} n\langle x_{(n)},y_{(-n)}\rangle(\zeta_1/\zeta_2)^n,$$

$$F = \Sigma_{m,n\in\mathbb{Z}}([x_{(m)},y_{(n)}]\otimes t^{m+n})\zeta_1^m\zeta_2^n$$
$$= \Sigma_{m,n}([x_{(m)},y_{(n-m)}]\otimes t^n)(\zeta_1/\zeta_2)^m\zeta_2^n$$
$$= \Sigma_{n\in\mathbb{Z}} f_n(\zeta_1/\zeta_2)\delta((\zeta_1/\zeta_2)^2)\zeta_2^n,$$

where

$$f_n(\zeta) = \Sigma_{i=0,1}([x_{(i)},y_{(n-i)}]\otimes t^n)\zeta^i.$$

For $p=0,1$,

$$f_n((-1)^p) = [\nu^p x, y]_{(n)} \otimes t^n,$$

so that by (3.4),

$$F = \frac{1}{2}\Sigma_{p=0,1}[\nu^p x, y](\zeta_2)\delta((-1)^p \zeta_1/\zeta_2).$$

On the other hand,

$$G = \frac{1}{2}cg(\zeta_1/\zeta_2),$$

where

$$\begin{aligned} g(\zeta) &= D\Sigma_{n\in\mathbb{Z}}\langle x_{(n)}, y\rangle \zeta^n \\ &= D((\Sigma_{i=0,1}\langle x_{(i)}, y\rangle \zeta^i)\delta(\zeta^2)) \\ &= \frac{1}{2}D\Sigma_{p=0,1}\langle \nu^p x, y\rangle \delta((-1)^p \zeta), \end{aligned}$$

again by (3.4). ■

The Lie algebra $\tilde{\mathfrak{g}}$ is spanned by $\tilde{\mathfrak{h}}$ (see (2.10)-(2.13)) and the coefficients of ζ^n ($n\in\mathbb{Z}$) in the periodic element $x_\beta(\zeta)$. By (3.6),

(3.9) $$x_{-\beta}(\zeta) = x_\beta(-\zeta).$$

Clearly,

(3.10) $$[\beta(m), x_\beta(n)] = 2x_\beta(m+n)$$

for $m\in 2\mathbb{Z}+1$, $n\in\mathbb{Z}$, so that

(3.11) $$[\beta(m), x_\beta(\zeta)] = 2\zeta^{-m}x_\beta(\zeta).$$

The bracket structure of $\tilde{\mathfrak{g}}$ is completely described by (2.12), (3.7), (3.10) or (3.11), and the following immediate consequence of (3.8):

$$(3.12)\quad [x_\beta(\zeta_1),x_\beta(\zeta_2)] = \frac{1}{2}\beta(\zeta_2)\delta(-\zeta_1/\zeta_2) - \frac{1}{4}c(D\delta)(-\zeta_1/\zeta_2).$$

This last formula is equivalent to its list of component formulas:

$$[x_\beta(m),x_\beta(n)] = \frac{1}{2}(-1)^m\beta(m+n) - \frac{1}{4}(-1)^m m\,\delta_{m+n,0}c$$

for $m,n\in\mathbb{Z}$; recall that $\beta(n)=0$ if $n\in 2\mathbb{Z}$.

Remark 3.2. The element $x_\beta(\zeta)$ is sometimes written $X(\beta,\zeta)$, as in [16], for example.

4. THE ALGEBRA $\mathcal{X}_V$.

The main objects of study and their elementary properties are described in this section.

For a $\tilde{\mathfrak{g}}$-module V (or more generally, a vector space V on which the element d of $\tilde{\mathfrak{g}}$ acts) and an element $z\in\mathbb{C}$, we set

$$(4.1)\qquad V_z = \{v\in V \mid d\cdot v = zv\},$$

and we call this the space of elements *homogeneous of degree* z. (This generalizes the notation $\tilde{\mathfrak{g}}_n$, $n\in\mathbb{Z}$; here d acts by ad d on $\tilde{\mathfrak{g}}$.) In particular, we say that $T\in\text{End}\ V$ is *homogeneous of degree* $y\in\mathbb{C}$ if

$$[d,T] = yT$$

(as elements of End V), and we denote by $(\text{End}\ V)_y$ the space of all endomorphisms homogeneous of degree y. Observe that if

$$V = \coprod_{z\in\mathbb{C}} V_z,$$

then $T\in\text{End}\ V$ has degree $y\in\mathbb{C}$ if and only if

$$TV_z \subset V_{y+z}$$

for all $z\in\mathbb{C}$.

Definition 4.1. For $k\in\mathbb{C}$, we say that a $\tilde{g}$-module (or, more generally, a vector space on which the element c of $\tilde{g}$ acts) has *level* k if c acts as the scalar k on it.

We now define the category $\mathcal{C}_k$:

Definition 4.2. For $k\in\mathbb{C}$, we denote by $\mathcal{C}_k$ the category of $\tilde{g}$-modules V such that:

(i) V has level k

(ii) $V = \coprod_{z\in\mathbb{C}} V_z$

(iii) For every $z\in\mathbb{C}$, there exists $n_0\geq 0$ such that for all $n>n_0$, $V_{z+n}=0$.

Throughout the rest of this section, let

$$k\in\mathbb{C}^* \text{ and } (V,\pi)\in\mathcal{C}_k, \tag{4.2}$$

$\pi\colon \tilde{g} \longrightarrow \mathrm{End}\, V$ being the representation associated with V.

Consider the element

$$\pi x_{\pm\beta}(\zeta) = \Sigma_{n\in\mathbb{Z}}\pi(x_{\pm\beta}(n))\zeta^n\in(\mathrm{End}\ V)\{\zeta\} \tag{4.3}$$

(see (3.1), (3.2), (3.5)), which we shall sometimes write as simply $x_{\pm\beta}(\zeta)$, and define the following elements of $(\mathrm{End}\ V)\{\zeta\}$:

$$E^{\pm}(\gamma,\zeta) = E^{\pm}(\gamma,\zeta,\pi) = \exp(2\Sigma_{n>0,n\ \mathrm{odd}}\pi(\gamma(\pm n))\zeta^{\pm n}/(\pm n)k) \tag{4.4}$$

for $\gamma\in\mathfrak{h}=\mathbb{C}\beta$. Here "exp" denotes the formal exponential series. Observe that the elements $E^{\pm}(\gamma,\zeta)$ are indeed well defined.

Note that if we designate any of the three elements (4.3), (4.4) by Y, then

$$[\pi(d),Y] = DY \tag{4.5}$$

in $(\mathrm{End}\ V)\{\zeta\}$. In other words, if we define the corresponding coefficients $Y_n \in \mathrm{End}\ V$ by the condition

$$Y = \Sigma_{n\in\mathbb{Z}} Y_n \zeta^n,$$

then Y_n is homogeneous of degree n, that is,

$$Y_n \in (\mathrm{End}\ V)_n.$$

Because the grading of $V\in\mathcal{C}_k$ is "truncated from above", it follows that

$$z_{\pm\beta}(\zeta)=z_{\pm\beta}(\zeta,\pi) = E^-(\pm\beta,\zeta)x_{\pm\beta}(\zeta)E^+(\pm\beta,\zeta) \tag{4.6}$$

is a well-defined element of $(\mathrm{End}\ V)\{\zeta\}$. By (4.5) and the derivation law, we see that

$$[\pi(d),z_{\pm\beta}(\zeta)] = Dz_{\pm\beta}(\zeta), \tag{4.7}$$

so that for all $n\in\mathbb{Z}$,

$$z_{\pm\beta}(n)\in(\mathrm{End}\ V)_n,$$

where $z_{\pm\beta}(n)$ is defined by the expansion

$$z_{\pm\beta}(\zeta) = \Sigma_{n\in\mathbb{Z}} z_{\pm\beta}(n)\zeta^n. \tag{4.8}$$

We can "solve (4.6) for $x_\beta(\zeta)$" to obtain

$$x_{\pm\beta}(\zeta) = E^-(\mp\beta,\zeta)z_{\pm\beta}(\zeta)E^+(\mp\beta,\zeta). \tag{4.9}$$

Using (3.9) we observe that

$$z_{-\beta}(\zeta) = z_{\beta}(-\zeta),$$

i.e., that

$$z_{-\beta}(n) = (-1)^n z_{\beta}(n)$$

for all $n\in\mathbb{Z}$.

Definition 4.3. Denote by $\mathcal{Z}_V'$ the subalgebra of End V generated by the operators $z_{\beta}(n)$, $n\in\mathbb{Z}$ (or equivalently, by the operators $z_{\pm\beta}(n)$, $n\in\mathbb{Z}$), and denote by $\mathcal{Z}_V$ the subalgebra of End V generated by these operators and $\pi(d)$.

Proposition 4.4. We have

$$[\pi(\hat{\mathfrak{h}}),\mathcal{Z}_V'] = 0,$$

that is, $z_{\beta}(n)$ commutes with the action of the principal Heisenberg subalgebra for all $n\in\mathbb{Z}$.

Proof. It suffices to show that

$$(4.10) \qquad [\pi(\beta(m)),\ z_{\beta}(\zeta)] = 0$$

for all $m\in 2\mathbb{Z}+1$. But if $m>0$, m odd, we have

$$[\pi(\beta(m)),\ 2\Sigma_{n>0,\ n\ \mathrm{odd}}\pi(\beta(-n))\zeta^{-n}/(-n)k] =$$

$$= 2m\pi(c)\zeta^{-m}/(-m)k$$

$$= -2\zeta^{-m}$$

by (2.12) and (4.2), and since this expression commutes with everything,

$$[\pi(\beta(m)),\ E^-(\beta,\zeta)] = -2\zeta^{-m}E^-(\beta,\zeta)$$

(see (4.4)). Thus

$$[\pi(\beta(m)),\ z_\beta(\zeta)] =$$

$$= [\pi(\beta(m)),\ E^-(\beta,\zeta)]x_\beta(\zeta)E^+(\beta,\zeta) +$$

$$E^-(\beta,\zeta)[\pi(\beta(m)), x_\beta(\zeta)]E^+(\beta,\zeta) +$$

$$E^-(\beta,\zeta)x_\beta(\zeta)[\pi(\beta(m)),\ E^+(\beta,\zeta)]$$

$$= -2\zeta^{-m}z_\beta(\zeta) + 2\zeta^{-m}z_\beta(\zeta) + 0$$

$$= 0,$$

using (3.11). For m < 0, m odd, (4.10) is proved similarly. ■

Define

$$\hat{\mathfrak{h}}_\pm = \mathfrak{h}\otimes t^{\pm 1}\mathbb{C}[t^{\pm 2}] = \hat{\mathfrak{h}}\cap\coprod_{n>0}\tilde{\mathfrak{g}}_{\pm n},$$

an abelian Lie subalgebra of $\hat{\mathfrak{h}}$.

Definition 4.5. Set

$$\Omega_V = \{w\in V \mid \hat{\mathfrak{h}}_+\cdot w=0\},$$

the "vacuum space" for the Heisenberg algebra $\hat{\mathfrak{h}}$ in V.

Using Proposition 4.4, we observe:

Proposition 4.6. The vacuum space Ω_V is stable under the algebra $\mathcal{Z}_V$. ■

The $\tilde{g}$-module V can in fact be reconstructed from the $\mathfrak{T}_V$-module Ω_V, as we now explain. Set

$$\tilde{\mathfrak{h}}^+ = \hat{\mathfrak{h}}_+ \oplus \mathbb{C}c \oplus \mathbb{C}d = \tilde{\mathfrak{h}} \cap \amalg_{n \geq 0} \tilde{g}_n,$$

a Lie subalgebra of $\tilde{\mathfrak{h}}$. For a $\mathbb{C}c\oplus\mathbb{C}d$-module W, define an $\tilde{\mathfrak{h}}$-module Ind W as follows: Extend the action of $\mathbb{C}c\oplus\mathbb{C}d$ on W to an action of $\tilde{\mathfrak{h}}^+$ by requiring that $\hat{\mathfrak{h}}_+ \cdot W = 0$. Take Ind W to be the induced $\tilde{\mathfrak{h}}$-module

$$\text{Ind } W = U(\tilde{\mathfrak{h}}) \otimes_{U(\tilde{\mathfrak{h}}^+)} W.$$

Standard properties of induced modules imply that there is a natural $\tilde{\mathfrak{h}}_-$-module isomorphism

(4.11) $$U(\tilde{\mathfrak{h}}_-) \otimes_{\mathbb{C}} W \xrightarrow{\sim} \text{Ind } W$$

$$u \otimes w \mapsto u \otimes w,$$

with $\tilde{\mathfrak{h}}_-$ acting on $U(\tilde{\mathfrak{h}}_-) \otimes W$ by left multiplication on the universal enveloping algebra $U(\tilde{\mathfrak{h}}_-)$.

For W the one-dimensional $\mathbb{C}c\oplus\mathbb{C}d$-module which is the space $\mathbb{C}$, with c acting as multiplication by the scalar k and d acting trivially, set

(4.12) $$K(k) = \text{Ind } W.$$

Then we have an $\tilde{\mathfrak{h}}_-$-module isomorphism

(4.13) $$S(\tilde{\mathfrak{h}}_-) = U(\tilde{\mathfrak{h}}_-) \xrightarrow{\sim} K(k)$$

$$u \mapsto u \otimes 1,$$

S denoting the symmetric algebra, and K(k) is a level k $\tilde{\mathfrak{h}}$-module on which d acts according to the adjoint action of d on $U(\tilde{\mathfrak{h}}_-)$, so that

(4.14) $$K(k) = \coprod_{n\geq 0} K(k)_{-n}.$$

Observe that $K(k) \simeq S(\tilde{\mathfrak{h}}_-)$ is isomorphic to the polynomial algebra on the infinitely many variables $\beta(-n)$ for $n > 0$, n odd. It follows from a standard elementary argument that $K(k)$ is irreducible as an $\tilde{\mathfrak{h}}$-module, and that for each $n>0$, n odd, $\beta(n)$ can be realized as the differential operator $n\partial/\partial\beta(-n)$ on

(4.15) $$K(k) = S(\hat{\mathfrak{h}}_-) = \mathbb{C}[\beta(-1),\ \beta(-3),\ \beta(-5),\ldots]$$

(the "uniqueness of the Heisenberg commutation relations (2.12)"). Also the vacuum space $\{x\in K(k) \mid \hat{\mathfrak{h}}_+\cdot x=0\}$ consists simply of the scalars in $S(\hat{\mathfrak{h}}_-)$.

Combining (4.11) and (4.13), we see that for a $\mathbb{C}c\oplus\mathbb{C}d$-module W of level k, we have a vector space isomorphism

(4.16) $$K(k)\otimes_{\mathbb{C}} W \xrightarrow{\sim} \text{Ind } W$$

$$(u\otimes 1)\otimes w \longmapsto u\otimes w$$

($u\in U(\tilde{\mathfrak{h}}_-)$, $w\in W$). This enables us to identify the $\tilde{\mathfrak{h}}$-module Ind W with the vector space $K(k)\otimes W$ equipped with the following action of $\tilde{\mathfrak{h}}$:

$$c \longmapsto k$$

(4.17) $$d \longmapsto d\otimes 1 + 1\otimes d$$

$$x \longmapsto x\otimes 1$$

for $x\in\hat{\mathfrak{h}}$.

We have the following analogue, for the Heisenberg algebra $\hat{\mathfrak{h}}$, of the classical Stone-Von Neumann theorem; The elementary self-contained proof is given in [14, Proposition 3.6]:

Proposition 4.7. The map

$$\text{(4.18)}\qquad \operatorname{Ind}\Omega_V = U(\tilde{\mathfrak{h}})\otimes_{U(\tilde{\mathfrak{h}}^+)}\Omega_V \longrightarrow V$$

$$u\otimes w \longmapsto u\cdot w$$

is an $\tilde{\mathfrak{h}}$-module isomorphism. In particular, regarded as an $\hat{\mathfrak{h}}$-module, V is a direct sum of copies of the irreducible $\hat{\mathfrak{h}}$-module K(k). ■

By (4.16)-(4.18), we now have:

Theorem 4.8. The correspondence

$$\text{(4.19)}\qquad K(k)\otimes_{\mathbb{C}}\Omega_V \longrightarrow V$$

$$(u\otimes 1)\otimes w \longmapsto u\cdot w$$

($u\in U(\tilde{\mathfrak{h}}_-)$, $w\in\Omega_V$; see (4.11), (4.12)) induces a well-defined linear isomorphism. The action of $\tilde{\mathfrak{g}}$ on V transports to the following action on $K(k)\otimes\Omega_V$: The subalgebra $\tilde{\mathfrak{h}}$ acts according to the formulas (4.17), and $x_\beta(n)$ ($n\in\mathbb{Z}$) acts according to the formula

$$\text{(4.20)}\qquad x_\beta(\zeta) \longmapsto E^-(-\beta,\zeta)E^+(-\beta,\zeta)\otimes z_\beta(\zeta). \quad ■$$

Hence, keeping in mind (4.6) and (4.20) (cf. (4.9)), we have:

Corollary 4.9. The correspondences

$$M \longmapsto U(\hat{\mathfrak{h}})\cdot M, \qquad N \longmapsto N\cap\Omega_V$$

define mutually inverse bijections beteween the set of all $\mathcal{Z}_V$-submodules M of Ω_V and the set of all $\tilde{\mathfrak{g}}$-submodules N of V. In particular, V is $\tilde{\mathfrak{g}}$-irreducible if and only if Ω_V is $\mathcal{Z}_V$-irreducible.

■

Thus we have completely reconstructed the $\tilde{\mathfrak{g}}$-module V from the $\mathcal{Z}_V$-module Ω_V. In later sections, we shall construct Ω_V (and therefore V) in several interesting cases.

Suppose that V is a vector space on which the element d of $\tilde{\mathfrak{g}}$ acts, such that

(4.21) $$V = \coprod_{n\geq 0} V_{-n}$$

(see (4.1)) and

(4.22) $$\dim V_{-n} < \infty \quad \text{for all } n \geq 0.$$

We define the *(principal) character* of V to be the formal power series in the indeterminate q given by

(4.23) $$\chi(V) = \Sigma_{n\geq 0}(\dim V_{-n})q^n.$$

It is clear from (4.14) and (4.15) that $\chi(K(k))$ is defined and

(4.24) $$\chi(K(k)) = \Pi_{n>0} (1-q^{2n-1})^{-1},$$

and so Theorem 4.8 implies:

Proposition 4.10. Suppose that V satisfies conditions (4.21) and (4.22). Then Ω_V does also, and

(4.25) $$\chi(V) = \chi(\Omega_V)\Pi_{n>0}(1-q^{2n-1})^{-1}.$$

Remark 4.11. In [16], the notations $Z(\beta,\zeta)$ and $Z_n(\beta)$ are used for $z_\beta(\zeta)$ and $z_\beta(n)$, respectively. The present notation makes the elements $z_\beta(n)$ appear analogous to the elements $x_\beta(n)$, as they should.

5. THE GENERALIZED COMMUTATION RELATIONS.

Here we establish the most interesting relations among the

generators of the algebras $\mathfrak{X}_V$.

Throughout this section, let $k \in \mathbb{C}^*$ *and* $(V,\pi) \in \mathcal{C}_k$.

Lemma 5.1. In the algebra of formal power series in ζ_1 and ζ_2^{-1} with coefficients in End V, we have

$$E^+(\beta,\zeta_1)E^-(\beta,\zeta_2) =$$

$$= E^-(\beta,\zeta_2)E^+(\beta,\zeta_1)\Pi_{p=0,1}(1-(-1)^p\zeta_1/\zeta_2)^{\langle\nu^p\beta,\beta\rangle/k}$$

$$= E^-(\beta,\zeta_2)E^+(\beta,\zeta_1)(1-\zeta_1/\zeta_2)^{2/k}(1+\zeta_1/\zeta_2)^{-2/k},$$

the last two factors being computed by means of the binomial expansion.

Proof. The bracket of the exponents in the exponentials $E^+(\beta,\zeta_1)$ and $E^-(\beta,\zeta_2)$ equals

$$-4\Sigma_{n>0,n\ \mathrm{odd}}(\zeta_1/\zeta_2)^n\pi(c)/nk^2 = -4\Sigma_{n>0,n\ \mathrm{odd}}(\zeta_1/\zeta_2)^n/nk$$

$$= 2k^{-1}\Sigma_{p=0,1}(-1)^p\log\,(1-(-1)^p\zeta_1/\zeta_2),$$

(using the logarithmic formal power series), and this expression commutes with everything. Since its exponential is

$$(1-\zeta_1/\zeta_2)^{2/k}(1+\zeta_1/\zeta_2)^{-2/k},$$

we obtain the desired result. ∎

We leave the next lemma as an exercise:

Lemma 5.2. As formal power series in ζ_1 (respectively, ζ_2^{-1}) and formal Laurent series in ζ_2 (respectively, ζ_1) with coefficients in End V,

$$E^+(\beta,\zeta_1)x_\beta(\zeta_2)=x_\beta(\zeta_2)E^+(\beta,\zeta_1)\Pi_{p=0,1}(1-(-1)^p\zeta_1/\zeta_2)^{-\langle\nu^p\beta,\beta\rangle/k}$$

$$x_\beta(\zeta_1)E^-(\beta,\zeta_2) =$$

$$= E^-(\beta,\zeta_2)x_\beta(\zeta_1)\Pi_{p=0,1}(1-(-1)^p\zeta_1/\zeta_2)^{-\langle\nu^p\beta,\beta\rangle/k}. \quad \blacksquare$$

Combining Lemma 5.1 and 5.2, we have:

Proposition 5.3. In (End V)$[\zeta_1,\zeta_2]$,

$$z_\beta(\zeta_1)z_\beta(\zeta_2) = (1-\zeta_1/\zeta_2)^{-2/k}(1+\zeta_1/\zeta_2)^{2/k} \cdot$$

$$\cdot\ E^-(\beta,\zeta_1)E^-(\beta,\zeta_2)x_\beta(\zeta_1)x_\beta(\zeta_2)E^+(\beta,\zeta_1)E^+(\beta,\zeta_2).$$

Now we multiply both sides by the inverse of the first factor on the right and then subtract the expression obtained by interchanging the roles of the subscripts 1 and 2. The result is:

Corollary 5.4. In (End V)$[\zeta_1,\zeta_2]$,

$$(1-\zeta_1/\zeta_2)^{2/k}(1+\zeta_1/\zeta_2)^{-2/k}z_\beta(\zeta_1)z_\beta(\zeta_2) -$$

(5.1) $$(1-\zeta_2/\zeta_1)^{2/k}(1+\zeta_2/\zeta_1)^{-2/k}z_\beta(\zeta_2)z_\beta(\zeta_1) =$$

$$= E^-(\beta,\zeta_1)E^-(\beta,\zeta_2)[x_\beta(\zeta_1),\ x_\beta(\zeta_2)]E^+(\beta,\zeta_1)E^+(\beta,\zeta_2).$$

We combine this with (3.12) and use the following extension of (3.3):

Lemma 5.5. Let V be a vector space and let

$$f(\zeta_1,\zeta_2) = \Sigma_{m,n\in\mathbf{Z}}v_{mn}\zeta_1^m\zeta_2^n\in V[\zeta_1,\zeta_2],$$

where each $v_{mn}\in V$, and suppose that for every $p\in\mathbf{Z}$, $v_{mn}=0$ for all but finitely many pairs (m,n) with m+n = p. Set

$$(D_\ell f)(\zeta_1,\zeta_2) = \zeta_\ell(\partial f/\partial\zeta_\ell)(\zeta_1,\zeta_2) \in V\{\zeta_1,\zeta_2\},$$

$\ell = 1,2$. Let $a\in\mathbb{C}^*$. Then in $V\{\zeta_1,\zeta_2\}$,

(5.2) $$\delta(a\zeta_1/\zeta_2)f(\zeta_1,\zeta_2) = \delta(a\zeta_1/\zeta_2)f(a^{-1}\zeta_2,\zeta_2),$$

(5.3) $$(D\delta)(a\zeta_1/\zeta_2)f(\zeta_1,\zeta_2) =$$

$$= (D\delta)(a\zeta_1/\zeta_2)f(a^{-1}\zeta_2,\zeta_2)-\delta(a\zeta_1/\zeta_2)(D_1f)(a^{-1}\zeta_2,\zeta_2).$$

<u>**Proof**</u>. Formula (5.2) is easily checked directly, and (5.3) follows by differentiating with respect to ζ_1 and then using (5.2). ■

By straightforward computation starting from (5.1), applying both sides to a typical vector in V to confirm the applicability of Lemma 5.5, we obtain the "generalized commutation relation":

<u>**Theorem**</u> **5.6.** Let $k\in\mathbb{C}^*$, $V\in\mathcal{C}_k$. Then in $(\mathrm{End}\ V)\{\zeta_1,\zeta_2\}$,

$$(1-\zeta_1/\zeta_2)^{2/k}(1+\zeta_1/\zeta_2)^{-2/k}z_\beta(\zeta_1)z_\beta(\zeta_2)$$

(5.4) $$-(1-\zeta_2/\zeta_1)^{2/k}(1+\zeta_2/\zeta_1)^{-2/k}z_\beta(\zeta_2)z_\beta(\zeta_1) =$$

$$= -\tfrac{1}{4}k(D\delta)(-\zeta_1/\zeta_2). \quad ■$$

To understand the significance of this formula, define the following expansion coefficients $a_n\in\mathbb{C}$:

(5.5) $$(1-\zeta)^{2/k}(1+\zeta)^{-2/k} = \Sigma_{n\geq 0}a_n\zeta^n.$$

We have $a_0=1$. Formula (5.4) is equivalent to the following infinite sequence of identities, obtained by equating the coefficients of $\zeta_1^m\zeta_2^n$ $(m,n\in\mathbb{Z})$ on the two sides:

(5.6) $$\Sigma_{i\geq 0}a_i(z_\beta(m-i)z_\beta(n+i) - z_\beta(n-i)z_\beta(m+i)) =$$

$$= -\tfrac{1}{4}k(-1)^m m\,\delta_{m+n,0}\ .$$

Notice that the formal infinite sum on the left is indeed a well-defined operator on V.

Remark 5.7. If (5.5) had been replaced by the formal Laurent series 1, then (5.4) would have been replaced by a sequence of commutation relations, and this is why we call relations of the type (5.6) (or their "generating function" (5.4)) "generalized commutation relations".

6. THE CATEGORY $\mathcal{D}_k$.

Here we demonstrate the foundational role played by the generalized commutation relations by indicating how they can be used to construct $\tilde{g}$-modules. *Throughout this section, we fix* $k\in\mathbb{C}^*$.

Definition 6.1. Let $\mathcal{S}$ be the disjoint union

$$\mathcal{S} = \{c\} \cup \{d\} \cup \mathbb{Z}.$$

By an $\mathcal{S}$-module we shall mean a vector space W equipped with actions of c, d and the integers. Denote the operator corresponding to $n\in\mathbb{Z}$ by $z_{\beta}(n)$ and set

$$z_{\beta}(\zeta) = \Sigma_{n\in\mathbb{Z}} z_{\beta}(n)\zeta^n \in (\text{End } W)\{\zeta\}.$$

We define $\mathcal{D}_k$ to be the category of $\mathcal{S}$-modules W such that conditions (i)–(iii) in Definition 4.2 hold, together with the following additional conditions:

$$[d,z_{\beta}(\zeta)] = Dz_{\beta}(\zeta),$$

i.e., $z_{\beta}(n)$ is an operator of degree n for each $n\in\mathbb{Z}$, and finally, the generalized commutation relations (5.4) hold.

Using Proposition 4.6, formula (4.7) and Theorem 5.6, we define a functor

(6.1) $$\Omega\colon \mathcal{C}_k \longrightarrow \mathcal{D}_k$$
$$V \longmapsto \Omega_V$$

which takes a map $V \longrightarrow W$ in $\mathcal{C}_k$ to its restriction $\Omega_V \longrightarrow \Omega_W$ to the vacuum spaces.

An argument very similar to that which proves the generalized commutation relations (5.4) enables us to start from these relations and conclude with the bracket structure of $\tilde{g}$ as formulated in (3.12):

Theorem 6.2. Given $W \in \mathcal{D}_k$, the space $K(k)\otimes_{\mathbb{C}} W$ (see (4.16)), equipped with the action of $\tilde{g}$ indicated in (4.17) and (4.20), becomes a $\tilde{g}$-module in the category $\mathcal{C}_k$. ■

Hence we can define the functor

(6.2) $$\Lambda\colon \mathcal{D}_k \longrightarrow \mathcal{C}_k$$
$$W \longmapsto K(k)\otimes W = \text{Ind } W$$

which takes a map $W \longrightarrow W'$ in $\mathcal{D}_k$ to its induced $\tilde{h}$-module map $\text{Ind } W \longrightarrow \text{Ind } W'$, which is also easily checked to be a $\tilde{g}$-module map.

Using Theorem 4.8, we now have:

Theorem 6.3. The functors Ω and Λ (see (6.1), (6.2)) are exact and they define equivalences between the categories $\mathcal{C}_k$ and $\mathcal{D}_k$.

Remark 6.4. In [16], we have introduced a universal algebra $\mathfrak{A}_k$ for which the relations (5.4) are rigorously taken as the main set of defining relations, and a category of $\mathfrak{A}_k$-modules isomorphic to the category $\mathcal{D}_k$ of $\mathcal{S}$-modules defined in Definition 6.1. Essentially, the algebra $\mathfrak{A}_k$ plays the role of a "universal preimage" of all the algebras

$\mathfrak{X}_V$ for $V \in \mathcal{C}_k$. Theorem 6.3 makes precise our viewpoint that the study of the $\tilde{g}$-module V can be reduced to the study of Ω_V (viewed as either an $\mathcal{S}$-module or an $\mathfrak{A}_k$-module).

7. THE PRINCIPALLY GRADED BASIC $\tilde{g}$-MODULES.

By Theorem 6.2, we can construct a $\tilde{g}$-module in $\mathcal{C}_k$ whenever we can produce an $\mathcal{S}$-module in $\mathcal{D}_k$. Inspection of the generalized commutation relations (5.4) quickly leads to the simplest possible ways of doing this:

Proposition 7.1. Let $k \in \mathbb{C}^*$ and let W be an $\mathcal{S}$-module of dimension 1 (see Definition 6.1). Then $W \in \mathcal{D}_k$ if and only if

(i) k=1, i.e., c acts as 1

(ii) d acts as a scalar

(iii) $z_\beta(\zeta) = \pm\frac{1}{4}(-1)^{1/2}$, i.e., $z_\beta(0) = \pm\frac{1}{4}(-1)^{1/2}$ and $z_\beta(n) = 0$ for $n \neq 0$. ■

From Theorem 6.2, we now have the original vertex operator construction of the basic $A_1^{(1)}$-modules in the principal picture [12], showing that $A_1^{(1)}$ can be realized by formal differential operators (cf. Remark 7.7). By Theorems 4.8 and 6.3, we also obtain a natural characterization of these modules:

Theorem 7.2. On the $\tilde{h}$-module

$$\text{Ind } W = K(1) \otimes W = S(\hat{h}_-) \otimes W \tag{7.1}$$

(see (4.11)-(4.17)), let $x_\beta(n)$ $(n \in \mathbb{Z})$ act as the coefficient of ζ^n in the formula

$$x_\beta(\zeta) = \pm\frac{1}{4}(-1)^{1/2} E^-(-\beta,\zeta) E^+(-\beta,\zeta) \otimes 1$$

$$(7.2) \qquad = \pm(-1)^{1/2}\exp(2\Sigma_{m>0,m\ \mathrm{odd}}\beta(-m)\zeta^{-m}/m)\ \cdot$$

$$\cdot\ \exp(-2\Sigma_{m>0,m\ \mathrm{odd}}\beta(m)\zeta^{m}/m)\otimes 1$$

(taking $k=1$, $\gamma=-\beta$ in (4.4)). Then Ind W becomes a $\tilde{\mathfrak{g}}$-module in the category $\mathcal{C}_1$. Conversely, if $k\in\mathbb{C}^*$ and the irreducible $\tilde{\mathfrak{g}}$-module V lies in $\mathcal{C}_k$ and remains irreducible under the principal Heisenberg subalgebra $\hat{\mathfrak{h}}$ (or the subalgebra $\tilde{\mathfrak{h}}$), then $k=1$ and V is isomorphic to one of the modules Ind W just described. ■

Definition 7.3. The $\tilde{\mathfrak{g}}$-modules characterized and explicitly described in Theorem 7.2 are called the *(principally graded) basic* $\tilde{\mathfrak{g}}$-modules.

Remark 7.4. By abuse of notation, we sometimes drop the "$\otimes 1$" in formula (7.2).

Remark 7.5. Under the subalgebra $\hat{\mathfrak{g}}$ of $\tilde{\mathfrak{g}}$ (see (2.5)), and even under the Heisenberg subalgebra $\hat{\mathfrak{h}}$, the basic modules remain irreducible. Two inequivalent $\hat{\mathfrak{g}}$-modules arise this way, corresponding to the two signs in (7.2). We sometimes call these modules the *basic* $\hat{\mathfrak{g}}$-modules.

The character (4.23) of a basic module is very important to consider. By formula (4.24), we have:

Proposition 7.6. Suppose that d acts as the scalar 0 on W (in the notation of Proposition 7.1 and Theorem 7.2). Then each of the two basic modules Ind W (see (7.1) and (7.2)) satisfies conditions (4.21) and (4.22), and

$$(7.3) \qquad \chi(\mathrm{Ind}\ W) = \Pi_{n>0}(1-q^{2n-1})^{-1}.$$

Remark 7.7. A significant aspect of Theorem 7.2 is that the following formal differential operators on $S(\hat{\mathfrak{h}}_-)$ (see (4.15)) span a Lie algebra (namely, $\hat{\mathfrak{g}}$):

$$1,\ \beta(-n)\ (n<0,\ n\ \text{odd}),\ \partial/\partial\beta(-n)\quad (n>0,\ n\ \text{odd})$$

and the coefficients of ζ^n ($n\in\mathbb{Z}$) in

$$\exp(2\textstyle\sum_{m>0,m\ \text{odd}}\beta(-m)\zeta^{-m}/m)\exp(-2\textstyle\sum_{m>0,m\ \text{odd}}(\partial/\partial\beta(-m))\zeta^{m}).$$

8. THE TENSOR PRODUCT OF TWO BASIC $\tilde{\mathfrak{g}}$-MODULES AND THE PRINCIPALLY GRADED STANDARD LEVEL 2 $\tilde{\mathfrak{g}}$-MODULES.

We define the (principally graded) standard level 2 $\tilde{\mathfrak{g}}$-modules to be the submodules generated by "highest d-eigenvectors" in tensor products of two basic modules (see Definition 8.6). In this section, we construct these modules explicitly.

Let $k\in\mathbb{C}^*$ and $(V,\pi)\in\mathcal{C}_k$. Also let $r\in\mathbb{C}^*$. We generalize the elements (4.4) as follows: Define

$$E^{\pm}(\gamma,\zeta,\pi,r) = \exp(2\textstyle\sum_{n>0,n\ \text{odd}}\pi(\gamma(\pm n))\zeta^{\pm n}/(\pm n)r)) \tag{8.1}$$

for $\gamma\in\mathfrak{h}$. Note that

$$E^{\pm}(\gamma,\zeta) = E^{\pm}(\gamma,\zeta,\pi,k).$$

We also generalize the elements (4.6):

$$z_{\pm\beta}(\zeta,\pi,r) = E^{-}(\pm\beta,\zeta,\pi,r)x_{\pm\beta}(\zeta)E^{+}(\pm\beta,\zeta,\pi,r), \tag{8.2}$$

so that

$$z_{\pm\beta}(\zeta) = z_{\pm\beta}(\zeta,\pi,k)$$

(see (4.6)).

For each $i=1,2$, let $(V_i,\pi_i)\in\mathcal{C}_1$ be a basic $\tilde{\mathfrak{g}}$-module (see

Definition 7.3). We want to study the tensor product $\tilde{g}$-module

$$(V,\pi) = (V_1 \otimes V_2, \pi_1 \otimes \pi_2).$$

We have

$$(V,\pi) \in \mathcal{C}_2,$$

and standard properties of tensor products give:

$$E^{\pm}(\gamma,\zeta,\pi,r) = E^{\pm}(\gamma,\zeta,\pi_1,r) \otimes E^{\pm}(\gamma,\zeta,\pi_2,r) \tag{8.3}$$

$$\pi x_{\pm\beta}(\zeta) = \pi_1 x_{\pm\beta}(\zeta) \otimes 1 + 1 \otimes \pi_2 x_{\pm\beta}(\zeta). \tag{8.4}$$

Using (8.3) and (8.4), we obtain:

$$z_{\pm\beta}(\zeta) = z_{\pm\beta}(\zeta,\pi) = z_{\pm\beta}^{(1)}(\zeta) + z_{\pm\beta}^{(2)}(\zeta), \tag{8.5}$$

where

$$z_{\pm\beta}^{(1)}(\zeta) = E^{-}(\pm\beta,\zeta,\pi,2)(\pi_1 x_{\pm\beta}(\zeta) \otimes 1)E^{+}(\pm\beta,\zeta,\pi,2) \tag{8.6}$$

$$= z_{\pm\beta}(\zeta,\pi_1,2) \otimes E^{-}(\pm\beta,\zeta,\pi_2,2)E^{+}(\pm\beta,\zeta,\pi_2,2),$$

$$z_{\pm\beta}^{(2)}(\zeta) = E^{-}(\pm\beta,\zeta,\pi,2)(1 \otimes \pi_2 x_{\pm\beta}(\zeta))E^{+}(\pm\beta,\zeta,\pi,2) \tag{8.7}$$

$$= E^{-}(\pm\beta,\zeta,\pi_1,2)E^{+}(\pm\beta,\zeta,\pi_1,2) \otimes z_{\pm\beta}(\zeta,\pi_2,2).$$

Note that

$$z_{-\beta}^{(i)}(\zeta) = z_{\beta}^{(i)}(-\zeta) \quad \text{for } i = 1,2. \tag{8.8}$$

Imitating the relevant arguments concerning $z_{\pm\beta}(\zeta)$ in Section 4, we have:

(8.9) $$[(\hat{\mathfrak{h}}), z^{(i)}_{\pm\beta}(\zeta)] = 0,$$

(8.10) $$[\pi(d), z^{(i)}_{\pm\beta}(\zeta)] = Dz^{(i)}_{\pm\beta}(\zeta)$$

in $(\mathrm{End}\ V)\{\zeta\}$, for $i = 1,2$.

Remark 8.1. Property (8.9) motivates the definition of $z^{(i)}_{\pm\beta}(\zeta)$. By suitably generalizing the definitions in Section 4, one can arrange for $z^{(i)}_{\pm\beta}(\zeta)$ ((8.6), (8.7)) and $z_{\pm\beta}(\zeta)$ ((4.6), (8.5)) to be special cases of a single construction. The corresponding general concept of "$\mathfrak{Z}$-algebra" is what is actually developed in [16].

Now we invoke the construction of the basic modules (Theorem 7.2) to obtain that for i=1,2,

(8.11) $$z_{\beta}(\zeta,\pi_i,2) = \pm\tfrac{1}{4}(-1)^{1/2}E^{-}(-\beta,\zeta,\pi_i,2)E^{+}(-\beta,\zeta,\pi_i,2)$$

(see Remark 7.4), where the sign depends on the choice of sign in (7.2).

Proposition 8.2. For i=1,2, we have

(8.12) $$z^{(i)}_{\beta}(\zeta_1)z^{(i)}_{\beta}(\zeta_2) + z^{(i)}_{\beta}(\zeta_2)z^{(i)}_{\beta}(\zeta_1) =$$
$$= -\tfrac{1}{8}\delta(-\zeta_1/\zeta_2)$$

in $(\mathrm{End}\ V)\{\zeta_1,\zeta_2\}$.

Proof. By (8.6), (8.7), (8.11) and a natural analogue of Lemma 5.1 for the elements $E^{+}(\pm\beta,\zeta,\pi_i,2)$ and $E^{-}(\pm\beta,\zeta,\pi_i,2)$, we see that the left-hand side of (8.12) equals

$$-\frac{1}{16}\left[(1-\zeta_1/\zeta_2)(1+\zeta_1/\zeta_2)^{-1} + (1-\zeta_2/\zeta_1)(1+\zeta_2/\zeta_1)^{-1}\right]Q =$$
$$= -\tfrac{1}{8}\delta(-\zeta_1/\zeta_2)Q,$$

where Q is the tensor product of

$$E^-(-\beta,\zeta_1,\pi_i,2)E^-(-\beta,\zeta_2,\pi_i,2)E^+(-\beta,\zeta_1,\pi_i,2)E^+(-\beta,\zeta_2,\pi_i,2)$$

in the ith tensorand and

$$E^-(\beta,\zeta_1,\pi_j,2)E^-(\beta,\zeta_2,\pi_j,2)E^+(\beta,\zeta_1,\pi_j,2)E^+(\beta,\zeta_2,\pi_j,2)$$

in the jth tensorand ($j\neq i$). The proposition now follows from (5.2). ■

Formulas (8.6), (8.7) and (8.11) give:

<u>Proposition</u> **8.3.** We have

$$z_{\beta}^{(1)}(\zeta_1) = \pm z_{-\beta}^{(2)}(\zeta_1), \tag{8.13}$$

where the plus sign is used if and only if the basic modules V_1, V_2 are equivalent when viewed as $\hat{g}$-modules (see Remark 7.5). ■

Recalling (4.8), we also define the operators $z_{\pm\beta}^{(i)}(n)\in\mathrm{End}\ V$ for $i=1,2$ and $n\in\mathbb{Z}$ by:

$$z_{\pm\beta}^{(i)}(\zeta) = \Sigma_{n\in\mathbb{Z}} z_{\pm\beta}^{(i)}(n)\zeta^n. \tag{8.14}$$

Now we extract the coefficient of $\zeta_1^m\zeta_2^n$ in (8.12) for $m,n\in\mathbb{Z}$ to obtain:

$$\{z_{\beta}^{(i)}(m),z_{\beta}^{(i)}(n)\} = -\tfrac{1}{8}(-1)^m\delta_{m+n,0}, \tag{8.15}$$

$\{\cdot,\cdot\}$ denoting anticommutator. Similarly, (8.13) and (8.14) give

$$z_{\beta}^{(2)}(n) = (-1)^n z_{\beta}^{(1)}(n) \tag{8.16}$$

if $V_1\cong V_2$ as $\hat{g}$-modules and

$$z_{\beta}^{(2)}(n) = (-1)^{n+1} z_{\beta}^{(1)}(n) \tag{8.17}$$

if $V_1 \not\simeq V_2$, for all $n \in \mathbf{Z}$. Since

$$z_{\beta}(n) = z_{\beta}^{(1)}(n) + z_{\beta}^{(2)}(n), \tag{8.18}$$

we have

$$\begin{aligned} z_{\beta}(n) &= 2z_{\beta}^{(1)}(n) && \text{for } n \in 2\mathbf{Z} \\ z_{\beta}(n) &= 0 && \text{for } n \in 2\mathbf{Z}+1 \end{aligned} \tag{8.19}$$

if $V_1 \simeq V_2$ and

$$\begin{aligned} z_{\beta}(n) &= 0 && \text{for } n \in 2\mathbf{Z} \\ z_{\beta}(n) &= 2z_{\beta}^{(1)}(n) && \text{for } n \in 2\mathbf{Z}+1 \end{aligned} \tag{8.20}$$

if $V_1 \not\simeq V_2$.

Definition 8.4. (cf. Definition 4.3). Denote by $\mathfrak{Z}'_{V_1,V_2}$ the subalgebra of End V generated by the operators $z_{\beta}^{(1)}(n)$ for $n \in \mathbf{Z}$ (or equivalently, in view of (8.8), (8.16) and (8.17), by the operators $z_{\pm\beta}^{(i)}(n)$ for $n \in \mathbf{Z}$ and $i=1,2$). Denote by $\mathfrak{Z}_{V_1,V_2}$ the subalgebra of End V generated by these operators together with $\pi(d) = \pi_1(d) \otimes 1 + 1 \otimes \pi_2(d)$.

Proposition 8.5. For $i=1,2$, let $v_0^{(i)}$ be a nonzero vector in the one-dimensional space Ω_{V_i}, and set

$$v_0 = v_0^{(1)} \otimes v_0^{(2)} \in V. \tag{8.21}$$

Then $v_0 \in \Omega_V$ and

(8.22) $$\Omega_V = \mathcal{Z}_{V_1,V_2}v_0 = \mathcal{Z}'_{V_1,V_2}v_0.$$

Proof. It is clear that $v_0 \in \Omega_V$. To prove (8.22), observe that under the action of the operators

$$x_\beta(n)\otimes 1 \text{ and } 1\otimes x_\beta(n) \text{ for } n\in\mathbb{Z},$$

the vector v_0 generates V. By (8.9), $\mathcal{Z}_{V_1,V_2}v_0 \subset \Omega_V$, and (8.22) now follows from Theorem 4.8 and the definitions (8.6), (8.7). ∎

Definition 8.6. The *(principally graded) standard* $\tilde{\mathfrak{g}}$*-module associated with the principally graded basic* $\tilde{\mathfrak{g}}$*-modules* V_1 *and* V_2 is the $\tilde{\mathfrak{g}}$-module

(8.23) $$L = U(\tilde{\mathfrak{g}})\cdot v_0$$

generated by v_0 in V. Note that L has level 2. A *(principally graded) standard level* 2 $\tilde{\mathfrak{g}}$*-module* is a $\tilde{\mathfrak{g}}$-module equivalent to such a module for some pair of basic modules.

Remark 8.7. This definition, natural for the present theory, is equivalent to the usual definition, in the theory of Kac-Moody algebras, of principally graded standard level 2 $\tilde{\mathfrak{g}}$-module.

Using Theorem 4.8, we clearly have:

Proposition 8.8. The $\tilde{\mathfrak{g}}$-module L (see (8.23)) lies in the category $\mathcal{C}_2$, and

(8.24) $$\Omega_L = \mathcal{Z}_V v_0 = \mathcal{Z}'_V v_0 = \mathcal{Z}_L v_0 = \mathcal{Z}'_L v_0.$$

In view of the anticommutation relations (8.15) and formulas (8.16)-(8.20), together with easy analogue for Clifford algebras of the "uniqueness of the Heisenberg commutation relations" (see the discussion surrounding (4.15)), we have:

Theorem **8.9.** (1) (Frenkel) The algebra $\mathcal{Z}'_{V_1, V_2}$ is a Clifford algebra generated by the space $\mathfrak{z}^V$ with basis $\{z_\beta^{(1)}(n) \mid n \in \mathbb{Z}\}$. Set

$$\mathfrak{z}_-^V = \text{span}\, \{z_\beta^{(1)}(n) \mid n<0\} \subset \mathfrak{z}^V.$$

Then

(8.25) $$\Omega_V \simeq \Lambda(\mathfrak{z}_-^V)$$

(see (8.22)), an exterior algebra with basis

$$\{z_\beta^{(1)}(n_1) \dots z_\beta^{(1)}(n_r)v_0 \mid r \geq 0,\ n_1, \dots, n_r \in \mathbb{Z},\ n_1 < \dots < n_r < 0\}.$$

(2) If $V_1 \simeq V_2$ (respectively, $V_1 \not\simeq V_2$) as $\hat{\mathfrak{g}}$-modules, then $\mathcal{Z}'_V \simeq \mathcal{Z}'_L$ is a Clifford algebra generated by the space $\mathfrak{z}^L$ with basis

$$\{z_\beta(n) = 2z_\beta^{(1)}(n) \mid n \in 2\mathbb{Z} \text{ (resp., } 2\mathbb{Z}+1)\}.$$

Set

$$\mathfrak{z}_-^L = \text{span}\{z_\beta(n) \mid n<0,\ n \in 2\mathbb{Z} \quad \text{(resp., } 2\mathbb{Z}+1)\} \subset \mathfrak{z}^L.$$

Then

(8.26) $$\Lambda_L \simeq \Lambda(\mathfrak{z}_-^L)$$

(see (8.24)), an exterior algebra with basis

$$\{z_\beta(n_1) \dots z_\beta(n_r)v_0 \mid r \geq 0,\ n_1, \dots, n_r \in 2\mathbb{Z} \text{ (resp., } 2\mathbb{Z}+1),\ n_1 < \dots < n_r < 0\}. \ \blacksquare$$

Remark **8.10.** Using Theorem 4.8 as in Theorem 7.2, we could now of course write down explicit constructions of the $\tilde{\mathfrak{g}}$-modules $V = V_1 \otimes V_2$ and L as tensor products of the symmetric algebra $S(\tilde{\mathfrak{h}}_-)$ with the exterior algebras $\Lambda(3_-^V)$ or $\Lambda(3_-^L)$ (see (8.25), (8.26)). Note that these constructions exhibit a "supersymmetry" phenomenon, since the affine Lie algebra $A_1^{(1)}$ acts naturally and irreducibly on these tensor products, mixing boson states (elements of the symmetric algebra) and fermion states (elements of the exterior algebra).

<u>Remark</u> **8.11.** We relate an interesting chain of events: At a certain stage (reflected in [15]) of our work on $\mathcal{Z}$-algebras, we had completed among other things the $\mathcal{Z}$-algebra construction of the standard $A_1^{(1)}$-modules of level 2 (see Theorem 8.9(2)) and level 3 (see Theorem 9.8), thereby interpreting the Rogers-Ramanujan identities, but we still needed to use these identities to prove the linear independence of the basis (9.8) of the vacuum space for level 3. Meanwhile, Date, Jimbo, Kashiwara and Miwa had just circulated their preprints, including [2], [3], which among other things related our vertex operator [12] and our earlier study of Rogers-Ramanujan [13], [14] to the infinitesimal Bäcklund transformation generating the soliton solutions of the KdV equation. Frenkel then used our $\mathcal{Z}$-algebras in [4] to illuminate further their interpretation of soliton equations (see [5] for related material). In particular, he proved Theorem 8.9(1). This result inspired our discovery of Theorem 9.6, which gave us the linear independence proof in Theorem 9.8. Thus our present linear independence proof for level 3 was based on a cycle of ideas filtering through soliton theory. We are indebted to all these people for their communications with us.

Concerning characters, we have from (8.26):

<u>Proposition</u> **8.12.** Suppose that each basic module V_1, V_2 has highest d-eigenvalue 0. Then the standard $\tilde{g}$-module L and its subspace Ω_L satisfy conditions (4.12) and (4.22). If $V_1 \simeq V_2$ as $\hat{g}$-modules (and hence as $\tilde{g}$-modules), then

$$(8.27) \qquad \chi(\Omega_L) = \Pi_{n>0}(1+q^{2n})$$

and if $V_1 \not\simeq V_2$ as $\hat{g}$-modules, then

(8.28) $$\chi(\Omega_L) = \Pi_{n>0}(1+q^{2n-1}). \qquad \blacksquare$$

<u>**Remark**</u> **8.13.** By (8.25), we also have

$$\chi(\Omega_V) = \Pi_{n>0}(1+q^n)^{-1}.$$

This result also follows from (4.25) and (7.3), since

$$\Pi_{n>0}(1-q^{2n-1})^{-1} = \Pi_{n>0}(1-q^{2n})/(1-q^n)$$

$$= \Pi_{n>0}(1+q^n).$$

9. THE PRINCIPALLY GRADED STANDARD LEVEL 3 $\tilde{\mathfrak{g}}$-MODULES.

The level 3 case, presented here, is the first "deep" special case of the theory of $\mathcal{Z}$-algebras, since the relevant generalized commutation and anticommutation relations (9.2)-(9.5) involve infinite series of operators, in contrast with say (8.12) and (8.15). In this section, the level 3 standard modules are constructed explicitly inside the tensor products of three basic modules.

For each i=1,2,3, let $(V_i,\pi_i)\in\mathcal{C}_1$ be a basic $\tilde{\mathfrak{g}}$-module. We shall study the tensor product

$$(V,\pi) = (V_1\otimes V_2\otimes V_3,\ \pi_1\otimes\pi_2\otimes\pi_3) \in \mathcal{C}_3.$$

Recalling (8.1) and (8.2), we have obvious analogues of (8.3)-(8.10) and of Remark 8.1. In particular, we have

(9.1) $$z_{\pm\beta}(\zeta) = z^{(1)}_{\pm\beta}(\zeta) + z^{(2)}_{\pm\beta}(\zeta) + z^{(3)}_{\pm\beta}(\zeta),$$

where

$$z^{(1)}_{\pm\beta}(\zeta) = E^-(\pm\beta,\zeta,\pi,3)(\pi_1 x_{\pm\beta}(\zeta)\otimes 1\otimes 1)E^+(\pm\beta,\zeta,\pi,3),$$

and similarly for $z^{(2)}_{\pm\beta}(\zeta)$, $z^{(3)}_{\pm\beta}(\zeta)$. The analogue of (8.11) states:

$$z_\beta(\zeta,\pi_j,3) = \pm\tfrac{1}{4}(-1)^{1/2}E^-(-\beta,\zeta,\pi_j,3)^2E^-(-\beta,\zeta,\pi_j,3)^2$$

for j=1,2,3, and the analogue of Proposition 8.2, proved the same way, asserts:

Proposition 9.1. For i=1,2,3, we have

$$(9.2)\quad \begin{aligned} &(1+\zeta_1/\zeta_2)^{1/3}(1-\zeta_1/\zeta_2)^{-1/3}z^{(i)}_\beta(\zeta_1)z^{(i)}_\beta(\zeta_2) + \\ &+ (1+\zeta_2/\zeta_1)^{1/3}(1-\zeta_2/\zeta_1)^{-1/3}z^{(i)}_\beta(\zeta_2)z^{(i)}_\beta(\zeta_1) = \\ &= -\tfrac{1}{8}\delta(-\zeta_1/\zeta_2). \quad \blacksquare \end{aligned}$$

Remark 9.2. In the same spirit that we call (5.4) a "generalized commutation relation", we sometimes call identities of type (9.2) "generalized anticommutation relations" (cf. Remark 5.7). (The analogous result (8.12) for the level 2 case is a strict anticommutation relation.)

There is an additional generalized anticommutation relation which is an analogue of Proposition 8.3. The proof is automatic:

Proposition 9.3. We have

$$(9.3)\quad \begin{aligned} &(1+\zeta_1/\zeta_2)^{1/3}(1-\zeta_1/\zeta_2)^{-1/3}z^{(1)}_\beta(\zeta_1)z^{(2)}_\beta(\zeta_2) + \\ &+ (1+\zeta_2/\zeta_1)^{1/3}(1-\zeta_2/\zeta_1)^{-1/3}z^{(2)}_\beta(\zeta_2)z^{(1)}_\beta(\zeta_1) = \\ &= \pm\tfrac{1}{2}(-1)^{1/2}\delta(\zeta_1/\zeta_2)z^{(3)}_{-\beta}(\zeta_1), \end{aligned}$$

the sign being the product of the signs in (7.2) for V_1, V_2 and V_3. More generally, the superscripts (1), (2), (3) in (9.3) may be replaced by $(\tau 1)$, $(\tau 2)$, $(\tau 3)$ for any permutation τ of (1,2,3). ■

Formulas (9.1)–(9.3) give:

Corollary 9.4. We have

$$(9.4)\qquad \begin{aligned}&(1+\zeta_1/\zeta_2)^{1/3}(1-\zeta_1/\zeta_2)^{-1/3}z_\beta(\zeta_1)z_\beta(\zeta_2)+\\ &+(1+\zeta_2/\zeta_1)^{1/3}(1-\zeta_2/\zeta_1)^{-1/3}z_\beta(\zeta_2)z_\beta(\zeta_1)=\\ &=-\tfrac{3}{8}\delta(-\zeta_1/\zeta_2)\pm(-1)^{1/2}\delta(\zeta_1/\zeta_2)z_{-\beta}(\zeta_1),\end{aligned}$$

the sign being the same as in (9.3). ■

The generalized commutation relation (5.4) will be needed below, together with one more such relation, proved either by imitating the proof of Theorem 5.6 or preferably by carrying out the theory of $\mathcal{Z}$-algebras in the generality of [16], as mentioned in Remark 8.1, and quoting the resulting generalization of Theorem 5.6:

Proposition 9.5. For $i,j\in\{1,2,3\}$ with $i\neq j$,

$$(9.5)\qquad \begin{aligned}&(1-\zeta_1/\zeta_2)^{2/3}(1+\zeta_1/\zeta_2)^{-2/3}z_\beta^{(i)}(\zeta_1)z_\beta^{(j)}(\zeta_2)\\ &-(1-\zeta_2/\zeta_1)^{2/3}(1+\zeta_2/\zeta_1)^{-2/3}z_\beta^{(j)}(\zeta_2)z_\beta^{(i)}(\zeta_1)=0.\end{aligned}$$ ■

Consider the components $z_\beta(n)$ and $z_\beta^{(i)}(n)$ $(n\in\mathbb{Z},\ i=1,2,3)$ of $z_\beta(\zeta)$ and $z_\beta^{(i)}(\zeta)$, respectively, defined by (4.8) and (8.14). By equating coefficients, we produce from each of the identities (9.2)-(9.5) and (5.4) an infinite sequence of identities of the type (5.6) involving these components. We shall use these identities to "straighten" monomials in the elements $z_\beta(n)$ or $z_\beta^{(i)}(n)$ applied to the vector

$$v_0=v_0^{(1)}\otimes v_0^{(2)}\otimes v_0^{(3)}\in\Omega_V\subset V$$

(cf. (8.21)), where $v_0^{(i)}$ is defined exactly as in Proposition 8.5. It is clear that monomials in these elements applied to v_0 lie in Ω_V. The following result is analogous to Theorem 8.9(1):

Theorem 9.6. The space Ω_V has basis consisting of the elements

$$z_\beta^{(i_1)}(n_1)\cdots z_\beta^{(i_r)}(n_r)v_0, \tag{9.6}$$

where $n_j \in \mathbb{Z}$, $i_j \in \{1,2\}$,

$$n_1 \leqslant \cdots \leqslant n_r < 0$$

and if $n_j = n_{j+1}$ then $i_j < i_{j+1}$. In particular

$$\chi(\Omega_V) = \Pi_{n>0}(1+q^n)^2. \tag{9.7}$$

Proof. Without loss of generality, we may assume that each basic module V_1, V_2, V_3 has highest d-eigenvalue 0. By (4.25) and (7.3),

$$\chi(\Omega_V) = \Pi_{n>0}(1-q^{2n-1})^{-2}$$

$$= (\Pi_{n>0}(1-q^{2n})/(1-q^n))^2$$

$$= \Pi_{n>0}(1+q^n)^2,$$

establishing (9.7) (cf. Remark 8.13). Thus it is sufficient to show that the elements (9.6) span Ω_V.

As in (8.22), Ω_V is spanned by elements of the type (9.6) with no restriction on the indices $n_j \in \mathbb{Z}$, $i_j \in \{1,2,3\}$. By (9.3), Ω_V is spanned by such elements with $i_j \in \{1,2\}$. An inductive "straightening" procedure using the identities which are the components of (9.2) and (9.5) now gives the result. We leave this as an exercise. ∎

Remark 9.7. This straightening procedure is developed in considerable detail and generality in [16].

By analogy with Definition 8.6, we define the *standard* $\tilde{\mathfrak{g}}$*-module associated with* V_1, V_2, V_3 to be the level 3 module

$$L = U(\tilde{g})\cdot v_0.$$

A *standard level 3 $\tilde{g}$-module* is defined to be a $\tilde{g}$-module equivalent to such a module. Formula (8.24) holds for L.

We now have the analogue of Theorem 8.9(2):

Theorem 9.8. If $V_1 \simeq V_2 \simeq V_3$ (respectively, V_1, V_2 and V_3 are not all equivalent) as $\hat{g}$-modules, then Ω_L has basis

$$\{z_\beta(n_1)\ldots z_\beta(n_r)v_0 \,\big|\, r \geq 0;\ n_1,\ldots,n_r \in \mathbb{Z};$$

(9.8) $$n_1 < \ldots < n_r < -1 \quad (\text{resp., } n_1 < \ldots < n_r < 0);$$

$$n_j + 2 \leq n_{j+1} \quad \text{for } 1 \leq j \leq r-1\}.$$

Proof. Without loss of generality, we assume that $V_1 \simeq V_3$ under $\hat{g}$. By equating the coefficients of $\zeta_1^0 \zeta_2^{-1}$ in (9.3) and using (9.5), we see that

$$z_\beta^{(3)}(-1)v_0 = -z_\beta^{(1)}(-1)v_0 - z_\beta^{(2)}(-1)v_0$$

if $V_1 \simeq V_1 \simeq V_3$ as $\hat{g}$-modules and that

(9.9) $$z_\beta^{(3)}(-1)v_0 = -z_\beta^{(1)}(-1)v_0 + z_\beta^{(2)}(-1)v_0$$

otherwise. Thus by (9.1) and Theorem 9.6, we have $z_\beta(-1)v_0 = 0$ if and only if $V_1 \simeq V_2 \simeq V_3$.

To prove that the indicated elements span Ω_V, we inductively apply the generalized commutation and anticommutation relations (5.4) and (9.4) to straighten monomials in the $z_\beta(n)$'s applied to v_0. To obtain the difference-two condition $n_j + 2 \leq n_{j+1}$, we use (9.4) to "straighten"

$$\tfrac{5}{3} z_\beta(n) z_\beta(n+1) + z_\beta(n+1) z_\beta(n)$$

and (5.4) to "straighten"

$$\tfrac{7}{3}z_{\beta}(n)z_{\beta}(n+1) + z_{\beta}(n+1)z_{\beta}(n),$$

and then we solve for $z_{\beta}(n)z_{\beta}(n+1)$ and $z_{\beta}(n+1)z_{\beta}(n)$. (As we have already remarked, the straightening arguments are discussed in detail in [16].)

For the linear independence of the proposed basis elements, we write each $z_{\beta}(m)$ ($m<-1$) according to (9.1). For $n<-1$, $z_{\beta}^{(3)}(2n+1)v_0$ can be written using (9.3) and (9.5) as a linear combination of expressions of the form $z_{\beta}^{(1)}(a)z_{\beta}^{(2)}(2n+1-a)v_0$ and $z_{\beta}^{(2)}(a)z_{\beta}^{(1)}(2n+1-a)v_0$, with "leading term" a nonzero multiple of $z_{\beta}^{(2)}(n)z_{\beta}^{(1)}(n+1)v_0$. Similarly, $z_{\beta}^{(3)}(2n)$ can be written analogously with "leading term" a nonzero multiple of $z_{\beta}^{(1)}(n)z_{\beta}^{(2)}(n)v_0$.

If V_1, V_2 and V_3 are not $\hat{g}$-equivalent, then by (9.1) and (9.9), $z_{\beta}(-1)v_0$ is a nonzero multiple of $z_{\beta}^{(2)}(-1)z_{\beta}^{(1)}(0)v_0$.

Using these comments, we see that the element $z_{\beta}(n_1)\ldots z_{\beta}(n_r)v_0$ in (9.8) (where $n_r<-1$ if V_1, V_2, V_3 are not all equivalent) can be written as a linear combination of expressions of the form

$$z_{\beta}^{(i_1)}(m_1)\ldots z_{\beta}^{(i_s)}(m_s)v_0, \tag{9.10}$$

where $s\leq 2r$, $m_j\in\mathbb{Z}$, $i_j\in\{1,2\}$, and where the "leading term" is a nonzero multiple of the expression (9.10) with the properties $s=2r$,

$$m_{2u-1} = [n_u/2],$$

$$m_{2u} = [(n_u+1)/2],$$

$$(i_{2u-1},i_{2u}) = \begin{cases} (1,2) & \text{if } n_u \text{ is even} \\ (2,1) & \text{if } n_u \text{ is odd} \end{cases}$$

for $1\leq u\leq r$. Since $n_u+2\leq n_{u+1}$ implies that

$$[n_u/2] < [n_{u+1}/2],$$

$$[(n_u+1)/2] < [(n_{u+1}+1)/2],$$

this "leading term" is one of the basis elements (9.6) of Ω_V, and distinct elements (9.8) give distinct such basis elements. Moreover, the "non-leading terms" in all these expressions can always be expressed by an inductive procedure as linear combinations of basis elements (9.6) not affecting the linear independence of the "leading terms". (Again, see [16] for a formalization of this inductive procedure.) This proves the linear independence of the elements (9.8). ∎

Remark 9.9 (cf. Remark 8.10). Using Theorem 4.8, we have explicit constructions of the $\tilde{g}$-modules $V=V_1\otimes V_2\otimes V_3$ and L as tensor products of the symmetric algebra $S(\tilde{h}_-)$ with the space Ω_V as described in Theorem 9.6 or the space Ω_L described in Theorem 9.8.

10. CHARACTERS, PARTITIONS AND THE ROGERS-RAMANUJAN IDENTITIES.

To describe the character of Ω_L, we recall that a *partition* of a nonnegative integer m is an expression of the form

$$m = m_1 + ... + m_n$$

where $n\geq 0$, $m_i \in \mathbb{Z}$ and

$$m_1 \geq ... \geq m_n > 0.$$

the m_i are called the *parts*. We clearly have:

Theorem 10.1. Suppose that each basic module V_1, V_2, V_3 has highest d-eigenvalue 0, so that the standard $\tilde{g}$-module L and its subspace Ω_L satisfy conditions (4.21) and (4.22). Then

$$\chi(\Omega_L) = \Sigma_{m\geq 0} a_m q^m$$

where a_m is the number of partitions in which consecutive parts have difference at least two, and in addition, if $V_1 \simeq V_2 \simeq V_3$, with minimum part at least two. ■

Definition 10.2. By the $\mathcal{Z}$-filtration of Ω_L we mean the filtration

$$0 = \Omega_L^{[-1]} \subset \Omega_L^{[0]} \subset \Omega_L^{[1]} \subset \ldots \subset \Omega_L$$

such that for all $n \geq 0$, $\Omega_L^{[n]}$ is the span of all the elements

$$z_\beta(m_1)\ldots z_\beta(m_i)v_0, \quad 0 \leq i \leq n, \quad m_j \in \mathbb{Z}.$$

Definition 10.3. The $\hat{\mathfrak{h}}$-*filtration of* L is the filtration

$$0 = L_{[-1]} \subset L_{[0]} \subset L_{[1]} \subset \ldots \subset L$$

such that for all $n \geq 0$, $L_{[n]}$ is the span of all the elements

$$x_1 \ldots x_i \cdot v_0, \quad i \geq 0,$$

where each $x_j \in \tilde{\mathfrak{g}}$ and at most n of the x_j lie outside the subalgebra $\tilde{\mathfrak{h}}$ of $\tilde{\mathfrak{g}}$. The $\hat{\mathfrak{h}}$-*filtration of* Ω_L is the filtration

$$0 = \Omega_{L,[-1]} \subset \Omega_{L,[0]} \subset \Omega_{L,[1]} \subset \ldots \subset \Omega_L$$

where

$$\Omega_{L,[n]} = \Omega_L \cap L_{[n]}$$

for all $n \geq -1$.

By (4.6), (4.19) and (4.20), we have:

Proposition 10.4. The $\mathcal{Z}$-filtration and the $\hat{\mathfrak{h}}$-filtration of Ω_L coincide, i.e.,

$$\Omega_L^{[n]} = \Omega_{L,[n]}$$

for all $n \geq -1$. ∎

Assume that V_1, V_2 and V_3 have highest d-eigenvalue 0. Then for all $n \geq 0$, the space $\Omega_{L,[n]}/\Omega_{L,[n-1]}$ satisfies conditions (4.21) and (4.22), and

$$(10.1) \quad \chi(\Omega_L) = \Sigma_{n\geq 0}\chi(\Omega_{L,[n]}/\Omega_{L,[n-1]}) = \Sigma_{n\geq 0}\chi(\Omega_L^{[n]}/\Omega_L^{[n-1]}).$$

The following refinement of Theorem 10.1 is clear, using Theorem 9.8:

Theorem 10.5. In the setting of Theorem 10.1, for all $n \geq 0$,

$$(10.2) \quad \chi(\Omega_{L,[n]}/\Omega_{L,[n-1]}) = \chi(\Omega_L^{[n]}/\Omega_L^{[n-1]}) = \Sigma_{j\geq 0}a_m^{[n]}q^m,$$

where $a_m^{[n]}$ is the number of partitions of m into exactly n parts, and, if $V_1 \simeq V_2 \simeq V_3$, with minimum part at least 2. ∎

A simple combinatorial exercise shows that

$$(10.3) \qquad \Sigma_{m\geq 0}a_m^{[n]}q^m = q^{n^2}/(1-q)(1-q^2)\ldots(1-q^n)$$

in the case where there is no restriction on the minimum part and

$$(10.4) \qquad = q^{n(n+1)}/(1-q)(1-q^2)\ldots(1-q^n)$$

if the minimum part is required to be at least 2. (Hint: Write

$$n^2 = 1 + 3 + 5 + \ldots + (2n-1),$$

$$n(n+1) = 2 + 4 + 6 + \ldots + 2n.)$$

The principal specialization [11] of Kac's generalization [6] of Weyl's character formula (see also [7], [8]), combined with (4.25), gives product expansions of $\chi(\Omega_L)$ for the two level 3 standard $\tilde{\mathfrak{g}}$-modules L discussed in Theorem 10.1. Formula (10.1) becomes formula (1.1) if V_1, V_2 and V_3 are not all $\hat{\mathfrak{g}}$-equivalent, and (1.2) if $V_1 \simeq V_2 \simeq V_3$ (see (10.2) - (10.4)). These are the two Rogers-Ramanujan identities (see [1] for a historical exposition).

Remark 10.6. The analogues of (1.1) and (1.2) for the level 2 standard $\tilde{\mathfrak{g}}$-modules are:

(10.5) $$\Pi_{n\geq 0}(1+q^{2n-1}) = \Sigma_{n\geq 0}q^{n^2}/(1-q^2)(1-q^4)\ldots(1-q^{2n})$$

if $V_1 \not\simeq V_2$ as $\hat{\mathfrak{g}}$-modules, in the setting of Proposition 8.12, and

(10.6) $$\Pi_{n\geq 1}(1+q^{2n}) = \Sigma_{n\geq 0}q^{n(n+1)}/(1-q^2)(1-q^4)\ldots(1-q^{2n})$$

if $V_1 \simeq V_2$ (see (8.27), (8.28)). In contrast with the situation for level 3, these two identities are easy to prove, the coefficient of q^m on either side counting the number of partitions of m into distinct odd parts for (10.5) and into distinct even parts for (10.6). The formulas (10.5) and (10.6) are due to Euler (c.f. [1]).

REFERENCES

1. G.E. Andrews, *The theory of partitions*, Encyclopedia of Mathematics and its Applications, Vol. 2, ed. G.-C. Rota, Addison-Wesley, Reading, MA 1976.

2. E. Date, M. Jimbo, M. Kashiwara and T. Miwa, Transformation groups for soliton equations - Euclidean Lie algebras and reduction of the KP hierarchy, Publ. Research Inst. for Math. Sciences, Kyoto University 18 (1982), 1077-1110.

3. E. Date. M. Kashiwara and T. Miwa, Vertex operators and τ functions - transformation groups for soliton equations II, Proc. Japan Acad. Ser. A Math. Sci. 57 (1981), 387-392.

4. I.B. Frenkel, Lectures on Respresentation Theory and

Mechanics, Yale University, fall, 1981.

5. I.B. Frenkel, Representations of affine Lie algebras, Hecke modular forms and Korteweg-deVries type equations, Proc. 1981 Rutgers Conference on Lie Algebras and Related Topics, Springer-Verlag Lectures Notes in Mathematics 933 (1982), 71-110.

6. V.G. Kac, Infinite-dimensional Lie algebras and Dedekind's η-function, Funkcional. Anal. i Prilozhen. 8 (1974), 77-78; English transl., Functional Anal. Appl. 8 (1974), 68-70.

7. J. Lepowsky, Lectures on Kac-Moody Lie algebras, Université Paris VI, spring, 1978.

8. J. Lepowsky, Application of the numerator formula to k-rowed plane partitions, Advances in Math. 35 (1980), 179-194.

9. J. Lepowsky, Affine Lie algebras and combinatorial identities, Proc. 1981 Rutgers Conference on Lie Algebras and Related Topics, Springer-Verlag Lecture Notes in Mathematics 933 (1982), 130-156.

10. J. Lepowsky, Some constructions of the affine Lie algebra $A_1^{(1)}$, Proc. 1982 Summer Seminar on Applications of Group Theory in Physics and Mathematical Physics, Lectures in Applied Math., Amer. Math. Soc. 21 (1984).

11. J. Lepowsky and S. Milne, Lie algebraic approaches to classical partition identities, Advances in Math. 29 (1978), 15-59,

12. J. Lepowsky and R.L. Wilson, Construction of the affine Lie algebra $A_1^{(1)}$, Comm. Math. Phys. 62 (1978), 43-53.

13. J. Lepowsky and R.L. Wilson, The Rogers-Ramanujan identities: Lie theoretic intepretation and proof, Proc. Nat. Acad. Sci. U.S.A. 78 (1981), 699-701.

14. J. Lepowsky and R.L. Wilson, A Lie theoretic intepretation and proof of the Rogers-Ramanujan identities, Advances in Math. 45 (1982), 21-72.

15. J. Lepowsky and R.L. Wilson, A new family of algebras

underlying the Rogers-Ramanujan identities and generalizations, Proc. Nat. Acad. Sci. U.S.A. 78 (1981), 7254-7258.

16. J. Lepowsky and R.L. Wilson, The structure of standard modules, I: Universal algebras and the Rogers-Ramanujan identities, Invent. Math. (1984).

17. J. Lepowsky and R.L. Wilson, The structure of standard modules, II: The case $A_1^{(1)}$, principal gradation, to appear.

Mathematical Sciences Research Institute

and

Rutgers University
New Brunswick, NJ 08903

STRUCTURE OF THE STANDARD MODULES FOR THE AFFINE LIE ALGEBRA $A_1^{(1)}$ IN THE HOMOGENEOUS PICTURE

J. Lepowsky† and M. Primc*

We announce the main results of our paper [5] on the structure of the standard modules for the affine Lie algebra $A_1^{(1)}$ in the "homogeneous picture". This study, already begun in [4], was stimulated by the investigation [6] of the structure of the standard $A_1^{(1)}$-modules in the "principal picture". For detailed background and references, these papers and [1], [2], [7] and [8] should be consulted.

1. NOTATION

Let $\mathfrak{g}$ be a complex simple Lie algebra and let $\mathfrak{h}$ be a Cartan subalgebra. Let $\langle\cdot,\cdot\rangle$ be a nonzero invariant symmetric bilinear form on $\mathfrak{g}$, and identify $\mathfrak{h}$ with its dual $\mathfrak{h}^*$ via $\langle\cdot,\cdot\rangle$. We normalize this form so that the square length of a long root of $\mathfrak{g}$ with respect to $\mathfrak{h}$ is 2.

Write $\mathbb{C}[t,t^{-1}]$ for the algebra of Laurent polynomials in an indeterminate t. For $x \in \mathfrak{g}$ and $n \in \mathbb{Z}$ set

$$x(n) = x\otimes t^n \in \mathfrak{g}\otimes\mathbb{C}[t,t^{-1}] .$$

The affine Kac-Moody Lie algebra $\tilde{\mathfrak{g}}$ associated with $\mathfrak{g}$ is the Lie algebra

†* Partially supported by the National Science Foundation through the Mathematical Sciences Research Institute.
† Partially supported by NSF grants MCS 80-03000 and MCS 83-01664.
* Partially supported by the Institute for Advanced Study and Samoupravna interesna zajednica za znanstveni rad SRH-SIZ VI.

Vertex Operators in Mathematics and Physics - Proceedings of a Conference November 10-17, 1983. Publications of the Mathematical Sciences Research Institute #3, Springer-Verlag, 1984.

(1.1) $$\tilde{\mathfrak{g}} = \mathfrak{g}\otimes\mathbb{C}[t,t^{-1}]\oplus\mathbb{C}c\oplus\mathbb{C}d,$$

where c is a nonzero central element of $\tilde{\mathfrak{g}}$, and the other commutation relations are given by

(1.2)
$$[x(m),y(n)] = [x,y](m+n) + \langle x,y\rangle m\,\delta_{m+n,0}c,$$
$$[d,x(m)] = mx(m)$$

for $x,y \in \mathfrak{g}$ and $m,n \in \mathbb{Z}$.

We shall identify $\mathfrak{g}$ with the subalgebra $\mathfrak{g}\otimes 1$ of $\tilde{\mathfrak{g}}$. Extend the bilinear form $\langle\cdot,\cdot\rangle$ on $\mathfrak{g}$ to an invariant symmetric bilinear form on $\tilde{\mathfrak{g}}$ by setting

$$\langle x(m),y(n)\rangle = \langle x,y\rangle\,\delta_{m+n,0},$$

$$\langle x(m),c\rangle = \langle x(m),d\rangle = 0,$$

$$\langle c,d\rangle = 1,\ \langle c,c\rangle = \langle d,d\rangle = 0.$$

We have the following $\mathbb{Z}$-gradation of $\tilde{\mathfrak{g}}$:

$$\tilde{\mathfrak{g}} = \coprod_{n\in\mathbb{Z}} \tilde{\mathfrak{g}}_n,$$

where $\tilde{\mathfrak{g}}_n$ is the ad d - eigenspace in $\tilde{\mathfrak{g}}$ with eigenvalue n. For a subspace $\mathfrak{b}$ of $\tilde{\mathfrak{g}}$ and $n \in \mathbb{Z}$, set

$$\mathfrak{b}_n = \mathfrak{b} \cap \tilde{\mathfrak{g}}_n.$$

Define the subalgebra $\tilde{\mathfrak{h}}$ of $\tilde{\mathfrak{g}}$ by analogy with (1.1). Then

$$\tilde{\mathfrak{h}}_0 = \mathfrak{h}\oplus\mathbb{C}c\oplus\mathbb{C}d,$$

an abelian subalgebra of $\tilde{\mathfrak{g}}$. The form $\langle\cdot,\cdot\rangle$ is nonsingular on $\tilde{\mathfrak{h}}_0$, and we identify $\tilde{\mathfrak{h}}_0$ with $(\tilde{\mathfrak{h}}_0)^*$ via $\langle\cdot,\cdot\rangle$.

Let $\Phi \subset \mathfrak{h}^* = \mathfrak{h}$ denote the set of roots of $\mathfrak{g}$ with respect to $\mathfrak{h}$, so that we have the root space decomposition

$$\mathfrak{g} = \mathfrak{h} \oplus \coprod_{\varphi \in \Phi} \mathfrak{g}^{\varphi}.$$

Also set $\mathfrak{g}^0 = \mathfrak{h}$. Let $\Delta \subset \tilde{\mathfrak{h}}_0$ denote the set of roots of $\tilde{\mathfrak{g}}$ with respect to $\tilde{\mathfrak{h}}_0$, so that

$$\tilde{\mathfrak{g}} = \tilde{\mathfrak{h}}_0 \oplus \coprod_{\varphi \in \Delta} \tilde{\mathfrak{g}}^{\varphi},$$

where $\tilde{\mathfrak{g}}^{\varphi}$ is the root space for φ. Then

$$\Delta = \{\varphi + nc \mid \varphi \in \Phi \cup \{0\},\ n \in \mathbf{Z}\} \setminus \{0\},$$

and for $\varphi + nc \in \Delta$ we have $\tilde{\mathfrak{g}}^{\varphi+nc} = \mathfrak{g}^{\varphi} \otimes t^n$.

Choose a system $\{\alpha_1, \ldots, \alpha_\ell\}$ of simple roots in Φ (ℓ = rank $\mathfrak{g}$), and let $\beta_0 \in \Phi$ be the highest root and $\alpha_0 = c - \beta_0 \in \Delta$. Then $\{\alpha_0, \alpha_1, \ldots, \alpha_\ell\}$ is the associated system of simple roots in Δ, with corresponding positive system

$$\Delta_+ = \{\varphi + nc \mid \text{either } n > 0 \text{ or } n = 0,\ \varphi \in \Phi_+\},$$

where Φ_+ is the positive system in Φ.

We say that $\lambda \in \tilde{\mathfrak{h}}_0$ is *dominant integral* if $\langle \lambda, h_i \rangle \in \mathbf{N}$ ($= \{0,1,2,\ldots\}$) for $i = 0, \ldots, \ell$, where $h_i = 2\alpha_i / \langle \alpha_i, \alpha_i \rangle$.

For a $\tilde{\mathfrak{g}}$-module V and $\mu \in \tilde{\mathfrak{h}}_0$, the corresponding *weight space* consists of those $v \in V$ such that $x \cdot v = \langle \mu, x \rangle v$ for all $x \in \tilde{\mathfrak{h}}_0$, and μ is a *weight* if this space is nonzero. Its nonzero elements are called the *μ-weight vectors*. The module V is a *weight module* if it is the direct sum of its weight spaces.

A *highest weight vector* in a $\tilde{\mathfrak{g}}$-module V is a weight vector $v \in V$ such that $\tilde{\mathfrak{g}}^{\varphi} \cdot v = (0)$ for all $\varphi \in \Delta_+$, and V is a *highest weight module* if it is generated by a highest weight vector. In this case, V is a weight module with finite-dimensional weight spaces, and the generating highest weight vector is uniquely determined up to multiplication by a nonzero scalar. Its weight is called the *highest weight* of V.

For a dominant integral $\lambda \in \tilde{\mathfrak{h}}_0$, the *standard* $\tilde{\mathfrak{g}}$*-module* V *with highest weight* λ is by definition the irreducible $\tilde{\mathfrak{g}}$-module with highest weight λ. In this case c acts on V as the scalar $\langle\lambda,c\rangle \in \mathbb{N}$.

For $k \in \mathbb{C}\setminus\{0\}$ we say that a $\tilde{\mathfrak{g}}$-module V has *level* k if c acts as the scalar k on it.

Let $\zeta, \zeta_1, \zeta_2, \ldots$ be commuting formal indeterminates. For $x \in \mathfrak{g}$, set

$$x(\zeta) = \Sigma_{n\in\mathbb{Z}}\, x(n)\zeta^n,$$

a formal Laurent series in ζ with coefficients in $\tilde{\mathfrak{g}}$. Then the commutation relations (1.2) can be written as follows: For $x,y \in \mathfrak{g}$,

$$(1.3)\quad [x(\zeta_1),y(\zeta_2)] = [x,y](\zeta_1)\delta(\zeta_1/\zeta_2) + \langle x,y\rangle c(D\delta)(\zeta_1/\zeta_2),$$

where $\delta(\zeta)$ and $(D\delta)(\zeta)$ denote the formal Laurent series

$$\delta(\zeta) = \Sigma_{n\in\mathbb{Z}}\, \zeta^n, \qquad D\delta(\zeta) = \Sigma_{n\in\mathbb{Z}} n\zeta^n.$$

(More generally, D is the operator $D = \zeta d/d\zeta$.)

In what follows we shall view formulas like (1.3) as "generating functions" for formulas like (1.2).

For each $\varphi \in \Phi$ we fix a root vector $x_\varphi \in \mathfrak{g}^\varphi$. For φ, ψ, $\varphi+\psi \in \Phi$, define $\epsilon(\varphi,\psi) \in \mathbb{C}^*$ by the relation

$$[x_\varphi,x_\psi] = \epsilon(\varphi,\psi)x_{\varphi+\psi}.$$

Now let $k \in \mathbb{C}^*$ and let V be a highest weight $\tilde{\mathfrak{g}}$-module of level k. Then each $x \in \tilde{\mathfrak{g}}$ is represented by an operator on V usually denoted again by x. For $x \in \mathfrak{g}$ we may form the formal Laurent series $x(\zeta)$ in ζ with coefficients in End (V). Then $x(\zeta)$ will be viewed as a "generating function" for the operators x(n).

Define the *homogeneous Heisenberg subalgebra*

$$\tilde{\mathfrak{h}}' = [\tilde{\mathfrak{h}},\tilde{\mathfrak{h}}] = (\amalg_{n\in\mathbb{Z}\setminus\{0\}}\mathfrak{h}\otimes t^n)\oplus\mathbb{C}c$$

of $\tilde{\mathfrak{g}}$, and define the commutative subalgebras

$$\tilde{\mathfrak{h}}_{\pm} = \amalg_{n>0}\mathfrak{h}\otimes t^{\pm n}$$

of $\tilde{\mathfrak{h}}'$. On the $\tilde{\mathfrak{g}}$-module V we have

$$[\varphi(m),\psi(n)] = \langle\varphi,\psi\rangle m\,\delta_{m+n,0}k$$

(1.4)

$$[d,\varphi(m)] = m\varphi(m)$$

for φ, $\psi \in \mathfrak{h}$ and $m,n \in \mathbb{Z}\setminus\{0\}$.

For an $\tilde{\mathfrak{h}}'$-module W, set

$$\Omega_W = \{w \in W \mid \tilde{\mathfrak{h}}_+\cdot w = (0)\},$$

and call Ω_W the *vacuum space* for $\tilde{\mathfrak{h}}'$ in W and its nonzero elements *vacuum vectors*. Denote by K(k) the unique (up to isomorphism) irreducible $\tilde{\mathfrak{h}}'$-module generated by a vacuum vector and such that c acts as the scalar k on K(k). Then (1.4) holds on K(k), and K(k) may be realized as the symmetric algebra $S(\tilde{\mathfrak{h}}_-)$, with $\tilde{\mathfrak{h}}_-$ acting by multiplication operators and $\tilde{\mathfrak{h}}_+$ as derivations. The natural action of d on $S(\tilde{\mathfrak{h}}_-)$ gives K(k) a nonpositive $\mathbb{Z}$-grading for which a vacuum vector has degree 0. The vacuum space Ω_V of the $\tilde{\mathfrak{g}}$-module V is invariant under the action of $\tilde{\mathfrak{h}}_0$, and V may be decomposed as

(1.5)
$$V = K(k)\otimes\Omega_V$$

with $\tilde{\mathfrak{h}}'$ acting as $\tilde{\mathfrak{h}}'\otimes 1$, $\mathfrak{h}$ acting as $1\otimes\mathfrak{h}$ and d acting as $d\otimes 1 + 1\otimes d$. We want to describe the action of the rest of $\tilde{\mathfrak{g}}$ on V in the same spirit.

2. THE Z-OPERATORS

For $\varphi \in \Phi$ let $E^{\pm}(\varphi,\zeta)$ be the formal Laurent series in ζ with coefficients in End V defined as the formal exponential series

$$(2.1)\qquad E^{\pm}(\varphi,\zeta) = E^{\pm}_{(k)}(\varphi,\zeta) = \exp\Sigma_{n>0}\ \varphi(\pm n)\zeta^{\pm n}/(\pm n)k.$$

The coefficients $z_{\varphi}(n)$ of the formal Laurent series

$$z_{\varphi}(\zeta) = \Sigma_{n\in\mathbb{Z}}\ z_{\varphi}(n)\zeta^{n}$$

defined by

$$(2.2)\qquad z_{\varphi}(\zeta) = E^{-}(\varphi,\zeta)x_{\varphi}(\zeta)E^{+}(\varphi,\zeta)$$

commute with the operators $\psi(n)$ for $\psi \in \mathfrak{h}$, $n \neq 0$, and hence they preserve Ω_V. We can "reverse" (2.2), and by using the identification (1.5) we see that on the space $V = K(k)\otimes\Omega_V$,

$$(2.3)\qquad x_{\varphi}(\zeta) = E^{-}(-\varphi,\zeta)E^{+}(-\varphi,\zeta)\otimes z_{\varphi}(\zeta).$$

Thus one knows the action of $x_{\varphi}(\zeta)$ (and hence all of $\tilde{g}$) on V if one knows the action of $z_{\varphi}(\zeta)$ on Ω_V. For the special case of a basic (i.e., level 1 standard) module V for an affine Lie algebra of type $A_{\ell}^{(1)}$, $D_{\ell}^{(1)}$ or $E_{\ell}^{(1)}$, $z_{\varphi}(\zeta)$ has a simple form (cf. [4]; for $A_1^{(1)}$, cf. below) so as to make $x_{\varphi}(\zeta)$ a vertex operator ([1], [2], [8]).

For $\varphi_1,\dots,\varphi_s \in \Phi$, define a family of operators

$$(2.4)\qquad z(\varphi_1,\dots,\varphi_s;\ n_1,\dots,n_s) \in \text{End } V$$

($n_1,\dots,n_s \in \mathbb{Z}$) by the conditions

$$z(\varphi_1,\dots,\varphi_s;\ \zeta_1,\dots,\zeta_s) =$$

$$= \Sigma_{n_1,\dots,n_s\in\mathbb{Z}}\ z(\varphi_1,\dots,\varphi_s;\ n_1,\dots,n_s)\zeta_1^{n_1}\dots\zeta_s^{n_s},$$

where

$$z(\varphi_1,\dots,\varphi_s;\ \zeta_1,\dots,\zeta_s) = z_{(s)}(\varphi_1,\dots,\varphi_s;\ \zeta_1,\dots,\zeta_s) =$$

(2.5)

$$= E^-(\varphi_1,\zeta_1)\dots E^-(\varphi_s,\zeta_s)x_{\varphi_1}(\zeta_1)\dots x_{\varphi_s}(\zeta_s)E^+(\varphi_1,\zeta_1)\dots E^+(\varphi_s,\zeta_s)$$

$$= \Pi_{1\leq i<j\leq s}(1-\zeta_i/\zeta_j)^{\langle\varphi_i,\varphi_j\rangle/k} z_{\varphi_1}(\zeta_1)\dots z_{\varphi_s}(\zeta_s).$$

The factors of the form $(1-\zeta_i/\zeta_j)^a$ are to be understood as formal power series in ζ_i/ζ_j computed by means of the binomial expansion. The equality of the two formal expressions on the right-hand side of (2.5) follows from formal manipulations with the expressions (2.1) familiar in quantum field theory. Using (2.5) we see that the operators (2.4) commute with the Heisenberg algebra $\tilde{h}'$.

Let $v_0 \in V$ be a generating highest weight vector for V with weight $\lambda \in \tilde{h}_0$. Note that $v_0 \in \Omega_V$. Using the same notation as in (2.4), define the vectors

(2.6) $$v(\varphi_1,\dots,\varphi_s;\ n_1,\dots,n_s) \in \Omega_V$$

by

$$v(\varphi_1,\dots,\varphi_s;\ \zeta_1,\dots,\zeta_s) =$$

$$= \Sigma_{n_1,\dots,n_s \in \mathbb{Z}}\ v(\varphi_1,\dots,\varphi_s;\ n_1,\dots,n_s)\zeta_1^{n_1}\dots\zeta_s^{n_s}$$

$$= z(\varphi_1,\dots,\varphi_s;\ \zeta_1,\dots,\zeta_s)v_0.$$

From the definitions, we obtain:

$$d\cdot v(\varphi_1,\dots,\varphi_s;\ \zeta_1,\dots,\zeta_s) =$$

$$= (\langle d,\lambda\rangle + D_1+\dots+D_s)v(\varphi_1,\dots,\varphi_s;\ \zeta_1,\dots,\zeta_s)$$

$$\varphi\cdot v(\varphi_1,\dots,\varphi_s;\ \zeta_1,\dots,\zeta_s) = \langle\varphi,\lambda+\varphi_1+\dots+\varphi_s\rangle v(\varphi_1,\dots,\varphi_s;\zeta_1,\dots,\zeta_s)$$

$$z_\varphi(\zeta)v(\varphi_1,\dots,\varphi_s;\ \zeta_1,\dots,\zeta_s) =$$

(2.7)

$$= \Pi_{i=1}^{s}(1-\zeta/\zeta_i)^{-\langle\varphi,\varphi_i\rangle/k}\ v(\varphi,\varphi_1,\dots,\varphi_s;\ \zeta,\zeta_1,\dots,\zeta_s)$$

for $\varphi \in \Phi$; here $D_j = \zeta_j \partial/\partial\zeta_j$ for $j = 1,\dots,s$. Observe that the coefficient of each monomial $\zeta^n \zeta_1^{n_1} \dots \zeta_s^{n_s}$ ($n, n_i \in \mathbb{Z}$) in the right-hand side of (2.7) is a formal infinite linear combination of vectors of the type (2.6), but only finitely many of these vectors are nonzero since the vector (2.6) is nonzero only if

$$n_s \leq 0,\ n_{s-1} + n_s \leq 0,\ \dots,\ n_1 + \dots + n_s \leq 0.$$

Starting from the commutation rules (1.3), we can prove:

Theorem 2.1. Let $\varphi_1,\dots,\varphi_s \in \Phi$ and $1 \leq r < s$. Then on V,

$$z(\varphi_1,\dots,\varphi_r,\ \varphi_{r+1},\dots,\varphi_s;\ \zeta_1,\dots,\zeta_r,\ \zeta_{r+1},\dots,\zeta_s)$$

$$-z(\varphi_1,\dots,\varphi_{r+1},\varphi_r,\dots,\varphi_s;\ \zeta_1,\dots,\zeta_{r+1},\zeta_r,\dots,\zeta_s) =$$

$$= \begin{cases} 0 & \text{if } \varphi_r + \varphi_{r+1} \notin \Phi\cup\{0\} \\ \epsilon(\varphi_r,\varphi_{r+1})z(\varphi_1,\dots,\varphi_{r-1},\varphi_r+\varphi_{r+1},\varphi_{r+2},\dots,\varphi_s; \\ \quad \zeta_1,\dots,\hat{\zeta}_{r+1},\dots,\zeta_s)\delta(\zeta_r/\zeta_{r+1}) & \\ & \text{if } \varphi_r + \varphi_{r+1} \in \Phi \\ \langle x_{\varphi_r}, x_{-\varphi_r}\rangle z(\varphi_1,\dots,\hat{\varphi}_r,\hat{\varphi}_{r+1},\dots,\varphi_s;\zeta_1,\dots,\hat{\zeta}_r,\hat{\zeta}_{r+1},\dots,\zeta_s)\ \cdot \\ \cdot [kD\delta(\zeta_r/\zeta_{r+1}) + (\langle\varphi_r,\varphi_{r+2}+\dots+\varphi_s\rangle+\varphi_r)\delta(\zeta_r/\zeta_{r+1}) \\ \quad - \Sigma_{i<r}\langle\varphi_i,\varphi_r\rangle(\zeta_i/\zeta_{r+1})(1-\zeta_i/\zeta_{r+1})^{-1}\delta(\zeta_r/\zeta_{r+1}) \\ \quad + \Sigma_{j>r+1}\langle\varphi_r,\varphi_j\rangle(\zeta_{r+1}/\zeta_j)(1-\zeta_{r+1}/\zeta_j)^{-1}\delta(\zeta_r/\zeta_{r+1})] \\ & \text{if } \varphi_r + \varphi_{r+1} = 0, \end{cases}$$

where $\hat{x}$ means that the symbol x is to be omitted.

By applying these formulas to v_0 and extracting components, we obtain relations among the vectors (2.6). For example, in the case that $\varphi_r+\varphi_{r+1} = 0$ and $n_1,\dots,n_s \in \mathbf{Z}$, $n_r + n_{r+1} < 0$, we have

$$v(\varphi_1,\dots,\varphi_s;\ n_1,\dots,n_s) - v(\varphi_1,\dots,\varphi_{r+1},\varphi_r,\dots,\varphi_s;$$

$$n_1,\dots,n_{r+1},n_r,\dots,n_s) =$$

$$= - \langle x_{\varphi_r},x_{-\varphi_r}\rangle \Sigma_{i<r} \langle \varphi_i,\varphi_r\rangle v(\varphi_1,\dots,\hat{\varphi}_r,\hat{\varphi}_{r+1},\dots,\varphi_s;$$

$$n_1,\dots,n_i + n_r + n_{r+1},\dots,\hat{n}_r,\hat{n}_{r+1},\dots,n_s).$$

The vectors (2.6) span Ω_V. Using Theorem 2.1, we can reduce to a smaller spanning set. In some easy cases we can thus obtain a basis of Ω_V. For example:

<u>**Theorem**</u> **2.2.** Let V be a highest weight module for $\tilde{g}$ of level $k \in \mathbb{C}^*$ and let $v_0 \in V$ be a generating highest weight vector. Suppose that $g \cdot v_0 = 0$ and that the map

$$U(\amalg_{n<0}\tilde{g}_n) \longrightarrow V$$

$$u \longmapsto u \cdot v_0$$

is a linear isomorphism. (Here U(•) denotes the universal enveloping algebra. The module V is an example of a "generalized Verma module".) Let < be an arbitrary total ordering on the set Δ. Then Ω_V has two bases as follows:

$$\{v(\varphi_1,\dots,\varphi_s;\ n_1,\dots,n_s) \mid s \geqslant 0,\ \varphi_1,\dots,\varphi_s \in \Phi,$$

$$n_1 \leqslant \dots \leqslant n_s < 0 \text{ and } \varphi_p \leqslant \varphi_{p+1} \text{ if } n_p = n_{p+1}\},$$

$$\{z_{\varphi_1}(n_1)\dots z_{\varphi_s}(n_s)v_0 \mid \text{(same conditions)}\}.$$

By means of Theorem 2.1, one can write any vector (2.6) as a linear combination of elements from the first basis. The action of the operators $z_\varphi(n)$ for $\varphi \in \Delta$ and $n \in \mathbb{Z}$ on the two bases is given by (2.7) and inductive procedures which follow from Theorem 2.1 and the definitions. By (1.5) and (2.3), we have a description of the $\tilde{\mathfrak{g}}$-module V. This relatively simple result foreshadows the much deeper analysis required to describe the standard $\tilde{\mathfrak{g}}$-modules.

3. THE STANDARD $A_1^{(1)}$-MODULES

From now on, we take

$$\mathfrak{g} = \mathfrak{sl}(2,\mathbb{C}),$$

so that $\tilde{\mathfrak{g}}$ is of type $A_1^{(1)}$.

Define $\Lambda_0, \Lambda_1 \in \tilde{\mathfrak{h}}_0$ as follows:

$$\Lambda_i \in \mathfrak{h}\oplus\mathbb{C}d, \qquad \langle\Lambda_i,h_j\rangle = \delta_{ij}$$

for i, j = 0, 1. For $k \in \mathbb{N}$, the dominant integral elements

$$\lambda = i\Lambda_0 + (k-i)\Lambda_1 + ac, \tag{3.1}$$

for $i = 0,\ldots,k$ and $a \in \mathbb{C}$ are the highest weights of all the standard $\tilde{\mathfrak{g}}$-modules of level k. Two standard $\tilde{\mathfrak{g}}$-modules whose highest weights differ by a multiple of c are equivalent as $\hat{\mathfrak{g}}$-modules, and so the scalar a in formula (3.1) will not be important for our purposes.

Fix $k \geqslant 1$ and λ as in (3.1), and let L be the standard $\tilde{\mathfrak{g}}$-module with highest weight λ.

Denote by α the positive root in Φ, so that

$$\Phi = \{\pm\alpha\}.$$

Set

$$Q = \mathbb{Z}\alpha \subset \mathfrak{h}, \tag{3.2}$$

the root lattice of $\mathfrak{g}$.

Let $\varphi \in \Phi$. Since the operators $x_\varphi(n)$ $(n \in \mathbb{Z})$ commute with one another, $x_\varphi(\zeta)^k$ is well defined on L. Using the notation (2.1) and (2.5), set

$$\begin{aligned} e_\varphi &= (1/k!)E^-(\varphi,\zeta)^k x_\varphi(\zeta)^k E^+(\varphi,\zeta)^k \zeta^{k+\varphi} \\ &= (1/k!)z(\varphi,\ldots,\varphi;\ \zeta,\ldots,\zeta)\zeta^{k+\varphi}, \end{aligned} \tag{3.3}$$

a formal Laurent series in ζ with coefficients in End L. Here

$$\zeta^{k+\varphi} = \zeta^k \zeta^\varphi,$$

and ζ^φ is defined by the condition

$$\zeta^\varphi \cdot x = \zeta^{\langle\varphi,\mu\rangle} x$$

for $x \in L$ a weight vector with weight $\mu \in \tilde{\mathfrak{h}}_0$; the quantity $\langle\varphi,\mu\rangle$ is necessarily an integer.

Using the construction of the basic $\tilde{\mathfrak{g}}$-modules ([1], [8]), the case $k = 1$, we can show that e_φ is independent of ζ, i.e.,

$$e_\varphi \in \text{End } L,$$

and that

$$e_\varphi e_{-\varphi} = 1$$

on L. Thus we can define a representation of the group Q on L by the operators

$$e_{n\varphi} = (e_\varphi)^n$$

for $n \geq 0$, $\varphi \in \Phi$. This action realizes the translation subgroup of the Weyl group of $\tilde{g}$; cf. [2], [8].

Proposition 3.1. For $\varphi, \psi \in Q$, we have:

$$e_\varphi e_\psi = e_{\varphi+\psi}, \quad e_0 = 1$$

$$[\beta, e_\varphi] = k\langle\beta,\varphi\rangle e_\varphi \quad \text{for } \beta \in \mathfrak{h}$$

$$[d, e_\varphi] = -e_\varphi(\varphi + k\langle\varphi,\varphi\rangle/2)$$

$$\zeta^\varphi e_\psi = e_\psi \zeta^\varphi \zeta^{k\langle\varphi,\psi\rangle}$$

$$x_\varphi(\zeta)e_\psi = e_\psi x_\varphi(\zeta)\zeta^{-\langle\varphi,\psi\rangle} \quad \text{if } \varphi = \pm\alpha$$

$$z_\varphi(\zeta)e_\psi = e_\psi z_\varphi(\zeta)\zeta^{-\langle\varphi,\psi\rangle} \quad \text{if } \varphi = \pm\alpha$$

$$[\tilde{\mathfrak{h}}', e_\varphi] = 0.$$

The following result generalizes the definition (3.3):

Theorem 3.2. Let $\varphi = \pm\alpha$. On L we have

$$x_\varphi(\zeta)^{k+1} = 0$$

and

$$(3.4) \quad \begin{aligned} &(1/p!)E^-(\varphi,\zeta)^p x_\varphi(\zeta)^p E^+(\varphi,\zeta)^p = \\ &= (1/q!)e_\varphi \zeta^{-k-\varphi} E^-(-\varphi,\zeta)^q x_{-\varphi}(\zeta)^q E^+(-\varphi,\zeta)^q \end{aligned}$$

for $p,q \geq 0$ with $p+q = k$. Equivalently,

$$E^-(\varphi,\zeta)^k \exp x_\varphi(\zeta) E^+(\varphi,\zeta)^k = \exp x_{-\varphi}(\zeta) e_\varphi \zeta^{-k-\varphi}.$$

Note that formula (3.4) may be written as:

$$(3.5)\qquad \begin{aligned}&(1/p!)z_{(p)}(\varphi,\ldots,\varphi;\ \zeta,\ldots,\zeta) =\\ &= (1/q!)e_{\varphi}\zeta^{-k-\varphi}z_{(q)}(-\varphi,\ldots,-\varphi;\ \zeta,\ldots,\zeta),\end{aligned}$$

using the notation $z_{(s)}$ of (2.5).

For $n_1,\ldots,n_s \in \mathbb{Z}$, it will be convenient to write

$$v(n_1,\ldots,n_s) = v(\alpha,\ldots,\alpha;\ n_1,\ldots,n_s) \in \Omega_L$$

(see (2.6)). We have already mentioned that Ω_L is spanned by the vectors (2.6). Using Theorem 3.2 in the case $\varphi = -\alpha$, $p = 1$, we can show that Ω_L is spanned by the vectors

$$(3.6)\qquad e_{\varphi}v(n_1,\ldots,n_s)$$

where $\varphi \in Q$ and $n_1,\ldots,n_s \in \mathbb{Z}$. By Theorem 2.1, the subscripts $1,\ldots,s$ in (3.6) may be permuted without changing the value of the vector (3.6). Hence in (3.6) we may assume that

$$n_1 \leqslant \ldots \leqslant n_s.$$

Our main result (Theorem 3.3) gives a basis of Ω_L in terms of vectors of the type (3.6).

For $m,n \in \mathbb{N}$, denote by $R_{k,i}(m,n)$ the set of all sequences $(b_1,\ldots,b_m) \in \mathbb{Z}^m$ such that

$$n = b_1 +\ldots+ b_m$$

$$b_j \geqslant b_{j+1} \qquad \text{for } j = 1,\ldots,m-1$$

$$b_j \geqslant 2+b_{j+k-1} \qquad \text{for } j = 1,\ldots,m-k+1$$

$$b_m \geqslant 1$$

$$b_{m-i} \geqslant 2$$

and set

$$\rho_{k,i}(m,n) = \# R_{k,i}(m,n).$$

Also set

$$R_{k,i,m} = \cup_{n \geq 0} R_{k,i}(m,n).$$

__Theorem__ **3.3.** For the standard level k $\tilde{g}$-module L with highest weight λ (see (3.1)), Ω_L has two bases as follows:

$$\{e_\varphi v(-b_1,\ldots,-b_m) \mid \varphi \in Q,\ m \geq 0,$$

$$(b_1,\ldots,b_m) \in R_{k,i,m}\},$$

$$\{e_\varphi z_\alpha(-b_1)\ldots z_\alpha(-b_m)v_0 \mid \varphi \in Q,\ m \geq 0,$$

$$(b_1,\ldots,b_m) \in R_{k,i,m}\}.$$

The difference-two condition comes from Theorem 3.2 in the case $\varphi = \alpha$, $p = k$.

The action of the operators $z_\varphi(n)$ for $\varphi = \pm\alpha$ and $n \in \mathbb{Z}$ on the two bases is given by inductive procedures which follow from the definitions, Proposition 3.1 and Theorem 3.2. Now (1.5) and (2.3) complete the description of the $\tilde{g}$-module L.

Note that for $k = 1$, the set $R_{k,i,m}$ is empty unless $m = 0$, in which case it has only 1 element. Then Theorem 3.3 asserts that

$$\{e_\varphi v_0 \mid \varphi \in Q\}$$

is a basis of Ω_L, and formula (3.3) says that

$$z_\varphi(\zeta) = e_\varphi \zeta^{-1-\varphi}$$

for $\varphi = \pm\alpha$. This recovers the construction of the level 1

$A_1^{(1)}$-modules of [1], [2], [8].

Theorem 3.3 leads to a number of interesting combinatorial formulas (see [5]). For instance, by equating the known principally specialized character of L (see [3]) with an expression for it given by Theorem 3.3, we obtain:

Theorem 3.4.

$$\Pi_{n\equiv 0}(1-q^n)\Pi_{n\equiv i+1}(1-q^n)\Pi_{n\equiv k-i+1}(1-q^n)\Pi(1+q^n)^2 =$$

$$= \Sigma_{p\in \mathbf{Z}}\Sigma_{m,n\geqslant 0}\rho_{k,i}(m,n)q^{2kp^2+p(k-2i+4m)-m+2n},$$

where the products range over $n \geqslant 1$ and the congruences are mod(k+2).

4. THE LEVEL 2 CASE

The structure of the level 2 standard $\tilde{g}$-modules is much simpler than for level k, $k \geqslant 3$. In this case $A_1^{(1)}$ appears naturally as a Lie algebra of operators acting irreducibly on a space of mixed boson-fermion states (a subspace of the tensor product of a polynomial algebra, a group algebra and an exterior algebra). Here we reformulate the explicit construction of such modules given in [4, Section 6]. This construction extends naturally to the case of the level 1 standard modules for the affine algebra $B_\ell^{(1)}$ ($\ell \geqslant 2$) [4, Section 7].

For the standard modules with highest weights $2\Lambda_0$ and $2\Lambda_1 - \frac{1}{2}c$, denote by $\mathcal{C}$ the Clifford algebra generated by $\{\kappa(n) \mid n \in \mathbf{Z} + \frac{1}{2}\} \cup \{1\}$ and defined by the anticommutation relations

$$\text{(4.1)} \qquad \{\kappa(m), \kappa(n)\} = -2\delta_{m+n,0}, \quad m,n \in \mathbf{Z}+\tfrac{1}{2}.$$

Set

$$\text{(4.2)} \qquad \mathcal{C}_- = \text{span}\,\{\kappa(n) \mid n < 0\}.$$

The standard irreducible representation of the Clifford algebra $\mathcal{C}$ on the exterior algebra $\Lambda(\mathcal{C}_-)$ is defined uniquely be the anticommutation relations (4.1) and the conditions

$$\varkappa(n)\cdot v = \varkappa(n)\wedge v \quad \text{for } n < 0,\ v \in \Lambda(\mathcal{C}_-)$$

(4.3)

$$\varkappa(n)\cdot 1 = 0 \quad \text{for } n > 0.$$

Define an action of $d \in \tilde{g}$ on $\Lambda(\mathcal{C}_-)$ by

$$d(\varkappa(n_1)\wedge\ldots\wedge\varkappa(n_s)) =$$

(4.4)

$$= (n_1+\ldots+n_s)\varkappa(n_1)\wedge\ldots\wedge\varkappa(n_s)$$

for $n_1,\ldots,n_s < 0$.

Denote by $\mathbb{C}[Q]$ the group algebra of the root lattice Q (see (3.2)). For $\varphi \in Q$ let e^{φ} be the corresponding element of $\mathbb{C}[Q]$, and define the formal Laurent series ζ^{φ} with coefficients in End $\mathbb{C}[Q]$ by

$$\zeta^{\varphi}\cdot e^{\mu} = \zeta^{<\varphi,\mu>}e^{\mu}$$

for all $\mu \in Q$. Also define $e_{\varphi} \in \text{End}\ \mathbb{C}[Q]$ by

$$e_{\varphi}\cdot e^{\mu} = e^{\varphi+\mu}$$

for $\mu \in Q$. Let $\mathfrak{h}$ act on $\mathbb{C}[Q]$ by

$$x\cdot e^{\mu} = <x,\mu>e^{\mu}, \qquad \mu \in Q,$$

and let d act by

(4.5) $$d\cdot e^{\mu} = -\tfrac{1}{4}<\mu,\mu>e^{\mu}, \qquad \mu \in Q.$$

Set

$$\Omega = \Lambda(\mathcal{C}_-)\otimes\mathbb{C}[Q]$$

and let $\mathfrak{h}$ act on Ω as $1\otimes\mathfrak{h}$ and d as $d\otimes 1+1\otimes d$. Then the operators

$$(4.6) \qquad x_{\varphi}(\zeta) = E^{-}_{(2)}(-\varphi,\zeta)E^{+}_{(2)}(-\varphi,\zeta)\otimes z_{\varphi}(\zeta),$$

$\varphi = \pm\alpha$ (see (2.1)), together with the action $\tilde{\mathfrak{h}}'\otimes 1$ of $\tilde{\mathfrak{h}}'$, the action $1\otimes\mathfrak{h}$ of $\mathfrak{h}$ and the action $d\otimes 1 + 1\otimes d$ of d, define a representation of $\tilde{\mathfrak{g}}$ on

$$(4.7) \qquad V = K(2)\otimes\Omega,$$

where

$$(4.8) \qquad z_{\varphi}(\zeta) = \epsilon(\Sigma_{n\in\mathbb{Z}+1/2}\kappa(n)\zeta^{n})\otimes e_{\varphi}\zeta^{-1/2-\varphi/2},$$

with $\epsilon = \pm 1$ for $\varphi = \pm\alpha$, respectively. Furthermore, V is the direct sum of the standard $\tilde{\mathfrak{g}}$-modules with highest weights $2\Lambda_0$ and $2\Lambda_1 - \frac{1}{2}c$.

The level 2 module with highest weight $\Lambda_0 + \Lambda_1$ can be constructed similarly: This time, let $\mathcal{C}$ be the Clifford algebra generated by $\{\kappa(n) \mid n \in \mathbb{Z}\} \cup \{1\}$ and defined by the relations

$$\{\kappa(m), \kappa(n)\} = 2\delta_{m+n,0}, \qquad m,n \in \mathbb{Z}.$$

Define $\mathcal{C}_{-}$ as in (4.2) and consider the standard irreducible representation of the Clifford subalgebra $\mathcal{C}_{*}$ of $\mathcal{C}$ generated by $\{\kappa(n) \mid n \neq 0\} \cup \{1\}$ on the space $\Lambda(\mathcal{C}_{-})$ determined as in (4.3). Then

$$M = \Lambda(\mathcal{C}_{-})\otimes\mathbb{C}^{2}$$

has a $\mathcal{C}$-module structure defined by the conditions that $\mathcal{C}_{*}$ act as $\mathcal{C}_{*}\otimes 1$ and that $\kappa(0)$ act on $1\otimes\mathbb{C}^{2}$ as $1\otimes\begin{pmatrix} 0 & 1 \\ 1 & 0 \end{pmatrix}$.

Define an action of d on $\Lambda(\mathcal{C}_{-})$ as in (4.4) and then on M by $d\otimes 1$. Also let d act on $e^{\alpha/2}\mathbb{C}[Q] \subset \mathbb{C}[\frac{1}{2}Q]$ (the group algebra of $\frac{1}{2}Q$) as in (4.5).

Set

$$\Omega = M \otimes e^{\alpha/2}\mathbb{C}[Q]$$

and let $\mathfrak{h}$ act as $1\otimes\mathfrak{h}$ and d as $d\otimes 1 + 1\otimes d$. Let

$$V = K(2)\otimes\Omega$$

(cf. (4.7)). Then we can define a representation of $\tilde{\mathfrak{g}}$ on V as above, using (4.6) and (4.8) (with $\mathbb{Z}+\frac{1}{2}$ replaced by $\mathbb{Z}$ and $\epsilon = 1$), and V is the direct sum of two copies of the standard $\tilde{\mathfrak{g}}$-module with highest weight $\Lambda_0 + \Lambda_1 - \frac{1}{8}c$.

Set $Z = \mathbb{Z}+\frac{1}{2}$ in the case of the half-integral Clifford algebra and $Z = \mathbb{Z}$ in the case of the integral Clifford algebra. Then in both cases,

(4.9)
$$\begin{aligned} &z(\varphi,\dots,\varphi;\ \zeta_1,\dots,\zeta_s) = \\ &= \epsilon^s \Sigma_{n_1,\dots,n_s\in Z}\{\Sigma_{w\in\mathcal{S}_s}\epsilon(w)\kappa(n_1+s-w(1)) \dots \\ &\kappa(n_s+s-w(s))\}\zeta_1^{n_1} \dots \zeta_s^{n_s} \otimes \\ &e_{s\varphi}(\zeta_1\dots\zeta_s)^{-1/2-\varphi/2}, \end{aligned}$$

where $\varphi = \pm\alpha$, $s \geq 1$, $\mathcal{S}_s$ denotes the permutation group of $\{1,\dots,s\}$ and $\epsilon(w)$ is the sign (± 1) of the permutation w. Note that the expression in braces on the right-hand side of (4.9) can be written as the following noncommutative determinant:

$$\det\begin{bmatrix} \kappa(n_1+s-1) & \dots & \kappa(n_s+s-1) \\ \kappa(n_1+s-2) & \dots & \kappa(n_s+s-2) \\ \vdots & & \vdots \\ \kappa(n_1) & \dots & \kappa(n_s) \end{bmatrix}.$$

Also, for $\varphi \in Q$,

(4.10) $$e_{\varphi} = -1 \otimes 1 \otimes e_{2\varphi}$$

on

$$V = K(2) \otimes \Lambda(\mathcal{C}_-) \otimes \mathbb{C}[Q]$$

when $Z = \mathbb{Z}+\frac{1}{2}$ and

(4.11) $$e_{\varphi} = 1 \otimes 1 \otimes e_{2\varphi}$$

on

$$V = K(2) \otimes M \otimes e^{\alpha/2}\mathbb{C}[Q]$$

when $Z = \mathbb{Z}$.

Using (4.9), (4.10) and (4.11), we can now write explicit expressions for the basis vectors given in Theorem 3.3 for each of the level 2 modules under consideration.

REFERENCES

[1] I.B. Frenkel, Two constructions of affine Lie algebra representations and the boson-fermion correspondence in quantum field theory, J. Functional Anal. 44 (1981), 259-327.

[2] I.B. Frenkel and V.G. Kac, Basic representations of affine Lie algebras and dual resonance models, Invent. Math. 62 (1980), 23-66.

[3] J. Lepowsky and S. Milne, Lie algebraic approaches to classical partition identities, Advances in Math. 29 (1978), 15-59.

[4] J. Lepowsky and M. Primc, Standard modules for type one affine Lie algebras, Number Theory, New York, 1982, Springer-Verlag Lecture Notes in Mathematics 1052 (1984), 194-251.

[5] J. Lepowsky and M. Primc, Structure of the standard modules for the affine Lie algebra $A_1^{(1)}$, to appear.

[6] J. Lepowsky and R.L. Wilson, A new family of algebras underlying the Rogers-Ramanujan identities and generalizations, Proc. Nat. Acad. Sci. U.S.A. 78 (1981), 7254-7258.

[7] J. Lepowsky and R.L. Wilson, The structure of standard modules, I: Universal algebras and the Rogers-Ramanujan identities, Invent. Math. (1984).

[8] G. Segal, Unitary representations of some infinite-dimensional groups, Comm. Math. Phys. 80 (1981), 301-342.

Rutgers University
New Brunswick, NJ 08903

University of Zagreb
P.O. Box 187
41001 Zagreb, Yugoslavia

Mathematical Sciences Research Institute
Berkeley, CA 94720

STANDARD REPRESENTATIONS OF SOME AFFINE LIE ALGEBRAS

Kailash C. Misra[†]

INTRODUCTION.

The theory of Kac-Moody Lie algebras was originally developed by Kac [5] and Moody [19]. These are Lie algebras $L = L(A)$ associated with *generalized Cartan matrices* (GCM) A - i.e. $A = (a_{ij})$ is an $n \times n$ integral matrix satisfying: (1) $a_{ii} = 2$, for all i, (2) $a_{ij} \leq 0$, for all $i \neq j$, and (3) $a_{ij} = 0 \iff a_{ji} = 0$ for all $i \neq j$. The algebra L(A) is not necessarily finite-dimensional. Two GCM's $A = (a_{ij})$ and $B = (b_{ij})$ are called *equivalent* if there is a permutation π of the indices such that $b_{ij} = a_{\pi i, \pi j}$, for all i,j. A GCM is *indecomposable* if it is not equivalent to a matrix in block form $\begin{bmatrix} * & 0 \\ 0 & * \end{bmatrix}$. The GCM A is *symmetrizable* if there exists a nonsingular diagonal matrix D such that DA is symmetric. The GCM A is said to be of *finite type* if it is the Cartan matrix of a finite-dimensional split semisimple Lie algebra. The GCM A is *Euclidean* if it is indecomposable, symmetrizable, singular and every principal submatrix is of finite type. The infinite-dimensional Kac-Moody Lie algebras associated with the Euclidean generalized Cartan matrices are called *affine Lie algebras*. These algebras are completely classified [5,20]. They are one among $A_n^{(1)}$, $B_n^{(1)}$, $C_n^{(1)}$, $D_n^{(1)}$, $E_6^{(1)}$, $E_7^{(1)}$, $E_8^{(1)}$, $F_4^{(1)}$, $G_2^{(1)}$, $A_{2n}^{(2)}$, $A_{2n-1}^{(2)}$, $D_{n+1}^{(2)}$, $E_6^{(2)}$ and $D_4^{(3)}$, where the superscripts denote the type of the affine Lie algebra. For a simple complex Lie algebra M, the corresponding affine Lie algebra $M^{(1)}$ is defined as the algebra associated with the extended Cartan matrix of M. It is known [5,20] that $M^{(1)}$ is isomorphic to a one-dimensional central extension of the complex Lie algebra $\tilde{M} = M \otimes_{\mathbb{C}} \mathbb{C}[t,t^{-1}]$, where t is an indeterminate and $\mathbb{C}[t,t^{-1}]$ denotes the algebra of Laurent polynomials in t.

[†] Partially supported by the National Science Foundation through the Mathematical Sciences Research Institute.

Vertex Operators in Mathematics and Physics - Proceedings of a Conference November 10-17, 1983. Publications of the Mathematical Sciences Research Institute #3, Springer-Verlag, 1984.

Let L be an affine Lie algebra over the field of complex numbers $\mathbb{C}$ with canonical generators e_i, f_i, h_i, $i = 0,1,2,\ldots,n-1$. The *principal gradation* of L is defined by the conditions

$$\deg e_i = 1 = -\deg f_i \qquad \text{and} \qquad \deg h_i = 0,$$

$i = 0,1,2,\ldots,n-1$. This induces in the natural way a gradation $U(L) = \oplus_{i\in\mathbb{Z}} U_i$ of the universal enveloping algebra U(L) of L. A *highest weight vector* in an L-module is a nonzero vector annihilated by $e_0, e_1, \ldots, e_{n-1}$, and whose span is preserved by $h_0, h_1, \ldots, h_{n-1}$. An L-module generated by a highest weight vector, say v_0, such that the sum $\sum_{j\leq 0} V_j$, where $V_j = U_j \cdot v_0$, is direct, is called a *graded highest weight L-module*. Clearly each V_j is finite-dimensional. A *standard L-module* V is a graded highest weight module with a highest weight vector v_0 say, such that for some positive integer n, $f_i^n \cdot v_0 = 0$ for all $i = 0,1,\ldots,n-1$. Thus the principal gradation of L induces a gradation (also called 'principal') on the standard module V and we have a direct sum decomposition

$$V = \bigoplus_{j\leq 0} V_j$$

with $\dim(V_j) < \infty$ for each j. Let q be an indeterminate, and consider the formal power series ring $\mathbb{Z}[[q]]$. The (principally specialized) character $\chi(V)$ of V is defined by

$$\chi(V) = \sum_{j\geq 0} (\dim V_{-j})q^j \in \mathbb{Z}[[q]].$$

This character has a known product expansion due to Kac and Lepowsky (cf. [17], Formula 1.1). It has recently been established that these characters give rise to many important power series identities (for example see, [2], [9], [11], [14], [17], and [18]). This observation has been crucial in the sense that it has brought two different branches of mathematics, namely, Kac-Moody Lie algebras and Combinatorics, closer. For an excellent survey on this the reader is referred to [10]. Using these connections one can hope to prove known or new power series identities by the help of the structure theory of affine Lie algebras. Conversely, one can study structure theory of affine algebras using power series indentities.

Lepowsky and Milne [11] first observed that the product sides of the Rogers-Ramanujan identities (cf. [1]) differ by a simple factor from the characters of certain modules for the affine Lie algebra $A_1^{(1)}$. In 1978 Lepowsky and Wilson [12] made an important discovery by giving an explicit construction of the basic representation of the affine Lie algebra $A_1^{(1)}$ in terms of differential operators acting on a certain polynomial algebra in infinitely many variables via an exponential generating function. Subsequently, they utilized this construction in [14] to give a Lie theoretic proof of the Rogers-Ramanujan identities. In [8] the basic module construction given by Lepowsky and Wilson was generalized for all the affine Lie algebras of ADE type. Other constructions of the basic representations for these affine Lie algebras were found by Frenkel and Kac [4] and Segal [21]. Recently, using the spin module, Frenkel [3] and Kac and Peterson [7] have given explicit constructions of certain representations of orthogonal Lie algebras.

More recently, starting from the viewpoint in [8], Lepowsky and Wilson [13] have introduced certain new algebras $\underline{Z}_V$ corresponding to a highest weight L-module V. In [15], they have done a more elaborate study of these new algebras corresponding to the highest weight representations of all affine Lie algebras. They have also used these algebras corresponding to certain representations of $A_1^{(1)}$ to give another proof of the Rogers-Ramanujan identities.

In [17] and [18] we used the $\underline{Z}_V$ algebras to give explicit constructions of some standard modules for the affine Lie algebras $A_n^{(1)}$ and $C_n^{(1)}$. We also gave new proofs of the Rogers-Ramanujan identities using certain representations of the affine Lie algebras $A_2^{(1)}$ and $C_3^{(1)}$. In this paper we intend to summarize the main results of [17] and [18]. To see more details of what follows the reader is referred to [17] and [18].

1. THE AFFINE LIE ALGEBRA $A_{n-1}^{(1)}$.

Consider the simple Lie algebra $\mathfrak{sl}(n,\mathbb{C})$. Let E_{ij} (i,j = 1,2,...,n) denote the n × n matrix which is 1 in the i,j - entry and 0 everywhere else. Define

$$E_0 = E_{n,1}, \quad E_i = E_{i,i+1}, \quad i = 1,2,\dots,n-1$$

$$F_0 = E_{1,n}, \quad F_i = E_{i+1,i}, \quad i = 1,2,\dots,n-1,$$

$$H_0 = E_{n,n} - E_{1,1}, \quad \text{and} \quad H_i = E_{i,i} - E_{i+1,i+1}, \; i = 1,2,\dots,n-1.$$

Note that $\{E_i, F_i, H_i \mid i = 1,2,\dots,n-1\}$ are the canonical generators for the Lie algebra $\mathfrak{sl}(n,\mathbb{C})$. Consider the Lie algebra

(1.1) $$\mathcal{L} = \mathfrak{sl}(n,\mathbb{C}) \otimes_{\mathbb{C}} \mathbb{C}[t,t^{-1}] \oplus \mathbb{C}\, c$$

with the bracket operation given by

(1.2) $$[x \otimes t^i, y \otimes t^j] = [x,y] \otimes t^{i+j} + i\delta_{i+j,0}\, n^{-1}\mathrm{tr}(xy)c$$

for all $x,y \in \mathfrak{sl}(n,\mathbb{C})$ and c central. Here $\mathbb{C}[t,t^{-1}]$ denotes the algebra of Laurent polynomials in the indeterminate t. Let

$$e_0 = E_0 \otimes t, \qquad f_0 = F_0 \otimes t^{-1}, \qquad h_0 = H_0 \otimes 1 \oplus n^{-1}c$$

and

$$e_i = E_i \otimes 1, \qquad f_i = F_i \otimes 1, \qquad h_i = H_i \otimes 1$$

for i = 1,2,...,n-1. Then (cf. [8]) $\mathcal{L}$ is isomorphic to the affine Lie algebra $A_{n-1}^{(1)}$, with e_i, f_i, h_i, i = 0,1,2,...,n-1, as the images of the corresponding canonical generators. Let ω be a primitive n^{th} root of unity. For m = 1,2,...,n-1, $i,j \in \mathbb{Z}$ and $i \not\equiv 0 \pmod n$ define

(1.3) $$X_j^{(m)} = D^m E^j \otimes t^j \quad \text{and} \quad B_i = E^i \otimes t^i$$

where $D = \mathrm{diag}(\omega,\omega^2,\ldots,\omega^n = 1)$, $E = (\delta_{i',j-1})_{i,j=1}$, where (') denotes reduction modulo n. Using (1.2), the following Lemma can be proved by direct computation.

__Lemma__ 1.1 [17]. For $i,j \in \mathbb{Z}$ and $m,\ell = 1,2,\ldots,n-1$,

(a) $[B_i, B_j] = \delta_{i+j,0}\, ic$, for $i,j \not\equiv 0 \pmod n$,

(b) $[B_i, X_j^{(m)}] = (\omega^{mi} - 1)X_{i+j}^{(m)}$, for $i \not\equiv 0 \pmod n$,

$(\omega^{\ell j} - \omega^{mi})\, X_{i+j}^{(m+\ell)}$, for $(m+\ell) \neq n$,

(c) $[X_j^{(m)}, X_i^{(\ell)}] = j\omega^{\ell j}\delta_{i+j,0}c$, for $m+\ell = n$, $i+j \equiv 0 \pmod n$

$(\omega^{\ell j} - \omega^{mi})\, B_{i+j}$, for $m+\ell = n$, $i+j \not\equiv 0 \pmod n$.

□

The set $\{c, B_i, X_j^{(m)} \mid i,j \in \mathbb{Z}, i \not\equiv 0 \pmod n, m = 1,2,\ldots,n-1\}$ forms a basis of the affine Lie algebra $\mathcal{L}$. The principal gradation of $\mathcal{L}$ is given by the conditions

$$\deg c = 0, \qquad \deg B_i = i, \text{ and } \deg X_j^{(m)} = j,$$

for all $i,j \in \mathbb{Z}$, $i \not\equiv 0 \pmod n$ and $m = 1,2,\ldots,n-1$. The resulting gradation $\mathcal{L} = \oplus_{j\in\mathbb{Z}} \mathcal{L}_j$ is a Lie algebra gradation. A subspace of $\mathcal{L}$ is graded if it is the direct sum of its intersections with $\mathcal{L}_j$. The subalgebra $\mathfrak{s}$ with basis $\{c, B_i \mid i \not\equiv 0 \pmod n\}$ is a (principally) graded Heisenberg subalgebra of $\mathcal{L}$. Set

$$\mathfrak{h} = \mathcal{L}_0 = \sum_{m=1}^{n-1} \mathbb{C}\, X_0^{(m)} \oplus \mathbb{C}\, c,$$

$$\mathfrak{n} = \oplus_{j>0} \mathcal{L}_j \quad \text{and} \quad \mathfrak{n}^- = \oplus_{j>0} \mathcal{L}_{-j}.$$

Then $\mathfrak{h}$, $\mathfrak{n}$ and $\mathfrak{n}^-$ are subalgebras of $\mathcal{L}$ and

$$\mathcal{L} = \mathfrak{n}^- \oplus \mathfrak{h} \oplus \mathfrak{n}.$$

Observe that, for $i = 0,1,2,\ldots,n-1$,

$$(1.4) \qquad h_i = n^{-1}\{c + \sum_{m=1}^{n-1} \omega^{-im}(1 - \omega^{-m})\, X_0^{(m)}\}$$

and $\{h_i \mid 0 \leq i \leq n-1\}$ form a basis of the Cartan subalgebra $\mathfrak{h}$ with $c = \sum_{i=0}^{n-1} h_i$. Let $\{h_i^* \mid 0 \leq i \leq n-1\}$ be the dual basis. A highest weight $\mathscr{L}$-module is said to be of level $r \in \mathbb{C}$ if the central element c acts on it as multiplication by the scalar r. Let V be a standard $\mathscr{L}$-module of level r with highest weight λ and highest weight vector v_0. Note that V is an irreducible $\mathscr{L}$-module (unique up to equivalence) and λ is dominant integral (i.e. $\lambda(h_i) \in \mathbb{N}$, $i = 0,1,\ldots,n-1$). Furthermore, observe that if $\lambda = \sum_{i=0}^{n-1} r_i h_i^*$, then $r = \sum_{i=0}^{n-1} r_i$.

Let $V = \oplus_{j\leq 0} V_j$ be the induced gradation on V. Define the completion $\hat{V}$ (resp. $\hat{\mathscr{L}}$) of V (resp. $\mathscr{L}$) to be the vector space $\prod_{j\leq 0} V_j$ (resp. $\prod_{j\in\mathbb{Z}} \mathscr{L}_j$). Observe that $\mathscr{L}$ acts on $\hat{V}$ and every element of $\hat{\mathscr{L}}$ can be viewed as a linear operator from V to $\hat{V}$. For $m = 1,2,\ldots,n-1$, define

$$(1.5) \qquad X^{(m)} = \sum_{j\in\mathbb{Z}} X_j^{(m)} \in \hat{\mathscr{L}}.$$

By Lemma 1.1. (b),

$$(1.6) \qquad [B_i, X^{(m)}] = (\omega^{mi} - 1)X^{(m)},$$

for $i \not\equiv 0 \pmod n$ and $m = 1,2,\ldots,n-1$. Assume that $r \in \mathbb{C}^*$. For $m = 1,2,\ldots,n-1$, define the following operators (cf. [17]):

$$(1.7) \qquad E^-_{(m)} = \exp(\Sigma(\frac{1-\omega^{mi}}{ri})B_{-i}) \in \mathrm{End}(\hat{V}),$$

$$(1.8)\qquad E^{+}_{(m)} = \exp\left(\Sigma\left(\frac{\omega^{-mi}-1}{ri}\right) B_i\right) \in \mathrm{End}(V),$$

where 'exp' denotes the formal exponential series, i runs through the positive integers $\not\equiv 0 \pmod{n}$, and

$$(1.9)\qquad Z^{(m)} = E^{-}_{(m)} X^{(m)} E^{+}_{(m)} \in \mathrm{Hom}(V,\hat{V}).$$

For each $j \in \mathbb{Z}$, let $Z_j^{(m)} \in \mathrm{End}(V)$ be the homogeneous component of degree j (in the obvious sense) of $Z^{(m)}$, so that

$$(1.10)\qquad Z^{(m)} = \sum_{j\in\mathbb{Z}} Z_j^{(m)}.$$

Denote by $\underline{Z}_V$ or $\underline{Z}$ the subalgebra of End(V) [13] generated by

$$(1.11)\qquad \{Z_j^{(m)} \mid j \in \mathbb{Z},\ m = 1,2,\ldots,n-1\}.$$

The algebra $\underline{Z}$ centralizes the action of $\mathfrak{s}$ on V and hence preserves the vacuum space $\Omega(V) = \{v \in V \mid \mathfrak{s}^{+}\cdot v = 0\}$ of V, where $\mathfrak{s}^{+}$ is the subalgebra $\oplus_{j>0}\ \mathfrak{s}_j$ of $\mathfrak{s}$. Let $\mathfrak{p}$ denote the subalgebra $\mathfrak{s}^{+} \oplus \mathbb{C}\, c$ of $\mathfrak{s}$.

<u>Proposition</u> 1.2. (cf. [13]) Suppose that $r \in \mathbb{C}^{*}$. Then the map

$$f\colon U(\mathfrak{s}) \otimes_{u(\mathfrak{p})} \Omega(V) \longrightarrow V$$

$$u \otimes w \longmapsto u\cdot w$$

($u \in U(\mathfrak{s})$, $w \in \Omega(V)$) is an $\mathfrak{s}$-module isomorphism. In particular,

$$(1.12)\qquad \chi(V) = F\ \chi(\Omega(V))$$

where

$$F = \prod_{\substack{i>0 \\ i\not\equiv 0 \pmod{n}}} (1-q^{i})^{-1} \qquad \square$$

Since V is $\mathcal{L}$-irreducible, hence $\Omega(V)$ is $\underline{Z}$-irreducible and $\Omega(V) = \underline{Z}\cdot v_0$ [13]. Define the $\underline{Z}$-filtration [13] of $\Omega = \Omega(V)$

$$0 = \Omega^{[-1]} \subseteq \Omega^{[0]} \subseteq \Omega^{[1]} \subseteq \cdots \subseteq \Omega$$

by the condition that for all $s \geq 0$, $\Omega^{[s]}$ is the span of all the elements $x_1 x_2 \cdots x_t \cdot v_0$, $0 \leq t \leq s$, where each x_t is one of the $Z_j^{(m)}$ ($j \in \mathbb{Z}$, $m = 1,2,\ldots,n-1$). Then

$$\chi(\Omega) = \sum_{s \geq 0} \chi(\Omega^{[s]} / \Omega^{[s-1]}). \tag{1.13}$$

Suppose now that V is of level one with highest weight h_i^* and highest weight vector v_0. Then V is also $\mathfrak{s}$-irreducible [8]. Hence $\Omega(V)$ is one dimensional and

$$X^{(m)} = c_m^{(i)} (E^-_{(m)})^{-1} (E^+_{(m)})^{-1} \tag{1.14}$$

where

$$c_m^{(i)} = h_i^*(D^m) = \omega^{m(i+1)}/(\omega-1), \tag{1.15}$$

for $m = 1,2,\ldots,n-1$.

2. SOME STANDARD REPRESENTATIONS OF $A_{n-1}^{(1)}$.

Let W_i denote the standard $\mathcal{L}$-module with highest weight h_i^*, $i = 0,1,2,\ldots,n-1$. Then $W_0 \otimes W_i$ is an $\mathcal{L}$-module of level two with highest weight $(h_0^* + h_i^*)$. Let V_i denote the submodule of $W_0 \otimes W_i$ generated by the tensor product of the highest weight vectors. Clearly, V_i is the standard $\mathcal{L}$-module of level two with highest weight $(h_0^* + h_i^*)$ and highest weight vector v_0, say. Here we want to study the structure of V_i for $i = 0,1,\ldots,[\frac{n}{2}]$. Other standard $\mathcal{L}$-modules of level two can be dealt with similarly.

Now, using Lemma 1.1 and by direct computations with formal power series involving the operators $Z_j^{(m)}$ acting on V_i, we can derive the so called *generalized commutator identities* ([17],

Theorem 2.11). Also from (1.9), (1.14) and (1.15) we can get an exponential representation of the operator $Z^{(m)}$, $m = 1,2,\dots,n-1$, acting on V_i. Using this, we can derive another set of identities, called *generalized anticommutator identities* ([17], Theorem 2.16).

<u>Case</u> I: $n = 3$.

Using the generalized commutator and anticommutator identities we can prove the following theorem.

<u>Theorem</u> 2.1. ([17], Theorem 3.1). For $i = 0,1$, the vacuum space $\Omega(V_i)$ of the standard module V_i with highest weight vector v_0 is spanned by

$$S = \{Z^{(1)}_{d_1} Z^{(1)}_{d_2} \cdots Z^{(1)}_{d_n} \cdot v_0\}$$

where $d_1 < d_2 < \cdots < d_n < 0$, $n \geqslant 0$, $d_k - d_{k-1} \geqslant 2$ and if $i = 0$, then $d_n < -1$. □

Now to prove that the set S in Theorem 2.1 is independent we observe that for $v \otimes w \in W_0 \otimes W_i$,

$$Z^{(m)}(v \otimes w) = (Z^{(m,1)} + Z^{(m,2)})(v \otimes w) \tag{2.1}$$

where

$$Z^{(m,1)} = E^-_{(m)} X^{(m)} E^+_{(m)} \otimes E^-_{(m)} E^+_{(m)} \tag{2.2}$$

and

$$Z^{(m,2)} = E^-_{(m)} E^+_{(m)} \otimes E^-_{(m)} X^{(m)} E^+_{(m)}, \tag{2.3}$$

for $m = 1,2$. Let $Z^{(m,u)}_j$ denote the jth homogeneous component of $Z^{(m,u)}$, $m,u = 1,2$, so that

(2.4) $$Z^{(m,u)} = \sum_{j\in\mathbf{Z}} Z_j^{(m,u)}.$$

Using (1.14) and (1.15) we get an exponential representation of the operator $Z^{(m,u)}$ $(m,u = 1,2)$, which gives

(2.5) $$\begin{aligned} Z_j^{(2,1)} &= (\omega+1)\omega^{2i+j}\, Z_j^{(1,2)} \\ Z_j^{(2,2)} &= (\omega+1)\omega^{2i+j}\, Z_j^{(1,1)}. \end{aligned}$$

Also using the exponential representation we can derive generalized commutator and anticommutator identities ([17] Proposition 3.3) involving the operators $Z_j^{(m,u)}$, $(m,u = 1,2)$. Let $\underline{Z}^*$ denote the algebra generated by the operators $Z_j^{(m,u)}$, $j \in \mathbf{Z}$, $m,u = 1,2$. Then $\underline{Z}^*$ centralizes the action of $\mathfrak{s}$ on $W_0 \otimes W_i$ and hence preserves the vacuum space $\Omega(W_0 \otimes W_i) = \{v \in W_0 \otimes W_i \mid \mathfrak{s}^+\cdot v = 0\}$. Furthermore, $\Omega(W_0 \otimes W_i) = \underline{Z}^*\cdot v_0$ (cf. [17], Proposition 1.5). Observe that

(2.6) $$\begin{aligned} \chi(\Omega(W_0 \otimes W_i)) &= \chi(W_0 \otimes W_i)/F \\ &= \prod_{\substack{j>0 \\ j\not\equiv 0 \pmod 3}} (1-q^j)^{-1} \end{aligned}$$

But (cf. [1])

(2.7) $$\begin{aligned} \prod_{\substack{j>0 \\ j\not\equiv 0 \pmod 3}} (1-q^j)^{-1} &= \prod_{j>0} (1-q^{3j})/(1-q^j) \\ &= \prod_{j>0} (1 + q^j + q^{2j}) \\ &= \sum_{\ell\geq 0} a(\ell)q^{\ell} \end{aligned}$$

where

$a(\ell)$ = number of partitions of ℓ in which each part appears at most twice.

Now using (2.5), (2.6), (2.7) and the generalized commutator and anticommutator identities involving the operators $Z_j^{(m,u)}$, $j \in \mathbf{Z}$, $m,u = 1,2$, the following theorem can be proved.

Theorem 2.2. ([17], Theorem 3.4). The vacuum space $\Omega(W_0 \otimes W_i)$, (i = 0,1) of $W_0 \otimes W_i$ with highest weight vector v_0 has the basis

$$T = \{Z_{i_1}^{(1,1)} Z_{i_2}^{(1,1)} \cdots Z_{i_n}^{(1,1)} \cdot v_0\}$$

where $i_1 \leqslant i_2 \leqslant \cdots \leqslant i_n < 0$, $n \geqslant 0$, and $i_{k-2} < i_k$. □

The next Theorem follows from Theorem 2.1 and Theorem 2.2.

Theorem 2.3. ([17], Theorem 3.5). The set S in Theorem 2.1 is a basis for the vacuum space $\Omega(V_i)$, $i = 0,1$. □

Thus combining Theorem 2.3 with Proposition 1.2 we obtain explicit constructions of the level two standard $A_2^{(1)}$-modules V_i, $i = 0,1$. Furthermore, since, for $i = 0,1$,

$$\chi(\Omega(V_i)) = \prod_{\substack{j>0 \\ j \not\equiv 0,\ \pm(i+1) \pmod 5}} (1 - q^j)^{-1} \tag{2.8}$$

which coincides with the product sides of the Rogers-Ramanujan identities (cf. [1]), Theorem 2.3 along with equation (1.13) gives a new Lie theoretic proof of these identities.

Case II: $n = 5$.

In this case $\mathcal{L} \cong A_4^{(1)}$. Consider the level two standard $\mathcal{L}$-modules V_i, $i = 0,1,2$. Observe that in this case

$$\chi(\Omega(V_i)) = \prod_{\substack{j>0 \\ j \not\equiv 0,\ \pm(i+1) \pmod 7}} (1 - q^j)^{-1} \tag{2.9}$$

But by the generalization of Rogers-Ramanujan identities due to Gordon (cf. [1]) we have, for $i = 0,1,2$,

$$(2.10)\qquad \prod_{\substack{j>0\\ j\not\equiv 0,\ \pm(i+1) \pmod 7}} (1-q^j)^{-1} = \sum_{\ell\geq 0} b_i(\ell)q^\ell$$

where

$b_i(\ell)$ = number of partitions of ℓ of the form $(d_1,d_2,\ldots,d_s)$ where $d_{j+2} - d_j \geq 2$ and at most i of the d_j equal to 1.

Now using (2.9), (2.10) and the generalized commutator and anticommutator identities we can prove the following theorem.

__Theorem__ **2.4.** ([17], Theorem 4.1). The vacuum space $\Omega(V_i)$ of the standard module V_i, $i = 0,1,2$, with highest weight vector v_0 has the basis

$$\{Z^{(1)}_{d_1} Z^{(1)}_{d_2} \cdots Z^{(1)}_{d_n} \cdot v_0\}$$

where $d_1 \leq d_2 \leq \cdots \leq d_n < 0$, $n \geq 0$ and $d_{k+2} - d_k \geq 2$. Furthermore, for $i = 0$, $d_n < -1$; and for $i = 1$, $d_{n-1} < -1$. □

3. SOME STANDARD REPRESENTATIONS OF $C_k^{(1)}$.

In order to be compatible with the notations in [18] we intend to use $\underline{Z}$, $Z^{(m)}$, $Z_j^{(m)}$ and Ω to denote the corresponding things for the affine Lie algebra $C_k^{(1)}$, $k \geq 2$.

In Section 1, set $n = 2k$ $(k \geq 2)$, so that $\mathcal{L}$ is isomorphic to $A^{(1)}_{2k-1}$. Consider the automorphism

$$\theta\colon \alpha \longmapsto -\beta\alpha^t\beta^{-1}$$

of $\underline{sl}(2k, \mathbb{C})$, where $\beta = ((-1)^{i+1}\delta_{i,2k-j+1})^{2k}_{i,j=1}$ and α^t denotes the transpose of α. Extend this to an automorphism of $\mathcal{L}$, again denoted by θ, by defining

$$\theta(x \otimes t^i) = \theta(x) \otimes t^i, \text{ for all } x \in \underline{sl}(2k, \mathbb{C})$$

and

$$\theta(c) = c.$$

Observe that

$$(3.1) \qquad \theta(B_i) = \begin{cases} B_i, & \text{for all } i \text{ odd} \\ -B_i, & \text{otherwise,} \end{cases}$$

and

$$(3.2) \qquad \theta(X_j^{(m)}) = -(-1)^j \omega^{-m(j-1)} X_j^{(-m)},$$

for $i, j \in \mathbb{Z}$, $i \not\equiv 0 \pmod{2k}$ and $m = 1,2,\dots,2k-1$, where $(-m)$ in the superscript denotes $(2k-m)$. Let

$$(3.3) \qquad \bar{X}_j^{(m)} = \{X_j^{(m)} + \theta(X_j^{(m)})\} / 2$$

for $j \in \mathbb{Z}$ and $m = 1,2,\dots,k$. Then the set (cf. [18])

$$\{c, B_i, \bar{X}_j^{(m)} \mid i,j \in \mathbb{Z}, i \text{ odd}, m = 1,2,\dots,k\}$$

is a basis for an affine Lie algebra $\mathfrak{g}$ which is isomorphic to the symplectic affine Lie algebra $C_k^{(1)}$. The principal gradation of $\mathfrak{g}$ is given by

$$\deg B_i = i, \qquad \deg c = 0 \qquad \text{and } \deg \bar{X}_j^{(m)} = j.$$

The subalgebra $\mathfrak{t}$ with basis $\{c, B_i \mid i \in 2\mathbb{Z}+1\}$ is a (principally) graded Heisenberg subalgebra of $\mathfrak{g}$. Let

$$e_0' = e_0, \qquad e_i' = e_i + e_{2k-i}, \qquad i = 1,2,\dots,k-1, \; e_k' = e_k,$$

$$h_0' = h_0, \qquad h_i' = h_i + h_{2k-i}, \qquad i = 1,2,\dots,k-1, \qquad h_k' = h_k,$$

$$f_0' = f_0, \qquad f_i' = f_i + f_{2k-i}, \qquad i = 1,2,\dots,k-1, \qquad f_k' = f_k.$$

Then e_i', f_i', h_i', $i = 0,1,2,\dots,k$ are canonical generators of $g \cong C_k^{(1)}$. Let $\{h_i'^*\}$ be the dual basis of $\{h_i'\}$ (i.e. $h_i'^*(h_j') = \delta_{i,j}$). Let W_i be the standard $\mathscr{L}$-module with highest weight h_i^*, $i = 1,2,\dots,k$, and highest weight vector v_0. Consider the g-module generated by v_0. Denote it by V_i. Since $f_i^m \cdot v_0 = 0$ for some m, and since $[f_i, f_{2k-i}] = 0$, for $i = 1,2,\dots,k-1$, it follows that $f_i'^r \cdot v_0 = 0$, $i = 0,1,2,\dots,k$, for some integer r. Also note that $h_i^*(h_j') = \delta_{i,j}$, for $i,j = 0,1,2,\dots,k$. Hence it is clear that V_i is the standard g-module with highest weight $h_i'^*$ and highest weight vector v_0. Let $\hat{g}$, $\hat{V}_i$ and $\hat{W}_i$ denote the completion of g, V_i and W_i respectively. For $m = 1,2,\dots,k$, define the operators

$$\bar{X}^{(m)} = \sum_{j\in\mathbb{Z}} \bar{X}_j^{(m)}. \tag{3.4}$$

It follows from (3.3), (3.4) and Lemma 1.1(b) that for $m = 1,2,\dots,k$ and $i \in 2\mathbb{Z}+1$,

$$[B_i, \bar{X}^{(m)}] = (\omega^{mi} - 1)\, \bar{X}^{(m)}, \tag{3.5}$$

where ω is a primitive $(2k)^{th}$ root of unity. Now define the operators

$$\underline{E}_{(m)}^- = \exp\Big(\sum_i \Big(\frac{1-\omega^{mi}}{i}\Big) B_{-i}\Big) \tag{3.6}$$

$$\underline{E}_{(m)}^+ = \exp\Big(\sum_i \Big(\frac{\omega^{-mi}-1}{i}\Big) B_i\Big) \tag{3.7}$$

where i runs through all odd positive integers and $m = 1,2,\dots,k$. Define

$$Z^{(m)} = \underline{E}_{(m)}^- \,\bar{X}^{(m)}\, \underline{E}_{(m)}^+ \tag{3.8}$$

for $m = 1,2,\dots,k$. Let

$$Z^{(m)} = \sum_{j\in\mathbb{Z}} Z_j^{(m)}$$

where $Z_j^{(m)}$ is the homogeneous component of degree j of $Z^{(m)}$. Let

$\underline{Z}$ denote the algebra generated by $\{Z_j^{(m)} \mid j \in \mathbf{Z},\ m = 1,2,\ldots,k\}$. Then, as before, the algebra $\underline{Z}$ centralizes the action of $\mathfrak{t}$ on V_i and hence preserves the vacuum space $\Omega(V_i) = \{v \in V_i \mid \mathfrak{t}^+ \cdot v = 0\}$. Also since V_i is $\mathfrak{g}$-irreducible, $\Omega(V_i)$ is $\underline{Z}$-irreducible and $\Omega(V_i) = \underline{Z} \cdot v_0$, [13]. Furthermore, V_i is isomorphic to $U(\mathfrak{t}) \otimes \Omega(V_i)$ and (cf. Proposition 1.2),

$$(3.9) \qquad \chi(V_i) = F_1\ \chi(\Omega(V_i))$$

where

$$(3.10) \qquad F_1 = \prod_{\substack{j>0 \\ j \in 2\mathbf{Z}+1}} (1-q^j)^{-1}.$$

Define the $\underline{Z}$-filtration [13] of $\Omega = \Omega(V_i)$,

$$0 = \Omega^{[-1]} \subseteq \Omega^{[0]} \subseteq \Omega^{[1]} \subseteq \cdots \subseteq \Omega$$

as before. Then

$$(3.11) \qquad \chi(\Omega) = \sum_{s \geq 0} \chi(\Omega^{[s]} / \Omega^{[s-1]}).$$

Now, using (3.2), (3.3), (3.8), Lemma 1.1 and by direct computations, we can derive a set of *generalized commutator identities* for $C_k^{(1)}$ acting on V_i ([18], Theorem 2.5). On the other hand, since $V_i \subset W_i$, $i = 0,1,2,\ldots,k$, by (3.3), (3.4), (3.8) and (1.14) we have an exponential representation of $Z^{(m)}$ on W_i hence on V_i. For $v \in W_i$, $m = 1,2,\ldots,k$,

$$(3.12) \qquad Z^{(m)} \cdot v = \underline{E}^-_{(m)} \{(X^{(m)} + \theta X^{(m)})/2\} \underline{E}^+_{(m)} \cdot v$$

$$= A\{\omega^{mi} \exp(-P_m) \exp(Q_m) + \omega^{-mi} \exp(P_m) \exp(-Q_m)\} \cdot v$$

where

$$A = \frac{1}{2}\, \omega^m/(\omega^m-1),$$

$$P_m = \sum_j (1-\omega^{mj})B_{-j}/j,$$

$$Q_m = \sum_j (1-\omega^{-mj})B_j/j,$$

with j varying over all even positive integers $\not\equiv 0 \pmod{2k}$. It now follows from (3.12) that on W_i, hence on V_i,

$$Z_j^{(m)} = 0, \text{ for all } j \text{ odd and } m = 1,2,\ldots,k,$$

(3.13) $$Z_j^{(k)} = 0, \text{ for all } j \neq 0 \text{ and } Z_0^{(k)} = (-1)^i/2,$$

$$Z_j^{(k-r)} = \left\{\frac{(-1)^i(\omega^r-1)}{\omega^r(\omega^r+1)}\right\}\omega^{rj}\, Z_j^{(r)},\ r = 1,2,\ldots,k-1.$$

Furthermore, using (3.12) we can derive the so called <u>generalized anticommutator identities</u> ([18], Theorem 2.6) for $C_k^{(1)}$ acting on W_i, hence on V_i.

<u>Case</u> I: k = 2.

Let V_i (i = 0,1,2) denote the standard $C_2^{(1)}$-modules with highest weight $h_i^{'*}$ and highest weight vector v_0. Let $\Omega(V_i)$ denote the vacuum space of V_i. Observe that (cf. [17], Formula 1.1),

(3.14) $$\chi(\Omega(V_0)) = \chi(\Omega(V_o)) = \prod_{j\geq 1} (1+q^{4j})$$

and

(3.15) $$\chi(\Omega(V_1)) = \prod_{j\geq 1} (1+q^{4j-2}).$$

The following theorem can now be proved using (3.13), (3.14), (3.15) and generalized commutator and anticommutator identities.

<u>Theorem</u> 3.1. ([18], Theorem 3.1). The vacuum space $\Omega(V_i)$ of the

standard $C_2^{(1)}$-module V_i (i=0,1,2) with highest weight vector v_0 has the basis

$$\{Z_{d_1}^{(1)} \cdots Z_{d_n}^{(1)} \cdot v_0 \mid n \geq 0,\ d_1 < d_2 < \cdots < d_n,\ 0,\ d_j \in \mathbf{Z}\}$$

with $d_j \equiv 0 \pmod 4$, when $i = 0,2$; and $d_j \equiv 2 \pmod 4$, when $i = 1$. □

Case II: $k = 3$.

Now let V_i ($i = 0,1,2,3$) denote the standard $C_3^{(1)}$-module with highest weight $h_i'^{*}$ and let $\Omega(V_i)$ denote its vacuum space. Using the generalized commutator and anticommutator identities and (3.13) we can prove the following theorem.

Theorem 3.2. ([18], Theorem 4.1). The vacuum space $\Omega(V_i)$ of the level one standard $C_3^{(1)}$-module V_i (i=0,1,2,3) with highest weight vector v_0 is spanned by

$$\mathcal{B} = \{Z_{2d_1}^{(1)} \cdots Z_{2d_n}^{(1)} \cdot v_0\}$$

where $d_1 < d_2 < \cdots < d_n < 0$, $n \geq 0$, $d_{j+1} - d_j \geq 2$ and if $i = 0,3$ then $d_n < -1$. □

To prove that the set $\mathcal{B}$ in the above theorem is independent we observe that on W_i, for $m = 1,2,3$,

$$Z^{(m)} = Z^{(m,1)} + Z^{(m,2)} \tag{3.16}$$

where

$$Z^{(m,1)} = \tfrac{1}{2}\, \underline{E}^{-}_{(m)}\, X^{(m)}\, \underline{E}^{+}_{(m)} \tag{3.17}$$

$$= A\, \omega^{mi} \exp(-P_m) \exp(Q_m)$$

and

$$(3.18) \qquad Z^{(m,2)} = \frac{1}{2}\, \underline{E}^{-}_{(m)}(\theta X^{(m)})\, \underline{E}^{+}_{(m)}$$

$$= A\, \omega^{-mi} \exp(P_m)\exp(-Q_m).$$

Let $Z_j^{(m,u)}$, (m=1,2,3; u = 1,2) denote the jth homogeneous component of $Z^{(m,u)}$. So that

$$(3.19) \qquad Z^{(m,u)} = \sum_{j\in\mathbb{Z}} Z_j^{(m,u)}$$

for m = 1,2,3; u = 1,2. Using (3.17) and (3.18) we can derive generalized commutator and anticommutator identities ([18], Proposition 4.3) involving the operators $Z_j^{(m,u)}$. Let $\underline{Z}^*$ denote the algebra generated by the operators $Z_j^{(m,u)}$, $j \in \mathbb{Z}$; m = 1,2,3; u = 1,2. Then $\underline{Z}^*$ centralizes the action of $\mathfrak{t}$ on W_i, hence preserves the vacuum space $\Omega(W_i) = \{v \in W_i \mid \mathfrak{t}^+\cdot v = 0\}$. Furthermore, $\Omega(W_i) = \underline{Z}^*\cdot v_0$ and

$$(3.20) \quad \chi(\Omega(W_i)) = \chi(W_i) \,/\, F_1,$$

$$= \prod_{\substack{j>0\\ j\not\equiv 0 \pmod 6}} (1-q^j)^{-1} \Big/ \prod_{\substack{j>0\\ j \text{ odd}}} (1-q^j)^{-1}$$

$$= \prod_{\substack{j>0\\ j\not\equiv 0 \pmod 3}} (1-q^{2j})^{-1}$$

$$= \sum_{\ell\geqslant 0} a(\ell)\, q^{2\ell}$$

where the coefficients $a(\ell)$ are defined in (2.7). Hence the following theorem which is analogous to Theorem 2.2 can be proved in a similar way.

__Theorem__ 3.3. ([18], Theorem 4.4). The vacuum space $\Omega(W_i)$, i = 0,1,2,3 of W_i with highest weight vector v_0 has the basis

$$\{Z_{2i_1}^{(1,1)} \cdots Z_{2i_n}^{(1,1)}\cdot v_0\}$$

where $i_1 \leqslant i_2 \leqslant \cdots \leqslant i_n < 0$, $n \geqslant 0$ and $i_{k-2} < i_k$. □

Now the next theorem follows from Theorems 3.2 and 3.3.

Theorem 3.4. ([18], Theorem 4.5). The set $\mathcal{B}$ in Theorem 3.2 is a basis for the vacuum space $\Omega(V_i)$, $i = 0,1,2,3$. □

Observe that

$$\chi(\Omega(V_0)) = \chi(\Omega(V_3)) = \prod_{\substack{j>0 \\ j\equiv\pm 2 (\mathrm{mod}\ 5)}} (1-q^{2j})^{-1}$$

and

$$\chi(\Omega(V_1)) = \chi(\Omega(V_2)) = \prod_{\substack{j>0 \\ j\equiv\pm 1 (\mathrm{mod}\ 5)}} (1-q^{2j})^{-1}.$$

Thus by proving Theorem 3.4, we not only have constructed the level one standard $C_3^{(1)}$-modules by giving explicit bases for their vacuum spaces, but we have also given another proof of the Rogers-Ramanujan identities (see (3.11)).

REFERENCES

[1] G.E. Andrews, The Theory of Partitions, Encyclopedia of Mathematics and its Applications, Vol. 2, ed. G.C. Rota, Addison-Wesley, Reading, Mass., (1976).

[2] A. Feingold and J. Lepowsky, The Weyl-Kac character formula and power series identities, Adv. in Math. 29 (1978).

[3] I.B. Frenkel, Spinor representations of affine Lie algebras, Proc. Natl. Acad. Sci. USA 77 (11) (1980).

[4] I.B. Frenkel and V.G. Kac, Basic representations of affine Lie algebras and dual resonance models, Inventiones Math. 62

(1980).

[5] V.G. Kac, Simple irreducible graded Lie algebras of finite growth, Math. USSR - Izv. 26 (1968).

[6] V.G. Kac, Infinite-dimensional Lie algebras and Dedekind's η-function, Funct. Anal. Appl. 8 (1974).

[7] V.G. Kac and D.H. Peterson, Spin and wedge representations of infinite-dimensional Lie algebras and groups, Proc. Natl. Acad. Sci. USA 78 (6) (1981).

[8] V.G. Kac, D.A. Kazhdan, J. Lepowsky and R.L. Wilson, Realization of the Basic Representations of the Euclidean Lie algebras, Adv. in Math. 42 (1981).

[9] J. Lepowsky, Macdonald-type identities, Adv. in Math. 27 (1978).

[10] J. Lepowsky, Affine Lie algebras and combinatorial identities, Proc. 1981 Rutgers Conference on Lie Algebras and Related Topics, Springer-Verlag Lecture Notes in Mathematics 933 (1982).

[11] J. Lepowsky and S. Milne, Lie algebraic approaches to classical partition identities, Adv. in Math. 29 (1978).

[12] J. Lepowsky and R.L. Wilson, Construction of the affine Lie algebra $A_1^{(1)}$, Comm. Math. Phys. 62 (1978).

[13] J. Lepowsky and R.L Wilson, A new family of algebras underlying the Rogers-Ramanujan identities and generalizations, Proc. Natl. Acad. Sci. USA 78 (12) (19(1).

[14] J. Lepowsky and R.L. Wilson, A Lie theoretic interpretation and proof of the Rogers-Ramanujan identities, Adv. in Math. 45 (1982).

[15] J. Lepowsky and R.L. Wilson, The structure of standard modules, I: Universal algebras and the Rogers-Ramanujan identities, Invent. Math. (1984).

[16] K.C. Misra, Structure of the standard modules for $A_n^{(1)}$ and $C_n^{(1)}$, Ph.D. thesis, Rutgers University (1982).

[17] K.C. Misra, Structure of certain standard modules for $A_n^{(1)}$ and the Rogers-Ramanujan identities, J. Algebra 88 (1984).

[18] K.C. Misra, Structure of some standard modules for $C_n^{(1)}$, J. Algebra 90 (1984).

[19] R.V. Moody, A new class of Lie algebras, J. Algebra 10 (1968).

[20] R.V. Moody, Euclidean Lie algebras, Canad. J. Math. 21 (1969).

[21] G. Segal, Unitary representations of some infinite-dimensional groups, Comm. Math. Phys. 80 (1981).

University of Virginia
Charlottesville, VA 22903
Current Address:
University of Wisconsin
Madison, WI 53706

SOME APPLICATIONS OF VERTEX OPERATORS TO KAC-MOODY ALGEBRAS

Alex J. Feingold†

1. INTRODUCTION

This is an account of some of my recent work [2,3] which has involved applications of vertex operators to Kac-Moody Lie theory. For V the basic $A_1^{(1)}$-module in the principal realization given by Lepowsky and Wilson [13], one may use vertex operators to describe the decomposition $V \otimes V = S(V) \oplus A(V)$ of $V \otimes V$ into symmetric tensors S(V) and antisymmetric tensors A(V). This turns out to be precisely the decomposition of $V \otimes V$ into two "strings" of level two standard $A_1^{(1)}$-modules which I found in [1]. This result has a remarkable application to the construction of the hyperbolic algebra F with Dynkin diagram  .

In [2] Frenkel and I gave a $\mathbb{Z}$-graded construction of F such that the 0, 1 and -1 graded pieces (levels) were $A_1^{(1)}$ extended by the derivation d, V, and its contragredient module V^*, respectively. The higher levels were graded pieces of quotients of free Lie algebras by a graded ideal. For level 2 these were precisely determined to be $V \wedge V \approx A(V)$ modulo a single irreducible component (the top module of the antisymmetric string), and similarly for level -2 using V^* in place of V. This gave the first precise formula for "higher level" hyperbolic root multiplicities beyond the general formula of Moody and Berman [17]. These multiplicities have a remarkable relationship with the values of the classical partition function which has led to conjectures concerning upper bounds for all hyperbolic root multiplicities [6]. Different ways of applying vertex operators to the construction of hyperbolic algebras will be discussed by others in this workshop, but, as of this writing, none has yet explained those root multiplicities for F which are known precisely.

† Partially supported by the National Science Foundation through the Mathematical Sciences Research Institute

Vertex Operators in Mathematics and Physics - Proceedings of a Conference November 10-17, 1983. Publications of the Mathematical Sciences Research Institute #3, Springer-Verlag, 1984.

In order to extend the results mentioned above to the hyperbolic algebra with Dynkin diagram  I have recently studied the decomposition of $V \otimes V$ where V is the basic $A_2^{(2)}$-module. The techniques of [1,2] give this decomposition into two strings of level 2 standard modules with outer multiplicities, remarkably, equal to the coefficients of the Rogers-Ramanujan identities. From numerical data sent to me by V.G. Kac, it appears that the second level of this hyperbolic algebra consists of those irreducible components of $V \otimes V$ whose highest weights have odd principal degree > 1 relative to $1 \otimes 1$. One expects this to follow as for the algebra F from the decomposition $V \otimes V = S(V) \oplus A(V)$ and from the identification of $A(V)$ modulo one irreducible component with the second level of the $\mathbb{Z}$-graded hyperbolic algebra. The point of interest to those at this workshop is how the vertex operator techniques used to find $V \otimes V = S(V) \oplus A(V)$ in the $A_1^{(1)}$ case can be modified for the $A_2^{(2)}$ case. The proof in the $A_1^{(1)}$ case depended on the introduction of an auxiliary vertex operator on $V \otimes V$ with components which form a Clifford algebra, which commute with the action of the principal Heisenberg subalgebra on $V \otimes V$, and which anticommute with the action of the real root vectors on $V \otimes V$. In the case of $A_2^{(2)}$ the components of the analogous auxiliary vertex operator have much more complicated relations with each other and with the real root vector action on $V \otimes V$. In fact, current joint work with J. Lepowsky shows that one is dealing here with Z-algebras [15-16]. One may hope to generalize these results to all affine algebras and apply the theory of Z-algebras to the decomposition of more general tensor products.

In [3] Frenkel and I were able to construct highest weight representations for all "classical" affine algebras and superalgebras. These consist of the orthogonal series ($D_\ell^{(1)}$, $B_\ell^{(1)}$, $D_{\ell+1}^{(2)}$), the symplectic series ($C_\ell^{(1)}$, $B^{(1)}(0,\ell)$, $C^{(2)}(\ell+1)$), and the general linear series ($A_{\ell-1}^{(1)}$, $A_{2\ell-1}^{(2)}$, $A_{2\ell}^{(2)}$, $A^{(2)}(0,2\ell-1)$, $A^{(4)}(0,2\ell)$). The representations are given by certain "normally ordered" quadratic expressions whose linear factors generate an infinite-dimensional Clifford or Weyl algebra. This provides representations of the affine algebras on exterior or symmetric algebras of polynomials, respectively, in perfect analogy with the spinor (D_ℓ, B_ℓ) and oscillator (C_ℓ, $B(0,\ell)$) representations. The linear generators from the Clifford or Weyl algebra play the role in quantum mechanics of creation and annihilation operators for particles obeying Fermi or Bose statistics, respectively. We, therefore, named the corresponding constructions of affine algebras as fermionic or bosonic. The orthogonal series then has fermonic representations which are the standard representations previously called spinor representations [4,5,9]. The symplectic series has bosonic representations which are non-standard highest weight representations whose existence, at least for $C_\ell^{(1)}$, was independently noticed by H. Garland and M. Primc. For the general linear series one has both fermionic (standard) and bosonic (non-standard) representations. For the orthogonal series the fermionic (spinor) construction was shown by Frenkel [5] to be isomorphic to the vertex construction, thus interpreting the boson-fermion correspondence of physics in the framework of Kac-Moody theory. The fermions are the linear Clifford generators and the bosons are the normally ordered quadratic elements which form a Lie algebra. Both are realized as vertex operators. Changing the Clifford generators to Weyl generators gives the bosonic (oscillator) construction of the symplectic series, but one no longer has an alternative description using vertex operators. A vertex construction does exist for type A affine algebras, and for type $A_{2\ell-1}^{(2)}$ it was instrumental in discovering the way to construct the general linear series by normally ordered quadratic elements with a "twisted" generating function.

2. A TENSOR PRODUCT DECOMPOSITION AND APPLICATIONS

Let $\mathfrak{s}$ be the infinite-dimensional Heisenberg algebra with generators $h(n)$, $n \in 2\mathbb{Z}+1$, and where

$$(2.1) \qquad [h(m), h(n)] = m\delta_{m,-n}.$$

Let V be the symmetric algebra of polynomials in $\{h(-n) \mid 0 < n \in 2\mathbb{Z}+1\}$. Then V is an irreducible $\mathfrak{s}$-module. Define the vertex operator

$$(2.2)\qquad X(2h,z) = -\tfrac{1}{2}\exp\Big(\sum_k \frac{z^k}{k}\, 2h(-k)\Big)\exp\Big(-\sum_k \frac{z^{-k}}{k}\, 2h(k)\Big)$$

where the summations are over $0 < k \in 2\mathbb{Z}+1$ (as they are throughout this section unless otherwise indicated). Let

$$(2.3)\qquad X(2h,z) = \sum_{n\in\mathbb{Z}} X_n(2h)\, z^{-n}$$

define the homogeneous components $X_n(2h)$ of $X(2h,z)$. Then the operators $X_n(2h)$ for $n \in \mathbb{Z}$ are well defined on V and satisfy commutation relations with each other and with $\mathfrak{s}$ so as to provide the basic representation of the affine Kac-Moody algebra $\mathfrak{g}$ of type $A_1^{(1)}$ in its principal realization [13]. In particular, we have

$$(2.4)\qquad [h(k), X_n(2h)] = 2\, X_{n+k}(2h),\ k \in 2\mathbb{Z}+1,\ n \in \mathbb{Z},$$

$$(2.5)\qquad [X_m(2h), X_n(2h)] = \begin{cases} m\delta_{m,-n} & \text{if } m,n \in 2\mathbb{Z} \\ -m\delta_{m,-n} & \text{if } m,n \in 2\mathbb{Z}+1 \\ 2h(m+n) & \text{if } m \in 2\mathbb{Z}+1,\ n \in 2\mathbb{Z}. \end{cases}$$

One way of doing such computations with vertex operators is to use contour integrals and normal ordering lemmas. For example, we define the normally ordered product of two vertex operators

$$(2.6)\qquad :X(2h,z)X(2h,w): = \tfrac{1}{4}\exp\Big(\sum \frac{z^k}{k}\, 2h(-k)\Big)\exp\Big(\sum \frac{w^k}{k}\, 2h(-k)\Big)$$
$$\cdot \exp\Big(-\sum \frac{z^{-k}}{k}\, 2h(k)\Big)\exp\Big(-\sum \frac{w^{-k}}{k}\, 2h(k)\Big)$$

so that all the annhilation operators are applied first, and then all the creation operators are applied. Then we have the following.

<u>**Lemma**</u> 1. For $|z| > |w|$,

$$X(2h,z)X(2h,w) = \Big(\frac{z-w}{z+w}\Big)^2 :X(2h,z)X(2h,w):.$$

Proof. Let $X(2h,z) = -\frac{1}{2}\exp(A)\exp(B)$ and $X(2h,w) = -\frac{1}{2}\exp(C)\exp(D)$, then

$$[B,C] = -4\sum_k \frac{(w/z)^k}{k} = 2\log(\frac{z-w}{z+w}) \text{ if } |z| > |w|.$$

This scalar commutes with the operators A, B, C, D, so by the Baker-Campbell-Hausdorff formula,

$$\exp(B)\exp(C) = \exp(C)\exp(B)\exp([B,C]).$$

Let us abbreviate $\frac{1}{2\pi i}\int_C f(z)\,dz$ by just $\int_C f(z)\,dz$. Then we can compute the bracket

(2.7)
$$[X_m(2h), X_n(2h)] =$$

$$[\int_{C_{R_1}} X(2h,z)\, z^{m-1}\, dz, \int_{C_{R_2}} X(2h,w)\, w^{n-1}\, dw]$$

$$= \int_{C_{R_1}} (\int_{C_R} X(2h,z)\, X(2h,w)\, z^{m-1}\, dz - \int_{C_r} X(2h,w)\, X(2h,z)$$

$$z^{m-1}\, dz)\, w^{n-1}\, dw$$

$$= \int_{C_{R_1}} \int_{C_R \setminus C_r} (\frac{z-w}{z+w})^2 : X(2h,z)\, X(2h,w) :\, z^{m-1} dz\, w^{n-1} dw$$

where the contours are circles about the origin having radii satisfying $r < R_1 < R$. The only residue of the inner integral coming from the pole $z = -w$ is

(2.8)
$$\frac{d}{dz}\left((z-w)^2 : X(2h,z)\, X(2h,w) : \right)_{z=-w} =$$

$$(-1)^m w^m (m - \sum_{k \in 2\mathbb{Z}+1} w^k 2h(-k))$$

so one gets

(2.9) $$\int_{C_{R_1}} (-1)^m \left(m - \sum_{k\in 2\mathbb{Z}+1} w^k 2h(-k)\right) w^{m+n-1}\, dw$$

$$= \begin{cases} (-1)^m\, m\, \delta_{m,-n} & \text{if } m+n \in 2\mathbb{Z} \\ (-1)^{m+1}\, 2h(m+n) & \text{if } m+n \in 2\mathbb{Z}+1. \end{cases}$$

Now consider the tensor product $V \otimes V$. Denote by $h^1(m)$ the action of $\mathfrak{s}$ on the first tensor factor and by $h^2(m)$ the action on the second factor. Then the action of $\mathfrak{g}$ on $V \otimes V$ is given by $h^1(m) + h^2(m)$, $m \in 2\mathbb{Z}+1$, $X_n(2h^1) + X_n(2h^2)$, $n \in \mathbb{Z}$. The central element which acted as the scalar 1 on V acts as 2 on $V \otimes V$. This is just the realization of $\mathfrak{g}$ as the diagonal subalgebra of $\mathfrak{g}^1 \times \mathfrak{g}^2$. Define the generating functions

(2.10)
$$E^+(h^i, z) = \exp\Big(\sum_k \frac{z^{-k}}{k}\, h^i(k) \Big) \quad \text{and}$$
$$E^-(h^i, z) = \exp\Big(-\sum_k \frac{z^{k}}{k}\, h^i(-k) \Big) \quad \text{for } i = 1,2.$$

Then one has directly the auxiliary operator

(2.11) $$E^-(h^1+h^2, z)\, X(2h^1, z)\, E^+(h^1+h^2, z) = -\tfrac{1}{2}\, X(h^1-h^2, z)$$

which obviously commutes with $\mathfrak{s}$ on $V \otimes V$. To find out what relations hold among the components $X_n(h^1 - h^2)$ of $X(h^1 - h^2, z)$ we need the following.

__Lemma__ 2. For $|z| > |w|$,

$$X(h^1 - h^2, z)\, X(h^1 - h^2, w) = \frac{z-w}{z+w} : X(h^1 - h^2, z)\, X(h^1 - h^2, w):.$$

__Proof__. Following the proof of Lemma 1, the only difference is that

$$[B,C] = -2 \sum_k \frac{(w/z)^k}{k} = \log\Big(\frac{z-w}{z+w} \Big).$$

The crucial difference between these operators and those

before which formed a Lie algebra, is that the normal ordering factor here is antisymmetric in z and w, whereas before it was symmetric. It means that in order to obtain the inner contour integral over $C_R \setminus C_r$ we must anticommute components of the auxiliary vertex operator. One easily obtains the following Clifford algebra.

<u>Corollary</u>. For all m, n $\in \mathbb{Z}$,

$$\{X_m(h^1 - h^2), X_n(h^1 - h^2)\} = 2(-1)^m \delta_{m,-n}. \tag{2.12}$$

To find the relations between components of $X(h^1 - h^2, z)$ and $X(2h^i, w)$, $i = 1,2$, we need the following.

<u>Lemma</u> **3.** For $|z| > |w|$ we have

$$X(h^1 - h^2, z)\, X(2h^1, w) = \frac{z-w}{z+w} : X(h^1 - h^2, z)\, X(2h^1, w):,$$

$$X(h^1 - h^2, z)\, X(2h^2, w) = \frac{z+w}{z-w} : X(h^1 - h^2, z)\, X(2h^2, w):$$

and for $|w| > |z|$ we have

$$X(2h^1, w)\, X(h^1 - h^2, z) = \frac{w-z}{w+z} : X(2h^1, w)\, X(h^1 - h^2, z):,$$

$$X(2h^2, w)\, X(h^1 - h^2, z) = \frac{w+z}{w-z} : X(2h^2, w)\, X(h^1 - h^2, z):.$$

<u>Corollary</u>. For all m, n $\in \mathbb{Z}$,

$$\{X_m(h^1 - h^2), X_n(2h^1)\} = 2\,(-1)^m\, X_{m+n}(h^1 + h^2), \tag{2.13}$$

$$\{X_m(h^1 - h^2), X_n(2h^2)\} = 2\, X_{m+n}(h^1 + h^2) \tag{2.14}$$

so that for $m \in 2\mathbb{Z}+1$,

$$\{X_m(h^1 - h^2), X_n(2h^1) + X_n(2h^2)\} = 0. \tag{2.15}$$

One may also see easily that

(2.16) $X_m(h^1-h^2)(1\otimes 1) = \int_C X(h^1-h^2, z)(1\otimes 1)z^{m-1}\,dz = 0$, for $m > 0$.

Now consider the collection of vectors in $V \otimes V$,

$$(2.17)\quad \{X_{-2n_1-1}(h^1-h^2)X_{-2n_2-1}(h^1-h^2)\cdots X_{-2n_k-1}(h^1-h^2)(1\otimes 1) \mid$$

$$k \geq 0,\quad n_1 > n_2 > \cdots > n_k \geq 0\}.$$

From (2.12), (2.15), (2.16) one finds these vectors are linearly independent and are killed by $X_n(2h^1) + X_n(2h^2)$ for $0 < n \in \mathbb{Z}$, and by $h^1(k) + h^2(k)$ for $0 < k \in 2\mathbb{Z}+1$ (which represent positive root vectors of $\mathfrak{g}$ which kill $1 \otimes 1$). The principally specialized character of the space of highest weight vectors having basis (2.17) is

$$(2.18)\quad e^{\lambda} \prod_{n\geq 1} (1 + u^{2n-1})$$

where λ is the weight of $1 \otimes 1$. But the decomposition of $V \otimes V$ into two "strings" of level 2 irreducible $\mathfrak{g}$-modules is known [1] to be

$$(2.19)\quad V \otimes V = \sum_{m\geq 0} (a_m V^{2\omega_2+m\omega_3} + b_m V^{2\omega_1+(m+1)\omega_3})$$

where the outer multiplicities a_m, b_m are defined by

$$(2.20)\quad \sum_{m\geq 0} (a_m x^{2m} + b_m x^{2m+1}) = \prod_{j\geq 1} (1 + x^{2j-1}).$$

The fundamental weights of $\mathfrak{g}$ are ω_1 and ω_2, dual to the simple roots α_1 and α_2, $V = V^{\omega_2}$, and $\omega_3 = -\alpha_1-\alpha_2$. This implies that the principally specialized character of the space Ω of highest weight vectors in $V \otimes V$ is exactly (2.18), so that (2.17) is a basis of Ω.

Since $X_m(h^1 - h^2) = (-1)^m X_m(h^2 - h^1)$ it is clear that the typical vector in (2.17) is symmetric if k is even and antisymmetric if k is odd. To determine which string that vector falls into, note that $1 \otimes 1$ has weight $\lambda = 2\omega_2$, $2\omega_1 = 2\omega_2 + \alpha_1$ and ω_3 has even principal degree. So when k is even we get a vector of weight $2\omega_2 + m\omega_3$ and when k is odd we get a vector of weight $2\omega_1 + (m+1)\omega_3$ for some $m \geqslant 0$. This gives the following.

Theorem 1 [2]. We have

$$S(V) = \sum_{m \geqslant 0} a_m V^{2\omega_2 + m\omega_3}$$

$$A(V) = \sum_{m \geqslant 0} b_m V^{2\omega_1 + (m+1)\omega_3}.$$

This result has a remarkable application to the determination of certain root multiplicities in the hyperbolic algebra F with Dynkin diagram  . This algebra has a $\mathbb{Z}$-graded construction [2]

(2.21) $$F = \sum_{n \in \mathbb{Z}} F_n$$

such that $F_0 = g + \mathbb{C}d$ is the usual extension of g by the derivation d, $F_1 = V^{\omega_2 - \omega_3} \approx V$ and $F_{-1} = V^{-\omega_2 + \omega_3} \approx V^*$ is the dual (contragredient) g-module. The higher "levels" of F are much more complicated, being the graded pieces of free Lie algebras generated by F_1 (for $n > 0$) or F_{-1} (for $n < 0$) modulo a graded ideal

(2.22) $$I = \sum_{n \in \mathbb{Z}} I_n.$$

It turns out that

(2.23) $$F_{\pm 2} \approx (F_{\pm 1} \wedge F_{\pm 1}) / I_{\pm 2}$$

can be precisely determined using Theorem 1 and its analog for $V^* \otimes V^* = S(V^*) \oplus A(V^*)$. One finds that

(2.24) $$F_1 \wedge F_1 = A(V^{\omega_2 - \omega_3}) = \sum_{m \geqslant 0} b_m V^{2\omega_1 + (m-1)\omega_3}$$

and

(2.25) $$I_2 = V^{2\omega_1 - \omega_3}.$$

This gives a precise formula for the hyperbolic root multiplicities on the second level of F which shows them to be closely related to the values of the classical partition function. Further details may be found in [2].

3. ANOTHER TENSOR PRODUCT DECOMPOSITION

Joint work is in progress with J. Lepowsky on the analogue of the results from section 2 in the case when $\mathfrak{g}$ is of type $A_2^{(2)}$. The decomposition $V \otimes V = S(V) \oplus A(V)$ for the basic $A_2^{(2)}$-module $V = V^{\omega_2}$ may be applied to the hyperbolic algebra $\bar{A}_2^{(2)}$ with Dynkin diagram  , simple roots $\alpha_1, \alpha_2, \alpha_3$ and fundamental weights $\omega_1, \omega_2, \omega_3$.

__Theorem__ 2. We have the decomposition

(3.1) $$V \otimes V = \sum_{m \geqslant 0} (a_m V^{2\omega_2 + 2m\omega_3} + b_m V^{\omega_1 + (2m+1)\omega_3})$$

into irreducible level 2 $\mathfrak{g}$-modules, where the outer multiplicities are given by

(3.2) $$\sum_{m \geqslant 0} a_m x^m = \prod_{n \geqslant 1} (1 - x^{5n-1})^{-1} (1 - x^{5n-4})^{-1}$$

(3.3) $$\sum_{m \geqslant 0} b_m x^m = \prod_{n \geqslant 1} (1 - x^{5n-2})^{-1} (1 - x^{5n-3})^{-1}.$$

These are the product sides of the famous Rogers-Ramanujan identities [10, 14-16], which provide two combinatorial descriptions of the coefficients. We have that a_m equals the number of partitions of m into parts $\equiv$ 1,4 mod 5 which equals the number of partitions of m into parts with difference at least 2. Also, b_m equals the number of partitions of m into parts $\equiv$ 2, 3 mod 5 which equals the number of partitions of m into parts with difference at least 2 and no part less than 2.

The principally specialized character of the space Ω of highest weight vectors in $V \otimes V$ is

$$\text{(3.4)} \qquad ch(\Omega) = e^{2\omega_2} \sum_{m \geq 0} (a_m u^{3m} + b_m u^{3m+1}).$$

Using some auxiliary vertex operator one expects to find a basis for Ω which explains the combinatorial descriptions of a_m and b_m. In fact, numerical data on the root muliplicities of the hyperbolic algebra $\bar{A}_2^{(2)}$ provided by V.G. Kac indicate that the second level of that algebra consists of those irreducible components of $V \otimes V$ having highest weight vector of odd principal degree greater than 1 relative to $1 \otimes 1$. This indicates that $A(V)$ consists of all components having highest weight vector of odd principal degree and that $I_2 \approx V^{\omega_1+\omega_3}$. Instead of two strings, one symmetric and one antisymmetric, here we have the following.

Theorem 3. If V is the basic $A_2^{(2)}$-module and $V \otimes V = S(V) \oplus A(V)$ then

$$\text{(3.5)} \quad S(V) = \sum_{m \geq 0} (a_{2m}\, V^{2\omega_2+4m\omega_3} + b_{2m+1}\, V^{\omega_1+(4m+3)\omega_3})$$

$$\text{(3.6)} \quad A(V) = \sum_{m \geq 0} (a_{2m+1}\, V^{2\omega_2+(4m+2)\omega_3} + b_{2m}\, V^{\omega_1+(4m+1)\omega_3}).$$

The basic module for $A_2^{(2)}$ is constructed as follows [8]. Let $\mathfrak{s}$ be the Heisenberg algebra with generators $h(n)$, $n \equiv \pm 1 \bmod 6$ such that

$$\text{(3.7)} \qquad [h(m), h(n)] = m\, \delta_{m,-n}.$$

V is the symmetric algebra of polynomials in $\{h(-n) \mid 0 < n \in \mathbb{Z},\ n \equiv \pm 1 \bmod 6\}$. Then V is an irreducible $\mathfrak{s}$-module. Let ϵ be a primitive 6th root of unity. Define the vertex operator

$$(3.8)\quad X(2h,z) = \exp\left(\sum_k \frac{z^k}{k}\,\frac{\epsilon^k+1}{2^{1/2}}\,2h(-k)\right)\exp\left(-\sum_k \frac{z^{-k}}{k}\,\frac{\epsilon^{-k}+1}{2^{1/2}}\,2h(k)\right)$$

where the summations are over $0 < k \in \mathbb{Z}$, $k \equiv \pm 1 \bmod 6$ (as they are throughout this section unless otherwise indicated). Then $\mathfrak{s}$ and the components $X_n(2h)$, $n \in \mathbb{Z}$, of $X(2h,z)$ provide the basic representation of $A_2^{(2)}$ on V in the principal realization. To find the bracket of two vertex components, for example, one uses the following.

<u>**Lemma**</u> **4.** For $|z| > |w|$,

$$X(2h,z)\,X(2h,w) = \left(\frac{z-w}{z+w}\right)^2\left(\frac{z-\epsilon w}{z+\epsilon w}\right)\left(\frac{z+\epsilon^2 w}{z-\epsilon^2 w}\right):X(2h,z)\,X(2h,w):.$$

The representation of $\mathfrak{g}$ on $V \otimes V$ is given by $h^1(m) + h^2(m)$, $m \equiv \pm 1 \bmod 6$ and $X_n(2h^1) + X_n(2h^2)$, $n \in \mathbb{Z}$. Defining generating functions

$$(3.9)\qquad E^+(h^i,z) = \exp\left(\sum_k \frac{z^{-k}}{k}\,\frac{\epsilon^{-k}+1}{2^{1/2}}\,h^i(k)\right)$$

$$(3.10)\qquad E^-(h^i, z) = \exp\left(-\sum_k \frac{z^k}{k}\,\frac{\epsilon^k+1}{2^{1/2}}\,h^i(-k)\right)$$

for $i=1,2$ one finds the auxiliary operator

$$(3.11)\qquad E^-(h^1 + h^2, z)\,X(2h^1, z)\,E^+(h^1 + h^2, z) = X(h^1 - h^2, z).$$

The components $X_n(h^1 - h^2)$ obviously commute with $\mathfrak{s}$ on $V \otimes V$. We need the following.

<u>**Lemma**</u> **5.** For $|z| > |w|$,

$$X(h^1 - h^2, z)\,X(h^1 - h^2, w) =$$

$$\left(\frac{z-w}{z+w}\right)\left(\frac{z-\epsilon w}{z+\epsilon w}\right)^{1/2}\left(\frac{z+\epsilon^2 w}{z-\epsilon^2 w}\right)^{1/2} : X(h^1-h^2, z)X(h^1-h^2, w):.$$

At this point the occurence of the fractional exponents makes the use of the contour integral technique much more difficult. However, the alternative technique of formal variables and "correction factors" [10, 12, 15–16] easily shows that the components of this auxiliary vertex operator satisfy so-called "generalized anticommutation relations" which give them the structure of a "Z-algebra". While much more complicated than (2.12), this structure does allow one to determine a set of monomials

$$(3.12) \qquad X_{-m_1}(h^1 - h^2)\, X_{-m_2}(h^1 - h^2) \cdots X_{-m_n}(h^1 - h^2)(1 \otimes 1)$$

which are a basis for the "vacuum space" of $\mathfrak{s}$ (vectors killed by $h^1(m) + h^2(m)$ for $0 < m \equiv \pm 1 \bmod 6$). It then remains to find the subspace of highest weight vectors also killed by $X_n(2h^1) + X_n(2h^2)$ for all $0 < n \in \mathbb{Z}$. That also requires the techniques of Z-algebras. Details will appear elsewhere.

4. THE VERTEX REPRESENTATION OF $\mathfrak{gl}^{(2)}(2\ell)$

In [3] constructions were given of all the "classical" affine algebras; the orthogonal series $D_\ell^{(1)}$, $B_\ell^{(1)}$, $D_{\ell+1}^{(2)}$, the symplectic series $C_\ell^{(1)}$, $B^{(1)}(0,\ell)$, $C^{(2)}(\ell+1)$, and the general linear series $A_{\ell-1}^{(1)}$, $A_{2\ell-1}^{(2)}$, $A_{2\ell}^{(2)}$, $A^{(2)}(0,2\ell-1)$, $A^{(4)}(0,2\ell)$. The representations of the orthogonal and symplectic series are quite analogous to the well-known spinor and oscillator representations, respectively, in the finite-dimensional theory. The representations are given by certain "normally ordered" quadratic expressions whose linear factors come from an infinite-dimensional Clifford or Weyl algebra. As was explained in the introduction, these are called fermonic or bosonic constructions, respectively. The fermionic constructions of the orthogonal series were previously known [4,5,9], as was the existence of a bosonic construction of $C_\ell^{(1)}$ [Garland and, independently, Primc, unpublished]. Essentially new was the possibility of having both fermionic and bosonic constructions of the general linear series. The homogeneous vertex representation [5,7] of type $A^{(2)}_{2\ell-1}$ (or more precisely, of $\mathfrak{gl}^{(2)}(2\ell)$) was essential to the discovery of these constructions. I will describe how this was done.

Consider the Lie algebra involution σ of $\mathfrak{gl}(2\ell)$ given by

$$(4.1)\qquad \sigma\begin{bmatrix} m & n \\ p & q \end{bmatrix} = \begin{bmatrix} -q^t & n^t \\ p^t & -m^t \end{bmatrix}.$$

Then $\mathfrak{gl}(2\ell) = \mathfrak{g}_0 \oplus \mathfrak{g}_1$ where $\mathfrak{g}_0 \approx \mathfrak{sp}(2\ell)$ is the fixed point set of σ and on $\mathfrak{g}_1$, σ acts as -1. Then we have

$$(4.2)\quad \mathfrak{gl}^{(2)}(2\ell) = \mathfrak{g}_0 \otimes \mathbb{C}[t^2, t^{-2}] + \mathfrak{g}_1 \otimes t\mathbb{C}[t^2, t^{-2}] + \mathbb{C}c$$

where c is central and

$$(4.3)\qquad [x \otimes t^i, y \otimes t^j] = [x,y] \otimes t^{i+j} + i\, \delta_{i,-j} \langle x,y\rangle c$$

with invariant bilinear form $\langle x,y\rangle = \mathrm{Tr}(xy)$ on $\mathfrak{gl}(2\ell)$. This is a subalgebra of

$$(4.4)\qquad \mathfrak{gl}^{(1)}(2\ell) = \mathfrak{gl}(2\ell) \otimes \mathbb{C}[t, t^{-1}] + \mathbb{C}c.$$

One usually extends these algebras by adjoining the derivation

$d = t\frac{d}{dt}$,

$$(4.5)\qquad \hat{\mathfrak{gl}}^{(i)}(2\ell) = \mathfrak{gl}^{(i)}(2\ell) + \mathbb{C}d, \quad i=1,2.$$

Let E_{ij}, $1 \leq i, j \leq 2\ell$ be the standard basis of $\mathfrak{gl}(2\ell)$, then $h_i = E_{ii}$, $1 \leq i \leq 2\ell$, is an orthonormal basis for the Cartan subalgebra $\mathfrak{h}$ of $\mathfrak{gl}(2\ell)$. Let $\mathfrak{h}_0$ be the Cartan subalgebra of $\mathfrak{g}_0$ with basis $\bar{h}_i = h_i - h_{i+\ell}$, $1 \leq i \leq \ell$, and let $\mathfrak{h} = \mathfrak{h}_0 \oplus \mathfrak{h}_1$ with $\mathfrak{h}_1 \subset \mathfrak{g}_1$. Using the form $\langle h_i, h_j\rangle = \delta_{i,j}$ to identify the dual space $\mathfrak{h}^*$ with $\mathfrak{h}$ we find the root system of $\mathfrak{gl}(2\ell)$ is

(4.6) $$\Phi = \{h_i - h_j \mid 1 \leq i \neq j \leq 2\ell\}.$$

For $x \in \mathfrak{gl}(2\ell)$ let $x^0 = \frac{1}{2}(x + \sigma x)$ and $x^1 = \frac{1}{2}(x - \sigma x)$ denote the projections of x onto $\mathfrak{g}_0$ and $\mathfrak{g}_1$, respectively. Then $\Phi^0 \subset \mathfrak{h}_0$ is the root system for $\mathfrak{g}_0$ of type C_ℓ. With respect to $\mathfrak{h}_0$ the root system of $\mathfrak{g}_1$ consists of the short roots of Φ^0.

For $x \in \mathfrak{gl}(2\ell)$, $n \in \mathbb{Z}$, let us denote

(4.7) $$x(n/2) = \begin{cases} x^0 \otimes t^n & \text{if } n \in 2\mathbb{Z} \\ x^1 \otimes t^n & \text{if } n \in 2\mathbb{Z}+1 \end{cases}$$

in $\mathfrak{gl}^{(2)}(2\ell)$. To get the homogeneous vertex representation of $\mathfrak{gl}^{(2)}(2\ell)$ we begin with the Heisenberg algebra $\mathfrak{h}_\sigma$ generated by

(4.8) $$\{\bar{h}_i(m),\ c \mid 1 \leq i \leq \ell,\ m \in \tfrac{1}{2}\mathbb{Z}\}$$

where

(4.9) $$[\bar{h}_i(m), \bar{h}_j(n)] = m\, \delta_{m,-n} \langle \bar{h}_i, \bar{h}_j \rangle\, c.$$

Let $S(\mathfrak{h}_\sigma^-)$ denote the symmetric algebra of polynomials in $\{\bar{h}_i(-m) \mid 1 \leq i \leq \ell,\ 0 < m \in \frac{1}{2}\mathbb{Z}\}$. Let $P = \sum_{i=1}^{2\ell} \mathbb{Z}h_i$ denote the weight lattice of $\mathfrak{gl}(2\ell)$ which projects to P^0, the weight lattice of $\mathfrak{g}_0$. We have the root lattice $Q = \sum_{i=1}^{2\ell-1} \mathbb{Z}\alpha_i$ of $\mathfrak{gl}(2\ell)$ (where $\alpha_i = h_i - h_{i+1}$, $1 \leq i \leq 2\ell-1$, are the simple roots) which projects onto the root lattice Q^0 of $\mathfrak{g}_0$. There are two cosets of Q^0 in P^0, $P^0 = Q^0 \cup Q^1$. Let $\mathbb{C}[P^0]$ denote the group algebra generated by formal exponentials e^μ, $\mu \in P^0$. Then one has the two representation spaces

(4.10) $$V_k^\sigma = S(\mathfrak{h}_\sigma^-) \otimes \mathbb{C}[Q^k],\ k=0,1$$

where the action of $\bar{h}(m)$, $m \neq 0$ is as usual on $S(h_{\sigma}^{-})$, but

$$\bar{h}(0)(v \otimes e^{\mu}) = \langle \bar{h}, \mu \rangle (v \otimes e^{\mu}). \tag{4.11}$$

The rest of $g\ell^{(2)}(2\ell)$ is represented by the action of the homogeneous components of the vertex operators

$$X^{\sigma\epsilon}(\mu,z) = \tag{4.12}$$

$$\exp\Big(2 \sum_{n\geq 1} \frac{z^n}{n} \mu(-n/2)\Big)\exp(2 \log(z)\mu(0)+\mu^0)$$

$$\cdot \exp\Big(-2 \sum_{n\geq 1} \frac{z^{-n}}{n} \mu(n/2)\Big)\epsilon^0_{\mu^0}$$

for $\mu \in \Phi$. Here ϵ^0: $P^0 \times P^0 \longrightarrow \{\pm 1\}$ is a bilinear function which must satisfy certain properties discussed below, and $\epsilon^0_{\mu^0}$ acts by

$$\epsilon^0_{\mu^0}(v \otimes e^{\lambda}) = \epsilon^0(\mu^0, \lambda)(v \otimes e^{\lambda}). \tag{4.13}$$

The components of $X^{\sigma\epsilon}(\mu,z)$ for $\mu \in P$ are defined by

$$X^{\sigma\epsilon}(\mu,z) = \sum_{m\in Z} X_m^{\sigma\epsilon}(\mu)\ z^{-m} \tag{4.14}$$

where $Z = \mathbb{Z}$ or $\mathbb{Z} + \frac{1}{2}$ depending on whether μ^0 is in Q^0 or Q^1. This is determined by the action of the middle exponential of (4.12) in $\mathbb{C}[Q^k]$,

$$\exp(2 \log(z)\mu(0)+\mu^0)e^{\lambda} = z^{2\langle\mu^0,\lambda^0\rangle+\langle\mu^0,\mu^0\rangle}\ e^{\lambda+\mu^0}. \tag{4.15}$$

Since $\Phi \subset Q$ the components representing $g\ell^{(2)}(2\ell)$ are all integral. In fact, V_0^{σ} is the basic module.

To compute brackets of vertex operator components one uses the following.

__Lemma__ 6. For $|z| > |w|$,

$$X^{\sigma\epsilon}(\lambda,z)\, X^{\sigma\epsilon}(\mu,w) = \frac{\epsilon^0(\lambda^0,\mu^0)(z-w)^{\langle\lambda,\mu\rangle}(z+w)^{\langle\lambda,\sigma\mu\rangle}}{(zw)^{\langle\lambda^0,\mu^0\rangle}} :X^{\sigma\epsilon}(\lambda,z)X^{\sigma\epsilon}(\mu,w):.$$

In order to find commutation relations one must have

$$\epsilon^0(\mu^0,\lambda^0)(-1)^{\langle\lambda,\mu\rangle} = \epsilon^0(\lambda^0,\mu^0) \tag{4.16}$$

for $\lambda,\ \mu \in \Phi$. Using the bilinearity of ϵ^0 it suffices to have condition (4.16) for the simple roots in Φ. This is easily obtained by setting $\epsilon^0(\alpha_i^0,\ \alpha_j^0) = 1$ for $i \leqslant j$, so the other values are determined by (4.16).

We wish to understand how to write

$$X^{\sigma\epsilon}(h_i - h_j,\ z) =: X^{\sigma\epsilon}(h_i,\ z)\ X^{\sigma\epsilon}(-h_j,\ z): \tag{4.17}$$

so that the components of $X^{\sigma\epsilon}(\pm h^i,\ z)$ form a Clifford algebra. It is necessary to extend the definition of ϵ^0 to $\epsilon\colon P \times P^0 \to \{\pm 1\}$ so as to obtain

$$\epsilon(\lambda,\mu^0) = -\ (-1)^{\langle\lambda,\mu\rangle}\ \epsilon(\mu,\lambda^0). \tag{4.18}$$

Define for $1 \leqslant i,\ j \leqslant \ell$,

$$\epsilon(h_i,\ h_j^0) = \begin{cases} +1 & \text{if } i \leqslant j \\ -1 & \text{if } i > j \end{cases} \tag{4.19}$$

and

$$\epsilon(h_{i+\ell},\ h_j^0)(-1)^{2\langle h_i^0,\ h_j^0\rangle} = \epsilon(h_i,\ h_j^0) \tag{4.20}$$

so that

$$(4.21)\qquad \epsilon(h_i, h_j^0) = \begin{cases} \epsilon(h_{i+\ell}, h_j^0) & \text{if } i \neq j \\ -\epsilon(h_{i+\ell}, h_j^0) & \text{if } i = j. \end{cases}$$

Using the bilinear function ϵ determined by these conditions in place of ϵ^0 in (4.12) we get anticommutation relations among the components $X_m^{\sigma\epsilon}(\pm h_i)$, for which $m \in \mathbb{Z} + \frac{1}{2}$, and we get the same commutation relations as before among the components $X_n^{\sigma\epsilon}(h_i - h_j)$, $n \in \mathbb{Z}$. We find that for m, $n \in \mathbb{Z} + \frac{1}{2}$

$$(4.22)\qquad [X_m^{\sigma\epsilon}(h_i), X_n^{\sigma\epsilon}(-h_j)] = \epsilon(h_i, -h_j^0)\, \delta_{i,j}\, \delta_{m,-n}$$

for $1 \leqslant i, j \leqslant 2\ell$, and

$$(4.23)\qquad [X_m^{\sigma\epsilon}(h_i), X_n^{\sigma\epsilon}(h_{j+\ell})] = \epsilon(h_i, -h_j^0)\, (-1)^{m-\frac{1}{2}}\, \delta_{i,j}\, \delta_{m,-n}$$

for $1 \leqslant i, j \leqslant \ell$.

Let us introduce the notation

$$(4.24)\qquad a_i(m) = X_m^{\sigma\epsilon}(h_i), \quad a_i^*(m) = X_m^{\sigma\epsilon}(-h_i)$$

for $1 \leqslant i \leqslant 2\ell$, $m \in \mathbb{Z}+\frac{1}{2}$. Then we have

$$(4.25)\qquad [a_i(m), a_j^*(n)] = \delta_{i,j}\, \delta_{m,-n}, \quad 1 \leqslant i, j \leqslant 2\ell$$

$$(4.26)\qquad [a_i(m), a_{j+\ell}(n)] = (-1)^{m-\frac{1}{2}}\, \delta_{i,j}\, \delta_{m,-n}, \quad 1 \leqslant i, j \leqslant \ell$$

$$(4.27)\qquad [a_i^*(m), a_{j+\ell}^*(n)] = (-1)^{m-\frac{1}{2}}\, \delta_{i,j}\, \delta_{m,-n}, \quad 1 \leqslant i, j \leqslant \ell.$$

As in the spinor construction of the orthogonal series, one would like to consider the normally ordered quadratic expressions

(4.28) $$\sum_{k \in \mathbb{Z}+\frac{1}{2}} :a_i(m-k)a_j^*(k):$$

for $m \in \mathbb{Z}$, as components of generating functions: $a_i(z)a_j^*(z)$: in fermonic normal order. However, there are relations among the generators

(4.29) $$a_i(z) = X^{\sigma\epsilon}(h_i, z), \quad a_i^*(z) = X^{\sigma\epsilon}(-h_i, z)$$

which can only be seen from the definitions of these vertex operators.

Lemma 7. For $1 \leqslant i \leqslant \ell$,

(4.30) $$X^{\sigma\epsilon}(h_{i+\ell}, z) = (-1)^{-\frac{1}{2}} X^{\sigma\epsilon}(-h_i, -z)$$

and

(4.31) $$X^{\sigma\epsilon}(-h_{i+\ell}, z) = (-1)^{-\frac{1}{2}} X^{\sigma\epsilon}(h_i, -z).$$

Proof. Writing out the definitions of $X^{\sigma\epsilon}(h_{i+\ell}, z)$ and $X^{\sigma\epsilon}(-h_i, -z)$ using (4.1) and (4.7) one sees that the first and third exponentials are equal. Consider the difference between the actions of their middle exponentials and their cocycles ϵ on e^{λ^0}. From (4.15) the first gives

$$z^{2\langle -h_i^0, \lambda^0\rangle + \langle h_i^0, h_i^0\rangle} \epsilon(h_{i+\ell}, \lambda^0)\, e^{\lambda^0 - h_i^0}$$

because $h_{i+\ell}^0 = -h_i^0$, and the second gives

$$(-z)^{2\langle -h_i^0, \lambda^0\rangle + \langle h_i^0, h_i^0\rangle} \epsilon(-h_i^0, \lambda^0)\, e^{\lambda^0 - h_i^0}.$$

But then from (4.20)

$$(-1)^{2<-h_i^0,\ \lambda^0>}\ \epsilon(-h_i^0, \lambda^0) = \epsilon(h_{i+\ell}, \lambda^0),$$

and $<h_i^0, h_i^0> = \frac{1}{2}$, so we get (4.30). The case of (4.31) is similar.

This shows that for $1 \leqslant i \leqslant \ell$,

$$(4.32)\qquad a_{i+\ell}(z) = (-1)^{-\frac{1}{2}}\ a_i^*(-z) \quad \text{and} \quad a_{i+\ell}^*(z) = (-1)^{-\frac{1}{2}}\ a_i(-z),$$

so that $g\ell^{(2)}(2\ell)$ can be generated by the components of

$$(4.33)\quad :a_i(z)a_j^*(z):,\quad :a_i(z)a_j(-z):,\quad :a_i^*(-z)\ a_j^*(z):\ \text{for } 1 \leqslant i,j \leqslant \ell.$$

Note how (4.32) gives (4.26) and (4.27) from (4.25).

This was the method by which the "twisted" constructions of the general linear series were found. It is remarkable that when the fermionic generators (4.25) are replaced by bosonic ones, or when the index set Z is changed from $\mathbb{Z}+\frac{1}{2}$ to $\mathbb{Z}$, the functions (4.33) still provide a representation of the same algebra. For further details see [3].

REFERENCES

[1] A.J. Feingold, Tensor products of certain modules for the generalized Cartan matrix Lie algebra $A_1^{(1)}$, Communications in Algebra, Vol. 9, No. 12 (1981), 1323-1341.

[2] A.J. Feingold, I. Frenkel, A hyperbolic Kac-Moody algebra and the theory of Siegel modular forms of genus 2, Math. Ann. 263 (1983), 87-144.

[3] A.J. Feingold, I. Frenkel, Classical affine algebras, Advances in Math, to appear.

[4] I. Frenkel, Spinor representations of affine Lie algebras, Proc. Natl. Acad. Sci. USA, Vol. 77, No. 11, (1980), 6303-6306.

[5] I. Frenkel, Two constructions of affine Lie algebra representations and boson-fermion correspondence in quantum field theory, J. of Funct. Anal., Vol. 44, No. 3 (1981), 259-327.

[6] I. Frenkel, Representations of Kac-Moody algebras and dual resonance models, Proceedings of the AMS-SIAM 1982 Summer Seminar on Applications of Group Theory in Physics and Mathematical Physics, July, 1982.

[7] I. Frenkel, V.G. Kac, Basic representations of affine Lie algebras and dual resonance models, Invent. Math. 62 (1980), 23-66.

[8] V.G. Kac, D.A. Kazhdan, J. Lepowsky, R.L. Wilson, Realization of the basic representations of the Euclidean Lie algebras, Advances in Math. 42, (1981), 83-112.

[9] V.G. Kac, D.H. Peterson, Spin and wedge representations of infinite-dimensional Lie algebras and groups, Proc. Natl. Acad. Sci. USA, Vol. 78, No. 6 (1981), 3308-3312.

[10] J. Lepowsky, Affine Lie algebras and combinatorial identities, Proc. 1981 Rutgers Conference on Lie algebras and Related Topics, Springer-Verlag Lecture Notes in Math. 933 (1982), 130-156.

[11] J. Lepowsky, Some constructions of the affine Lie algebra $A_1^{(1)}$, Proceedings of the AMS-SIAM 1982 Summer Seminar on Applications of Group Theory in Physics and Mathematical Physics, July, 1982.

[12] J. Lepowsky, M. Primc, Standard modules for type one affine Lie algebras, Number Theory, New York, 1982, Springer-Verlag Lecture Notes in Mathematics 1052 (1984) 194-251.

[13] J. Lepowsky, R.L. Wilson, Construction of the affine Lie algebra $A_1^{(1)}$, Comm. in Math. Phys. 62 (1978), 43-53.

[14] J. Lepowsky, R.L. Wilson, A Lie theoretic interpretation and proof of the Rogers-Ramanujan identities, Advances in Math. 45 (1982), 21-72.

[15] J. Lepowsky, R.L. Wilson, A new family of algebras underlying the Rogers-Ramanujan identities and generalizations, Proc. Natl. Acad. Sci. USA, Vol. 78 (1981), 7254-7258.

[16] J. Lepowsky, R.L. Wilson, The structure of standard modules, I: Universal algebras and the Rogers-Ramanujan identities, preprint, 1982, Invent. Math., (1984).

[17] R.V. Moody, S. Berman, Lie algebra multiplicities, Proceedings of the American Mathematical Society 76 (1979), 223-228.

Department of Mathematics
The State University of New York
Binghamton, N.Y. 13901

ON A DUALITY OF BRANCHING COEFFICIENTS

Michio Jimbo[†] and Tetsuji Miwa[†]

§1. PROBLEM.

Consider a pair of Lie algebra and its subalgebra

$$\tilde{g} \supset g. \tag{1.1}$$

Throughout this paper we shall deal with Kac-Moody Lie algebras of affine or finite type. Let $\tilde{\Lambda}$ be a dominant integral weight of $\tilde{g}$ and let $L(\tilde{\Lambda})$ be an irreducible $\tilde{g}$-module with the highest weight $\tilde{\Lambda}$. Suppose that $L(\tilde{\Lambda})$ splits into irreducible g-modules

$$L(\tilde{\Lambda}) = \bigoplus_{\Lambda} (L(\Lambda) \oplus \cdots \oplus L(\Lambda)). \tag{1.2}$$

We call the multiplicity of $L(\Lambda)$ in the decomposition (1.2) a branching coefficient and denote it by $(\tilde{\Lambda} : \Lambda)$. The aim of this paper is to report a kind of duality of branching coefficients ([1],[2]).

§2. CHARACTER AND WEYL-KAC FORMULA

Let $\mathfrak{h}$ be a Cartan subalgebra of g. The character of an irreducible g-module $L(\Lambda)$ is the following formal sum.

$$ch_{L(\Lambda)} = \sum_{\lambda \in \mathfrak{h}^*} \dim\{x \in L(\Lambda) \mid h \cdot x = \lambda(h)x,\ \forall h \in \mathfrak{h}\} e^{\lambda}. \tag{2.1}$$

The Weyl-Kac character formula is well-known:

---

† Partially supported by the National Science Foundation through the Mathematical Sciences Research Institute.

Vertex Operators in Mathematics and Physics - Proceedings of a Conference November 10-17, 1983. Publications of the Mathematical Sciences Research Institute #3, Springer-Verlag, 1984.

$$\text{(2.2)}\qquad ch_{L(\Lambda)} = \frac{\sum_{w\in W} \det w \; e^{w(\Lambda+\rho)}}{\sum_{w\in W} \det w \; e^{w(\rho)}} .$$

If g is of finite type, the character is essentially a polynomial in $z_1 = e^{-\alpha_1},\ldots,z_\ell = e^{-\alpha_\ell}$ where $\alpha_1,\ldots,\alpha_\ell$ are the simple roots. If g is affine, it is written as a ratio of alternating sums of suitable theta functions in $z_1,\ldots,z_\ell$ and $q = e^{-\delta}$ ([3],[4]) (δ: the null root).

Notations.

$$\eta(\tau) = q^{\frac{1}{24}} \prod_{n=0}^{\infty} (1-q^{n+1}),$$

$$\theta(\tau,u) = z^{\frac{1}{2}} q^{\frac{1}{8}} \prod_{n=0}^{\infty} (1-q^{n+1})(1-zq^{n+1})(1-z^{-1}q^{n}),$$

$$\theta_{\lambda,N}(\tau,u) = \sum_{n\in\mathbb{Z}} (-)^{Nn}\, e[\tfrac{N}{2}(n+\tfrac{\lambda}{N})^2\tau + (Nn+\lambda)u],$$

where $e[*] = \exp(2\pi i *)$.

Example 1. $g\ell(\ell,\mathbb{C})$.

The irreducible representation is specified by a set of integers $\lambda = (\lambda_1,\ldots,\lambda_\ell)$ s.t. $\lambda_1 > \cdots > \lambda_\ell$. The corresponding character reads as

$$\text{(2.3)}\qquad \chi_\lambda = \det(z_i^{\lambda_j})_{i,j=1,\ldots,\ell} \Big/ \prod_{1\leq i<j\leq\ell} (z_i - z_j).$$

Take an integer $e \leq \lambda_\ell$, set $f_1 = \lambda_1 - \ell + 1 - e$, $f_2 = \lambda_2 - \ell + 2 - e,\ldots, f_\ell = \lambda_\ell - e$, and denote by Y the Young diagram of signature $(f_1,\ldots,f_\ell)$.

(2.4) Y: 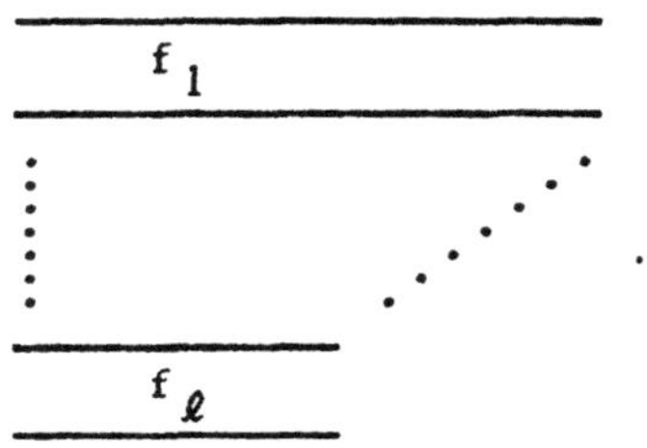

We define the character corresponding to the diagram Y by

$$(2.5)\qquad \chi_Y^{g\ell(\ell,\mathbb{C})} = (z_1 \cdots z_\ell)^{-e}\chi_\lambda .$$

__Example 2.__ $sp(2\ell,\mathbb{C})$

The irreducible representation is specified by a set of integers $\lambda = (\lambda_1,\ldots,\lambda_\ell)$ s.t. $\lambda_1 > \cdots > \lambda_\ell > 0$. The corresponding character reads as

$$(2.6)\qquad \frac{\chi_\lambda = \det(z_i^{\lambda_j} - z_i^{-\lambda_j})_{i,j=1,\ldots,\ell}}{\prod_{i=1}^{\ell}(z_i - z_i^{-1}) \prod_{1\leq i<j\leq \ell}(z_i - z_j)(1 - z_i^{-1}z_j^{-1})} .$$

We set $f_1 = \lambda_1 - \ell$, $f_2 = \lambda_2 - \ell + 1, \ldots, f_\ell = \lambda_\ell - 1$ and denote by Y the Young diagram of signature $(f_1,\ldots,f_\ell)$. We also write $\chi_Y^{sp(2\ell,\mathbb{C})}$ to mean χ_λ.

__Example 3.__ $A_\ell^{(1)}$

A level m irreducible representation is specified by $\lambda = (\lambda_1,\ldots,\lambda_{\ell+1})$ s.t. $\lambda_1 > \cdots > \lambda_{\ell+1}$, $\lambda_i - \lambda_j \in \mathbb{Z}$, $\sum_{i=1}^{\ell+1} \lambda_i = 0$ and $\ell + m \geqslant \lambda_1 - \lambda_{\ell+1}$. Consider an $\ell+1$ dimensional orthogonal space E with orthonormal basis $\epsilon_1,\ldots,\epsilon_{\ell+1}$. We denote by M the lattice $\{\sum_{i=1}^{\ell+1} \nu_i\epsilon_i \mid \nu_i \in \mathbb{Z}, \sum_{i=1}^{\ell+1} \nu_i = 0\}$. We introduce a coordinate $(\tau, u_1,\ldots,u_{\ell+1})$ so that $q = e[\tau]$, $z_1 = e[u_2 - u_1], \ldots, z_\ell = e[u_{\ell+1} - u_\ell]$. Then the character corresponding to (m,λ) is given (up to a fractional power of q) by

$$\chi_{\lambda,m} =$$

(2.7)

$$\frac{\sum_{w\in S_{\ell+1}} \det w \sum_{\gamma \in M + \sum_{i=1}^{\ell+1} \lambda_{w(i)}\epsilon_i/(\ell+m+1)} e[\frac{\tau}{2}(\ell+m+1)|\gamma|^2+(\lambda+m+1)(\gamma, \sum_{i=1}^{\ell+1} u_i\epsilon_i)]}{\eta(\tau)^{\frac{1}{2}\ell(1-\ell)} \prod_{1\leqslant i<j\leqslant \ell+1} \theta(\tau, u_i-u_j)}.$$

Note that $\chi_{\lambda,m}$ is unchanged by a shift $u_i \longmapsto u_i + c$, and hence is unambiguously defined as a function of $(q,z_1,\ldots,z_\ell)$. We set $f_1 = \lambda_1 - \lambda_{\ell+1} - \ell,\ldots, f_\ell = \lambda_\ell - \lambda_{\ell+1} - 1$, and denote by Y the Young diagram of signature $(f_1,\ldots,f_\ell)$. Note that Y is contained in a rectangle of size $\ell \times m$. For a fixed m we also write $\chi_Y^{A_\ell^{(1)}}$ (or $\chi_\Lambda^{A_\ell^{(1)}}$ where Λ is the corresponding highest weight) to mean $\chi_{\lambda,m}$.

Example 4. $C_\ell^{(1)}$

A level m irreducible representation is specified by a set of integers $\lambda = (\lambda_1,\ldots,\lambda_\ell)$ s.t. $\ell+m \geqslant \lambda_1 > \cdots > \lambda_\ell \geqslant 1$. We introduce a coordinate $(\tau,u_1,\ldots,u_\ell)$ so that $q = e[\tau]$, $z_1 = e[u_2 - u_1],\ldots, z_{\ell-1} = e[u_\ell - u_{\ell-1}]$, $z_\ell = e[2u_\ell]$. Then the character corresponding to (m,λ) is given (up to a fractional power of q) by

$$\chi_{\lambda,m} =$$

(2.8)

$$\frac{\det(\theta_{\lambda_i,2m}(\tau,u_j)-\theta_{-\lambda_i,2m}(\tau,u_j))_{i,j=1,\ldots,\ell}}{\eta(\tau)^{\ell(1-\ell)} \prod_{i=1}^{\ell} \theta(\tau,2u_i) \prod_{1\leqslant i<h\leqslant \ell} \theta(\tau,u_i+u_j)\theta(\tau,u_i-u_j)}.$$

We set $f_1 = \lambda_1-\ell,\ldots, f_\ell = \lambda_\ell - 1$, and denote by Y the Young diagram of signature $(f_1,..,f_\ell)$, which is contained in a rectangle of size $\ell \times m$. For a fixed m we also write $\chi_Y^{C_\ell^{(1)}}$ (or $\chi_Y^{C_\ell^{(1)}}$ where Λ is the corresponding highest weight) to mean $\chi_{\lambda,m}$.

Remark. If $\Lambda' \equiv \Lambda \bmod \mathbb{C}\delta$, $ch_{L(\Lambda')}$ differs from $ch_{L(\Lambda)}$ only by a fixed power of q. In (2.7) or (2.8) this ambiguity is fixed by requiring the expression in terms of theta functions.

§3. CHARACTER IDENTITY AND BRANCHING COEFFICIENTS.

The irreducible decomposition (1.2) implies a character identity of the form

$$ch_{L(\tilde{\Lambda})}\,|_{\mathfrak{g}} = \sum_{\Lambda} (\tilde{\Lambda}:\Lambda) ch_{L(\Lambda)}. \tag{3.1}$$

Here $|_{\mathfrak{g}}$ means the restriction to the Cartan subalgebra of $\mathfrak{g}$.

Example 1. $\mathfrak{gl}(2\ell,\mathbb{C}) \supset sp(2\ell,\mathbb{C})$

Let $\tilde{Y}_j$ (resp. Y_k) denote the Young diagram of signature

$$(\underbrace{1,\ldots,1}_{j},\underbrace{0,\ldots,0}_{2\ell-j}) \quad (\text{resp. } (\underbrace{1,\ldots,1}_{k},\underbrace{0,\ldots,0}_{\ell-k})).$$

We define q_ν $(\nu=0,\ldots,2\ell)$ by

$$\sum_{\nu=0}^{2\ell} q_\nu t^\nu = \prod_{i=1}^{\ell} (1+z_i t)(1+z_i^{-1}t). \tag{3.2}$$

Then we have

$$\chi_{\tilde{Y}_j}^{\mathfrak{gl}(2\ell,\mathbb{C})}\,|_{\mathfrak{g}} = q_j \tag{3.3}$$

and

$$(3.4)\qquad \chi_{Y_k}^{sp(2\ell,\mathbb{C})} = q_k - q_{k-2}.$$

Thus the character identity reads as

$$(3.5)\qquad q_j = (q_j - q_{j-2}) + (q_{j-2} - q_{j-4}) + \cdots,$$

and the branching coefficients are

$$(\tilde{Y}_j : Y_k) = \begin{cases} 1 & (\text{if } k \leqslant j \leqslant 2\ell-k,\ k \equiv j \bmod 2). \\ 0 & \text{otherwise.} \end{cases}$$

A specific feature of the decomposition (1.2) in the affine case is that the infinite sum over Λ can be separated into a finite sum over Λ mod $\mathbb{C}\delta$ and a 'string-like' sum over $\Lambda - n\delta$ ($n \in \mathbb{Z}$). Introducing the generating function

$$(3.6)\qquad E_{\tilde{\Lambda}\Lambda}(q) = \sum_{n\in\mathbb{Z}} (\tilde{\Lambda} : \Lambda - n\delta) q^n$$

the character identity reads as

$$(3.7)\qquad \mathrm{ch}_{L(\tilde{\Lambda})}\,|_{\mathfrak{g}} = \sum_{\Lambda \bmod \mathbb{C}\delta} E_{\tilde{\Lambda}\Lambda}(q)\,\mathrm{ch}_{L(\Lambda)}.$$

If the level of $\tilde{\Lambda}\,|_{\mathfrak{g}}$ is m in (3.7) the sum contains only such Λ with level m.

Take the pair $A_{2\ell-1}^{(1)} \supset C_{\ell}^{(1)}$ as an example. Then (3.7) combined with (2.7) and (2.8) gives a theta function identity of the form

$$(3.8)\qquad \chi_{\tilde{Y}}^{A_{2\ell-1}^{(1)}}\,|_{\mathfrak{g}}\,(\tau,u_1,\ldots,u_\ell) = \sum_{Y} e_{\tilde{Y}Y}(\tau)\,\chi_{Y}^{C_{\ell}^{(1)}}(\tau,u_1,\ldots,u_\ell).$$

Here $e_{\tilde{Y}Y}(\tau)$ ($\underset{\mathrm{def}}{=} e_{\tilde{\Lambda}\Lambda}(\tau)$) differs from $E_{\tilde{\Lambda}\Lambda}(\tau)$ in (3.7) by a suitable fractional power of q. As a consequence $\{e_{\tilde{\Lambda}\Lambda}(\tau)\}$ enjoy the automorphic property: For $\begin{pmatrix} a & b \\ c & d \end{pmatrix} \in SL(2,\mathbb{Z})$ $e_{\tilde{\Lambda}\Lambda}\left(\frac{a\tau+b}{c\tau+d}\right)$ is a linear combination of $e_{\tilde{\Lambda}'\Lambda'}(\tau)$'s with the same level m. In fact, as we shall see below, if $m = 1$, $\{e_{\tilde{\Lambda}\Lambda}(\tau)\}$ are given by Hecke indefinite modular forms.

§4. DUALITY.

In principle, we can compute $e_{\tilde{\Lambda}\Lambda}(\tau)$ by using (3.8) (Residue calculus). In practice, the computation is complicated especially for a large ℓ. Nevertheless, the result shows the following remarkable large ℓ behaviour: Consider the case $m = 1$. Then $\tilde{\Lambda}$ (resp. Λ) is one of fundamental weights $\tilde{\Lambda}_j$ (resp. Λ_k) of $A^{(1)}_{2\ell-1}$ (resp. $C^{(1)}_{\ell}$). The fact is for $\ell \longrightarrow \infty$

$$(4.1) \qquad E_{\tilde{\Lambda}_j\Lambda_k}(q) = F^{(0)}_{jk}(q) + q^{\ell}F^{(1)}_{jk}(q) + q^{2\ell}F^{(2)}_{jk}(q) + \cdots$$

where $\{F^{(\nu)}_{jk}(q)\}$ are independent of ℓ.

Guided by this fact we found a duality of branching coefficients for arbitrary rank ℓ and level m.

Let us state our duality first in the finite dimensional case; $gl(2\ell,\mathbb{C}) \supset sp(2\ell,\mathbb{C})$.

We choose m and consider a Young diagram $\tilde{Y}$ contained in a rectangle of size $2\ell \times m$. The character identity reads as

$$(4.2) \qquad \chi_{\tilde{Y}}^{gl(2\ell,\mathbb{C})}(z_1,\dots,z_\ell) = \sum_{Y} (\tilde{Y}:Y)\chi_Y^{sp(2\ell,\mathbb{C})}(z_1,\dots,z_\ell).$$

Here $(\tilde{Y}:Y)$ vanishes unless Y is contained in a rectangle of size $\ell \times$ m. Consider the second pair $gl(m;\mathbb{C}) \subset sp(m,\mathbb{C})$, and a Young diagram $Y^{\dagger}$ contained in a rectangle of size $m \times \ell$. the character identity which represents the irreducible decomposition of the irreducible $sp(m,\mathbb{C})$-module corresponding to $Y^{\dagger}$ can be written as

$$(4.3) \qquad \sum_{\tilde{Y}^{\dagger}} (w_1\dots w_m)^{-\ell}\ \chi_{\tilde{Y}^{\dagger}}^{gl(m,E)}(w_1,\dots,w_m)\cdot(Y^{\dagger}:\tilde{Y}^{\dagger})$$

$$= \chi_{Y^{\dagger}}^{sp(m,\mathbb{C})}(w_1,\dots,w_m).$$

Then $(Y^{\dagger}:\tilde{Y}^{\dagger})$ vanishes unless $\tilde{Y}^{\dagger}$ is contained in a rectangle of size m×2ℓ.

We say $\tilde{Y}$ and $\tilde{Y}^{\dagger}$ are complementary if $\tilde{Y}^{\dagger}$ is obtained by (upper-left ⟷ lower-right) transposition of the complement of $\tilde{Y}$ in the 2ℓ×m rectangle. Similarly, we define the complementary diagram $Y^{\dagger}$ of Y with respect to the ℓ×m rectangle. The duality reads as:

Theorem. $(\tilde{Y}:Y) = (Y^{\dagger}:\tilde{Y}^{\dagger})$ provided the pairs $(\tilde{Y},\tilde{Y}^{\dagger})$ and $(Y,Y^{\dagger})$ are both mutually complementary in the sense above.

Next, consider the affine case $A^{(1)}_{2\ell-1} \supset C^{(1)}_{\ell}$. Fix the level m. In this case, the dual pair is $A^{(1)}_{m-1} \supset C^{(1)}_{m}$. Let us prepare some notations. Let $\tilde{Y}^{\dagger}$ be a Young diagram in a rectangle of size m×2ℓ. Denote by $|\tilde{Y}^{\dagger}|$ the number of cells in $\tilde{Y}^{\dagger}$. Let $\tilde{\Lambda}^{\dagger}$ be a weight of level 2ℓ corresponding to $\tilde{Y}^{\dagger}$. Note that $\tilde{\Lambda}^{\dagger}$ be a weight of level 2ℓ corresponding to $\tilde{Y}^{\dagger}$. Note that $\tilde{\Lambda}^{\dagger}$ is unique up to $\mathbb{C}\delta$. We denote by σ the isomorphism of the Dynkin diagram for $A^{(1)}_{m-1}$ which rotates the diagram so that the i-th node goes to the (i+1)-th node. It induces an isomorphism on the weight space, which we also denote by σ. We define $\hat{\chi}^{A^{(1)}_{m-1}}_{\tilde{Y}^{\dagger}}(\tau,v_1,\ldots,v_m)$ to be the following positive linear combination of the characters of $A^{(1)}_{m-1}$.

$$
(4.4) \qquad \hat{\chi}^{A^{(1)}_{m-1}}_{\tilde{Y}^{\dagger}}(\tau,v_1,\ldots,v_m)
$$

$$
= \frac{1}{\eta(\tau)} \sum_{q=0}^{m-1} \theta_{2\ell q-\ell m+|\tilde{Y}^{\dagger}|,\, 2\ell m}\Big(gt, \frac{v_1+\cdots+v_m}{m} \Big)
$$

$$
\times \qquad \chi^{A^{(1)}_{m-1}}_{\sigma^q(\tilde{\Lambda}^{\dagger})} (\tau,v_1,\ldots,v_m).
$$

The character identity for the pair $A^{(1)}_{m-1} \subset C^{(1)}_{m}$ can be written as

$$(4.5)\qquad \sum_{\tilde{Y}^{\dagger}} x_{\tilde{Y}^{\dagger}}^{\hat{A}_{m-1}^{(1)}}(\tau,v_1,\dots,v_m)\, e_{\tilde{Y}^{\dagger}Y^{\dagger}}^{\dagger}(\tau) = x_{Y^{\dagger}}^{C_m^{(1)}}(\tau,v_1,\dots,v_m).$$

The duality reads as:

Theorem. $e_{\tilde{Y}Y}(\tau) = e_{\tilde{Y}^{\dagger}Y^{\dagger}}^{\dagger}(\tau)$ provided the pairs $(\tilde{Y},\tilde{Y}^{\dagger})$ and $(Y,Y^{\dagger})$ are both mutually complementary.

§5. HECKE INDEFINITE FORMS.

The duality theorem interchanges the rank and the level. In other words, it enables us to rewrite the theta function identity in ℓ variables into another theta function identity in m variables. If m = 1, this is a great simplification in the computation of branching coefficients. As a result, $e_{\tilde{Y}Y}(\tau)$'s for the pair $A_{2\ell-1}^{(1)} \supset C_{\ell}^{(1)}$ turns out to be expressible in terms of Hecke indefinite modular forms.

For level 1, $\tilde{Y}$ (resp. Y) corresponds to some fundamental weight $\tilde{\Lambda}_j$ (resp. Λ_k). Hence we write $e_{jk}(\tau)$ to mean $e_{\tilde{Y}Y}(\tau)$. Consider the following quadratic form

$$(5.1)\qquad B(\tbinom{x}{y},\tbinom{x}{y}) = 2(\ell+2)x^2 - 2\ell y^2.$$

We set $L = \mathbb{Z}^2$ and $L^* = \frac{\mathbb{Z}}{2(\ell+2)} \oplus \frac{\mathbb{Z}}{2\ell}$. We denote by G_0 the discrete group generated by $\begin{pmatrix}\ell+1 & \ell\\ \ell+2 & \ell+1\end{pmatrix}^2$. For $\gamma = \binom{x}{y}$ we set sgn γ = sgn x. For $\mu \in L^*$, we define

$$(5.2)\qquad \theta_{\mu}^{B}(\tau) = \sum_{\substack{\gamma\in G_0\backslash(L+\mu)\\ B(\gamma,\gamma)>0}} \operatorname{sgn}\gamma\; e[\tfrac{\tau}{2}B(\gamma,\gamma)].$$

Then we have:

Theorem. $\eta(\tau)^2 e_{jk}(\tau) = \theta^B_\mu(\tau)$ for $\mu = \begin{bmatrix} \frac{k+1}{2(\ell+2)} \\ \frac{j}{2\ell} \end{bmatrix}$.

Remark. Kac–Peterson ([3],[4]) computed the string functions for $A_1^{(1)}$ of arbitrary level m. Their result is

$$(5.3)\quad \eta(\tau)^3\, c^{(m-k)\Lambda_0+k\Lambda_1}_{(m-j)\Lambda_0+j\Lambda_1}(\tau) = \theta^B_\mu(\tau) \text{ for } \mu = \begin{bmatrix} \frac{k+1}{2(m+2)} \\ \frac{j}{2m} \end{bmatrix}$$

Thus the formula in the theorem above happens to be exactly the same as (5.3) with rank ℓ in place of level m.

REFERENCES

[1] Jimbo, M. and Miwa, T., Lett. Math. Phys. 6 (1982), 463.

[2] ______________, RIMS preprint 434, 453, Kyoto University, (1983), to appear in Adv. Stud. in Pure Math.

[3] Kac, V.G. and Peterson, D.H., Bull. Amer. Math. Soc. 3 (1980) 1057.

[4] ______________, Adv. of Math. 53 (1984).

R.I.M.S.
Kyoto University
Kyoto 606, JAPAN

A BRIEF INTRODUCTION TO THE FINITE SIMPLE GROUPS

Robert L. Griess, Jr.†

TABLE OF CONTENTS

1. INTRODUCTION

At this conference on vertex operators, we see more evidence that the theory of finite simple groups and the theory of vertex operators have a serious mathematical relationship. Because many conference participants have not worked in abstract algebra, a few brief statements about the current state of things in finite simple group theory seems in order. A number of recent articles [Gr4] [Gr5] and the book of Gorenstein [Go2] go into this subject in detail. Accordingly, I will limit myself to a few items which might be of greatest interest to those who work on vertex operator theory. I hope that workers in vertex operator theory become familiar with the finite simple groups. My feeling is that more interaction between these two subjects is forthcoming.

Some basic references for modern finite groups theory are [Go1], [Hu1].

---

† Partially supported by the National Science Foundation through the Mathematical Sciences Research Institute.

Vertex Operators in Mathematics and Physics - Proceedings of a Conference November 10-17, 1983. Publications of the Mathematical Sciences Research Institute #3, Springer-Verlag, 1984.

2. STATEMENT OF CLASSIFICATION.

The finite group theory community has, in an effort concentrated in the last 25 years or so, completed a program to classify the finite simple groups. The grand theorem says that every finite simple group is one of those on "the list," which we give in the table below, taken from [Gor2]. For a group of Lie type, the order of the "simply connected" version is given. The "adjoint version" (simple in all but a few cases) is obtained via factoring by a central subgroup of order d, given in the right column.

In addition, we make the following points:

(1) The groups $B_n(q)$ and $C_n(q)$ are isomorphic for all even prime powers q (but not for odd q). Note also that the lower bounds on n for the families B, C and D are needed to avoid duplication of isomorphism types.

(2) The groups on the list which are nonsimple are $A_1(2)$ (isomorphic to the symmetric group of degree 3), $A_1(3)$ (the alternating group of degree 4), $B_2(2)$ (the symmetric group of degree 6), $G_2(2)$ (the unitary group PSU(3,3) extended by the field automorphism of order 2), ${}^2A_2(2)$ (solvable of order 72), ${}^2B_2(2)$ (solvable of order 20), ${}^2G_2(3)$ (Aut(PSL(2,8))), ${}^2F_4(2)$ (it has a subgroup of index 2, the Tits simple group).

(3) There are a few exceptional isomorphisms, namely, $A_1(4) \cong A_1(5)$, $A_1(7) \cong A_2(2)$, $A_1(9) \cong A_6$, $A_3(2) \cong A_8$, ${}^2A_3(2) \cong B_2(3)$.

(4) The order of McLaughlin's group is given incorrectly in the table of [Gor2].

(5) The order of M(24)' is given incorrectly in the table of [Gor2].

The alternating groups are the most familiar. A bit less so are the groups of Lie type over finite fields. Roughly speaking, one could call them the classical groups over finite fields, although E_8 is not usually thought of as a classical group. There are groups on the above list which owe their existence to, among other things, properties of certain finite fields and have no analogue over the real or complex numbers. In the other direction, it is not true that distinct types of groups in characteristic zero correspond to distinct types of groups in prime characteristic. For example, given positive integers n, p and q such that n=p+q, there is a unitary group over the complex numbers associated to a Hermitian form of signature (p,q). However, given a dimension and a pair of finite fields, one quadratic over the other, there is only one nondegenerate Hermitian form (up to equivalence) and so only one unitary group. Thus, the finite groups of Lie type are analogues of the Lie groups, but there is nothing close to a precise bijection between the families of groups on both lists. See [Ca2] and [St1] for systematic treatments; the survey article [Ca1] is helpful for a first exposure.

As we see things now, the sporadic groups do not have any analogue among the Lie groups or the Weyl groups or any other families of groups one encounters in Lie theory. They have come to us in a number of ways, as automorphisms of combinatorial structures or of lattices and as solutions to certain problems posed abstractly in the classification theory, notably "centralizer of involution" problems. The sporadic groups are described systematically in [Go2]. See also [Ti1], [Ti3]. I have included a very brief discussion of the sporadic groups in Section 5 of this article.

The largest sporadic group, denoted F_1, is often called the Monster or the Friendly Giant. It was discovered by Bernd Fischer and myself in 1973 and was constructed by me in 1980 [Gr1], [Gr3] as automorphisms of a certain commutative nonassociative algebra of dimension 196884. See the recent article [Ti2].

3. MODULAR FORMS AND SPORADIC GROUPS.

Several years ago, McKay produced the remarkable equation 196884 = 1 + 196883, which related the coefficient of q in the elliptic module function $J = q^{-1} + 744 + 196884q + 21493760q^2 + \ldots$ to the two smallest degrees of representations of the monster

[Gr3[Co-No1]. Thompson [Th1] suggested that the higher coefficients be replaced by representations of the monster which they "suggest" (as it stands, the suggestion is not precise, since a positive integer may be writeable as a sum of degrees in several different ways). Conway and Norton [Co-No1] found a correspondence between rational conjugacy classes of the monster (two group elements are rationally conjugate if they generate conjugate cyclic subgroups) and a list of genus zero function fields with a number of remarkable algebraic properties. They defined a formal object $q^{-1} + \sum_{n\geq 1} H_n q^n$ where the H_n are in the ring of class functions of F_1 with the property that its evaluation on an element of F_1 gave the modular form associated to the aformentioned genus zero field corresponding to that element. With this idea, they were able to fulfill Thompson's suggestion in the following manner: they conjectured that the H_n are characters. A proof of this was given by Atkin, Fong and Smith [Fo1]. Using some reductions of Thompson, they proved that the H_n satisfied the conditions of Brauer's characterization of characters.

More recently, there has been an extension of the moonshine game involving M_{24} and a collection of eta functions with Euler product expansion; see [Ma1], [Ma2], [Ma3]. See also the survey [Br1].

It should be emphasized that the meaning of the "genus zero" property has not yet been made clear.

4. WHY VERTEX OPERATORS?

It has seemed desirable to have the graded module which gives the characters. A concretely given module, with accompanying algebraic structures, promises to give more insight than the graded character. Since the spaces vertex operators act on give (essentially) modular forms as their characters, it seemed reasonable to try to make something with them; indeed, there has not been a serious alternative to this method. For one thing, one can define a space and vertex operator based on any lattice, in particular the Leech lattice. However, the most straightforward such constructs [Ka1], [Le1], while certainly interesting, do not seem to have the right properties. The definition of such a module has been announced at this conference by Frenkel, Lepowsky and Meurman. It was an outgrowth of their work [F-L-M1].

5. A QUICK TOUR OF THE SPORADIC GROUPS.

One requires the classification of finite simple groups to say that there are twenty six sporadic groups. The five Mathieu groups were discovered in the 1860's, though their existence was not put on firm ground until the 1930's [Wi1]. The other twenty one were discovered in the ten year period 1965 to 1975. They came one at a time or in small clusters, and proof of existence has varied in style from the blitz of D. Higman and C. Sims (both discovery and construction in one day!) to, for example, extended work with computers, notably the work of C. Sims.

Two quick definitions. A *section* of a group is a quotient of a subgroup. We say that a group is *involved* in another group if it is isomorphic to a section of the other group.

A variety of ideas have been involved with the discovery and construction of the sporadic groups. At this time, no single, simply stated idea "captures" them all, though there is one theme which applies to twenty of them, the *Happy Family* (those sporadic groups which are involved in the largest one). Briefly, it goes as follows:

Step 1. The Steiner system S(5,8,24) and the related Golay code. Their automorphism group is M_{24}.

Step 2. The Leech lattice (in 24-dimensional Euclidean space). Its automorphism group is the finite group .0, whose central quotient is the simple group .1.

Step 3. The 196884-dimensional algebra B. Its automorphism group is the monster.

Each step makes essential use of the preceeding one. Also, once one has the automorphism group at each step, one can rather trivially derive existence of other sporadic groups, to wit:

Step 1. M_{11}, M_{12}, M_{22}, M_{23}, M_{24}

Step 2. HJ, Suz, HiS, McL, $\cdot 3$, $\cdot 2$, $\cdot 1$

Step 3. F_{22}, F_{23}, F'_{24}, Held, F_3, F_5, F_2, F_1

For more details on this theme, see [Gr1], [Gr2], and my survey articles [Gr4], [Gr5].

I should mention that I sometimes call the largest sporadic the "Friendly Giant" instead of the popular term "monster". So far, we have been unable to settle whether the smallest and first twentieth century sporadic group J_1 is a member of the Happy Family (its order is only 175,560). Probably in reference to its stubborness, J. Tits [Ti3] has named this group the "Wicked Dwarf" (le Mechant Nain).

I call the sporadic groups not involved in the monster the *Pariahs*. They include the groups of Rudvalis, Lyons, O'Nan, the third and fourth groups of Janko and possibly the first Janko group J_1. We can not yet capture these groups by a theme in the above sense. Instead, I shall just mention a few interesting points:

J_1 is a subgroup of $G_2(11)$ and is the fixed points of an outer automorphism of O'Nan's group.

Lyon's group contains $G_2(5)$.

J_3 involves no other sporadic group and is involved in no other sporadic group.

J_4 has a representation of degree 112 over the field of 2 elements.

J_4 contains M_{24} and involves M_{22} in two essentially different ways, one as a subgroup of M_{24} and one as the 6-fold cover of M_{22} which lies in the centralizer of an involution.

Ru has a projective representation of dimension 28 over the field $\mathbb{Q}(-1)^{1/2}$.

We reproduce a slightly updated version of the table of involvements of one sporadic group in another [Gr2]. It may help give a feeling for how the sporadic groups are related.

One of the real difficulties with the sporadic groups is that one has to use ad hoc methods to answer many questions about them. This is in contrast to the situation with finite groups of Lie type, where one can make rather uniform arguments by using the (B,N)-structure, the root systems of the underlying Lie algebras, etc. Thus, one of the attractive prospects at this time is for the vertex operator methodology to develop a context for a theory of sporadic groups. Other efforts to develop theories of finite groups which include the sporadic groups are in progress. I mention the work of finite geometers [Bu1], [Ro-Sm1] and the work on groups as automorphism groups of nonassociative algebras [Al1], [Al2], [Fr1], [Ha1], [Ha2], [Ha3], [No1], [Sm1]. In time, we shall see the merits of these approaches.

REFERENCES

[Al1] Allen, Harry P., Non-associative algebras associated with doubly transitive permutation groups: S_{n+1}, preprint.

[Al2] Allen, Harry P., A note on automorphism groups of nonassociative algebras associated with doubly transitive groups, preprint.

[Br1] Broue, Michel, Groupes finis, series formelles et fonctions modulaires, Seminaire sur les groupes finis, Tome 1, Publications Mathematique de l'Universite Paris VII.

[Bu1] Buekenhout, Francis, Diagrams for geometries and groups,

Journal of Combinatorial Theory 27 (1979) 121-151.

[Ca1] Carter, Roger, Simple groups and simple Lie algebras (survey article), Jour. London Math. Soc. 40(1965), 193-240.

[Ca2] Carter, Roger, Simple Groups of Lie Type, Wiley-Interscience, New York, 1972.

[Co-No1] Conway, John and Norton, Simon, Monstrous Moonshine, Bull. London Math., Soc. 11(1979), 308-339.

[Fo1] Fong, Paul, Characters arising in the monster-modular connection, article in the Santa Cruz Conference on Finite Groups, American Mathematical Society, Providence, 1980.

[F-L-M1] Frenkel, Igor B., Lepowsky, James and Meurman, Arne, An E_8-approach to F_1, to appear in the proceedings of the Montreal Conference on finite groups, 1982.

[Fr1] Frohardt, Daniel, A trilinear form for J_3, to appear in Journal of Algebra.

[Go1] Gorenstein, Daniel, Finite Groups, Harper and Row, New York, 1968; 2nd ed. Chelsea, New York, 1980.

[Go2] Gorenstein, Daniel, Finite Simple Groups, Plenum Press, New York, 1981.

[Gr1] Griess, Rober L., Jr., A construction of F_1 as automorphisms of a 196883 dimensional algebra, Proc. Nat. Acad. Sci. USA 78, 689-691 (1981).

[Gr2] Griess, Robert L., Jr., The Friendly Giant, Inventiones Math., 69, 1-102 (1982).

[Gr3] Griess, Robert L., Jr., The structure of the "monster" simple group, in W. Scott and F. Gross, Proceedings of the Conference on Finite Groups, Academic Press, New York, 1976, pp. 113-118.

[Gr4] Griess, Robert L., Jr., The monster and its nonassociative algebra, to appear in the proceedings of the Montreal Conference on finite groups, 1982.

[Gr5] Griess, Robert L., Jr., The sporadic simple groups and the construction of the monster, to appear in the proceedings of the International Congress of Mathematicians, 1983.

[Ha1] Harada, Koichiro, On a commutative nonassociative algebra associated with a multiply transitive group, J. Fac. Sci. Univ. Tokyo, vol. 28(1982), 843-849.

[Ha2] Harada, Koichiro, On a commutative nonassociative algebra associated with a doubly transitive group, preprint.

[Ha3] Harada, Koichiro, On commutative nonassociative algebras associated with the doubly transitive permutation groups PSL(m,q), m≥3, preprint.

[Hu1] Huppert, Bertram, Endliche Gruppen, I, Springer-Verlag, Berlin and New York, 1967.

[Ka1] Kac, Victor G., An elucidation of "Infinite dimensional algebras...and the very strange formula," $E_8^{(1)}$ and the cube root of the modular invariant j, Advances in Mathematics 3(1980) 264-273.

[Le1] Lepowsky, James, Euclidean lie algebras and the modular function j, in The Santa Cruz Conference on Finite Groups, Proc. Symp. Pure Math. 37, American Mathematical Society, Providence, 1980.

[Ma1] Mason, Geoffrey, M_{24} and certain automorphic forms, to appear in the proceedings of the Montreal group theory conference, 1982.

[Ma2] Mason, Geoffrey, Modular forms and the theory of Thompson series, to appear in the proceedings of the Rugers group theory year 1983.

[Ma3] Mason, Geoffrey, Frame shapes and rational characters of finite groups, to appear in Journal of Algebra.

[No1] Norton, Simon, unpublished work [on "Norton algebras"].

[Ro-Sm1] Ronan, Mark and Smith, Steven D., 2-local geometries for finite groups, article in The Santa Cruz Conference on Finite Groups, American Mathematical Society, 1980, Providence.

[Sm1] Smith, Stephen D., Nonassociative commutative algebras for triple covers of 3-transposition groups 24(1977), 273-289.

[St1] Steinberg, Robert, Lectures on Chevalley Groups, Yale Lecture Notes.

[Th1] Thompson, John G., Some numerology between the Fischer-Griess monster and the elliptic modular function, bull. London Math. Soc. 11 (1979), 352-353.

[Ti1] Tits, Jacques, Groups finis simple sporadique, Seminaire Bourbaki, No. 375, Febrier 1970.

[Ti2] Tits, Jacques, Le monstre (d'apres R. Griess, B. Fischer, et al., Seminaire Bourbaki, No. 620, Novembre 1983.

[Ti3] Tits, Jacques, Resume des cours de 1976-77, extrait de l'annuaire du College de France.

[Wi1] Witt, E., Die 5-fach transitiven Gruppen von Mathieu, Abn. Math. Hamburg 12(1983), 256-264.

Department of Mathematics
University of Michigan
Ann Arbor, Michigan 48109

Table 1: *Known Finite Simple Groups*

G	order of G	d
$A_n(q)$	$q^{n(n+1)/2} \prod_{i=1}^{n} (q^{i+1}-1)$	$(n+1,q-1)$
$B_n(q), n>1$	$q^{n^2} \prod_{i=1}^{n} (q^{2i}-1)$	$(2,q-1)$
$C_n(q), n>2$	$q^{n^2} \prod_{i=1}^{n} (q^{2i}-1)$	$(2,q-1)$
$D_n(q), n>3$	$q^{n(n-1)}(q^n-1) \prod_{i=1}^{n-1} (q^{2i}-1)$	$(4,q^n-1)$
$G_2(q)$	$q^6(q^6-1)(q^2-1)$	1
$F_4(q)$	$q^{24}(q^{12}-1)(q^8-1)(q^6-1)(q^2-1)$	1
$E_6(q)$	$q^{36}(q^{12}-1)(q^9-1)q^8-1)(q^6-1)(q^5-1)(q^2-1)$	$(3,q-1)$
$E_7(q)$	$q^{63}(q^{18}-1)(q^{14}-1)(q^{12}-1)(q^{10}-1)(q^8-1)(q^6-1)(q^2-1)$	$(2,q-1)$
$E_8(q)$	$q^{120}(q^{30}-1)(q^{24}-1)(q^{20}-1)(q^{18}-1)(q^{14}-1)(q^{12}-1)(q^8-1)(q^2-1)$	1
${}^2A_n(q), n>1$	$q^{n(n+1)/2} \prod_{i=1}^{n} (q^{i+1}-(-1)^{i+1})$	$(n+1,q+1)$
${}^2B_2(q), q=2^{2m+1}$	$q^2(q^2+1)(q-1)$	1
${}^2D_n(q), n>3$	$q^{n(n-1)}(q^n+1) \prod_{i=1}^{n-1} (q^{2i}-1)$	$(4,q^n+1)$
${}^3D_4(q)$	$q^{12}(q^8+q^4+1)(q^6-1)(q^2-1)$	1
${}^2G_2(q), q=3^{2m+1}$	$q^3(q^3+1)(q-1)$	1
${}^2F_4(q), q=2^{2m+1}$	$q^{12}(q^6+1)(q^4-1)(q^3+1)(q-1)$	1
${}^2E_6(q)$	$q^{36}(q^{12}-1)(q^9+1)(q^8-1)(q^6-1)(q^5+1)(q^2-1)$	$(3,q+1)$

(continued)

Table 1 (continued)

Alternating groups

$A_n, n \geq 5.$	$\frac{1}{2}(n!)$

Sporadic Groups

M_{11}	$7920=2^4 \cdot 3^2 \cdot 5 \cdot 11$
M_{12}	$95040=2^6 \cdot 3^3 \cdot 5 \cdot 11$
M_{22}	$443520=2^7 \cdot 3^2 \cdot 5 \cdot 7 \cdot 11$
M_{23}	$10200960=2^7 \cdot 3^2 \cdot 5 \cdot 7 \cdot 11 \cdot 23$
M_{24}	$244823040=2^{10} \cdot 3^3 \cdot 5 \cdot 7 \cdot 11 \cdot 23$
J_1	$175560=2^3 \cdot 3 \cdot 5 \cdot 7 \cdot 11 \cdot 19$
J_2	$2^7 \cdot 3^3 \cdot 5^2 \cdot 7$
J_3	$2^7 \cdot 3^5 \cdot 5 \cdot 17 \cdot 19$
J_4	$2^{21} \cdot 3^3 \cdot 5 \cdot 7 \cdot 11^3 \cdot 23 \cdot 29 \cdot 31 \cdot 37 \cdot 43$
HS	$2^9 3^2 5^3 \cdot 7 \cdot 11$
Mc	$2^7 3^6 5^3 \cdot 7 \cdot 11$
Suz	$2^{13} 3^7 5^2 7 \cdot 11 \cdot 23$
Ru	$2^{14} 3^3 5^3 7 \cdot 13 \cdot 29$
He	$2^{10} 3^3 5^2 \cdot 7^3 \cdot 17$
Ly	$2^8 3^7 5^6 7 \cdot 11 \cdot 31 \cdot 37 \cdot 67$
ON	$2^9 3^4 5 \cdot 7^3 \cdot 11 \cdot 19 \cdot 31$
$\cdot 1$	$2^{21} 3^9 5^4 7^2 11 \cdot 13 \cdot 23$
$\cdot 2$	$2^{18} 3^6 5^3 7 \cdot 11 \cdot 23$
$\cdot 3$	$2^{10} 3^7 5^3 7 \cdot 11 \cdot 23$
M(22)	$2^{17} 3^9 5^2 7 \cdot 11 \cdot 23$
M(23)	$2^{18} 3^{13} 5^2 7 \cdot 11 \cdot 13 \cdot 17 \cdot 23$
M(24)'	$2^{21} 3^{16} 5^2 7^3 \cdot 11 \cdot 13 \cdot 17 \cdot 23 \cdot 29$
F_5	$2^{15} 3^{10} 5^3 7^2 13 \cdot 19 \cdot 31$
F_3	$2^{14} 3^6 5^6 \cdot 7 \cdot 11 \cdot 19$
F_2	$2^{41} 3^{13} 5^6 7^2 11 \cdot 13 \cdot 17 \cdot 19 \cdot 23 \cdot 31 \cdot 47$
F_1	$2^{46} 3^{20} 5^9 \cdot 7^6 \cdot 11^2 \cdot 13^3 \cdot 17 \cdot 19 \cdot$ $23 \cdot 29 \cdot 31 \cdot 41 \cdot 47 \cdot 59 \cdot 71.$

Table 2: *Involvement of One Sporadic Group in Another*

(* = yes, · = no, ? = unsettled)

	M_{11}	M_{12}	M_{22}	M_{23}	M_{24}	J_1	J_2=HJ	J_3	Held	HiS	McL	Suz	·1	·2	·3	F_{22}	F_{23}	F'_{24}	LyS	Ru	O'S	F_2	F_1	F_3	F_5	J_4
M_{11}	*	*	·	*	*	·	·	·	·	*	*	*	*	*	*	*	*	*	*	·	*	*	*	·	*	*
M_{12}	·	*	·	·	*	·	·	·	·	·	·	*	*	·	*	*	*	*	·	·	·	*	*	·	*	*
M_{22}	·	·	*	*	*	·	·	·	·	*	*	·	*	*	*	*	*	*	*	·	·	*	*	·	*	*
M_{23}	·	·	·	*	*	·	·	·	·	·	·	·	*	*	*	·	*	*	·	·	·	*	*	·	·	*
M_{24}	·	·	·	·	*	·	·	·	·	·	·	·	*	·	·	·	·	*	·	·	·	?	*	·	·	*
J_1	·	·	·	·	·	*	·	·	·	·	·	·	·	·	·	·	·	·	·	·	*	?	?	·	·	·
J_2=HJ	·	·	·	·	·	·	*	·	·	·	·	*	*	·	·	·	?	?	·	·	·	?	*	·	·	·
J_3	·	·	·	·	·	·	·	*	·	·	·	·	·	·	·	·	·	·	·	·	·	·	·	·	·	·
Held	·	·	·	·	·	·	·	·	*	·	·	·	·	·	·	·	·	*	·	·	·	·	*	·	·	·
HiS	·	·	·	·	·	·	·	·	·	*	·	·	*	*	*	·	·	·	·	·	·	*	*	·	*	·
McL	·	·	·	·	·	·	·	·	·	·	*	·	*	*	*	·	·	·	*	·	·	*	*	·	·	·
Suz	·	·	·	·	·	·	·	·	·	·	·	*	*	·	·	·	·	·	·	·	·	·	*	·	·	·
·1	·	·	·	·	·	·	·	·	·	·	·	·	*	·	·	·	·	·	·	·	·	·	*	·	·	·
·2	·	·	·	·	·	·	·	·	·	·	·	·	*	*	·	·	·	·	·	·	·	*	*	·	·	·
·3	·	·	·	·	·	·	·	·	·	·	·	·	*	·	*	·	·	·	·	·	·	·	*	·	·	·
F_{22}	·	·	·	·	·	·	·	·	·	·	·	·	·	·	·	*	*	*	·	·	·	*	*	·	·	·
F_{23}	·	·	·	·	·	·	·	·	·	·	·	·	·	·	·	·	*	*	·	·	·	*	*	·	·	·
F'_{24}	·	·	·	·	·	·	·	·	·	·	·	·	·	·	·	·	·	*	·	·	·	·	*	·	·	·
LyS	·	·	·	·	·	·	·	·	·	·	·	·	·	·	·	·	·	·	*	·	·	·	·	·	·	·
Ru	·	·	·	·	·	·	·	·	·	·	·	·	·	·	·	·	·	·	·	*	·	·	·	·	·	·
O'S	·	·	·	·	·	·	·	·	·	·	·	·	·	·	·	·	·	·	·	·	*	·	·	·	·	·
F_2	·	·	·	·	·	·	·	·	·	·	·	·	·	·	·	·	·	·	·	·	·	*	*	·	·	·
F_1	·	·	·	·	·	·	·	·	·	·	·	·	·	·	·	·	·	·	·	·	·	·	*	·	·	·
F_3	·	·	·	·	·	·	·	·	·	·	·	·	·	·	·	·	·	·	·	·	·	*	*	*	·	·
F_5	·	·	·	·	·	·	·	·	·	·	·	·	·	·	·	·	·	·	·	·	·	*	*	·	*	·
J_4	·	·	·	·	·	·	·	·	·	·	·	·	·	·	·	·	·	·	·	·	·	·	·	·	·	*

A MOONSHINE MODULE FOR THE MONSTER

Igor B. Frenkel,[+] James Lepowsky[+]
and Arne Meurman[+]

0. INTRODUCTION

The theory of finite simple groups was for a long time a rather isolated and unusual branch of mathematics. It achieved its goal in 1981 when a proof of the classification theorem was completed. This unique proof, comprising several thousand pages of published articles and preprints, leads to the following list of finite simple groups (see [12] for the details): the groups of Lie type, the alternating groups and the 26 sporadic groups. While each of the first two classes has a uniform description, the groups in the third class still have quite different constructions. The largest sporadic group, called the Monster and denoted F_1, was predicted independently by B. Fischer and R. Griess in 1973. It contains 20 or 21 of the sporadic groups and has order $>8 \cdot 10^{53}$. This group gave rise to many mysteries even before its actual appearance, promising deep connections with different areas of mathematics.

Many amazing discoveries about the Fischer-Griess Monster were collected in the highly unusual paper "Monstrous Moonshine" by J. Conway and S. Norton [4]. Most of the discoveries were based on the existence of an irreducible F_1-module of dimension 196883. B. Fischer, D. Livingstone and M. Thorne computed the entire character table using this assumption, thus providing the numbers for the "Monstrous game". This started with J. McKay's observation that the modular function j, in its expansion

[+] Partially supported by the National Science Foundation through the Mathematical Sciences Research Institute.

Vertex Operators in Mathematics and Physics - Proceedings of a Conference November 10-17, 1983. Publications of the Mathematical Sciences Research Institute #3, Springer-Verlag, 1984.

$$(0.1)\quad j(q) = q^{-1} + 744 + 196884q + 21493760q^2 + \dots = \sum_{n\geqslant -1} a_n q^n,$$

has a coefficient 196884, which exceeds by only 1 the dimension of the minimal conjectured nontrivial F_1-module. Then Thompson observed [24] that the first few coefficients a_n (for $n \neq 0$!) are also simple linear combinations of the dimensions d_n of irreducible representations of the Monster, e.g., $a_{-1} = d_1$, $a_1 = d_1 + d_2$, $a_2 = d_1 + d_2 + d_3$, $a_3 = 2d_1 + 2d_2 + d_3 + d_4$, etc. The coefficient 744 is inessential for the modular property, so one can consider instead the normalized modular function

$$(0.2)\quad J(q) = j(q) - 744.$$

Thompson also proposed replacing the coefficients a_n in the q-series for J by the representations V_n of F_1 that they suggested and considering the series

$$(0.3)\quad T_m = q^{-1} + 0 + \operatorname{tr} m|_{V_1} q + \operatorname{tr} m|_{V_2} q^2 + \dots$$

for arbitrary elements m of the Monster (not just m = 1). A great deal of evidence concerning these "Thompson series" was then collected and they appeared to be the normalized generators of function fields of genus 0. Conway and Norton wrote down a list of such series -- one for each conjugacy class in F_1 -- and conjectured that there exist representations V_n of F_1 compatible with this list via (0.3). A. O. L. Atkin, P. Fong and S. Smith (see [23]), using a computer, produced overwhelming evidence for this conjecture, supporting the extreme likelihood that there should exist a natural graded module for F_1 with the J-function as character.

About a year after these observations, Griess constructed the 196883-dimensional representation of F_1 ([14], [15]). However, instead of resolving the mysteries of "Monstrous Moonshine" he added a new one: He gave a construction of the Monster as an automorphism group of a peculiar commutative nonassociative algebra, say B_0, of dimension 196883.

Although very far from a Lie algebra, Griess's algebra appeared to us to be tantalizingly close to a Lie algebra, especially the Lie algebra E_8, in some as yet unknown presentation. Also, $j(q)^{1/3}$ had already appeared as the character of the basic representation of the affine Lie algebra $\hat{E}_8$ ([16], [18]). Our work started as an attempt to understand the appearance of J(q) and B_0 using vertex operators and the representation theory of affine Lie algebras, and it now includes the construction of a natural module for the Monster with character J (see [7], [8]).

These notes serve as an introduction to the main ideas and steps of our construction, which is purely conceptual. We do not include detailed proofs, which will appear in a separate publication. Griess's construction was very important during our work but we can now obtain major results of his as corollaries of ours. In spite of the fact that our module is infinite-dimensional it is the most natural and simplest representation of the Monster.

We hope that it will now be possible to develop a unified theory of the finite simple groups, based on lattices and vertex operators. In such a theory, H. Garland's arithmetic theory of loop algebras and groups ([9], [10]), as simplified and generalized in [20], [26], would probably be important.

1. LIE ALGEBRAS OF TYPE A, D, E

We first recall several facts about the simple Lie algebras $\mathfrak{g}$ of type A, D, E (that is, type A_ℓ, D_ℓ, E_6, E_7 or E_8). Such an algebra $\mathfrak{g}$ has a root space decomposition with respect to a Cartan subalgebra $\mathfrak{h}$:

$$(1.1)\qquad \mathfrak{g} = \mathfrak{h} \oplus \bigoplus_{\alpha\in\Delta} \mathbb{C}x_\alpha.$$

Let $\langle\cdot,\cdot\rangle$ be a nonsingular symmetric invariant bilinear form on $\mathfrak{g}$. We identify $\mathfrak{g}$ with its dual $\mathfrak{g}^*$ and $\mathfrak{h}$ with $\mathfrak{h}^*$ by means of this form and we normalize the form so that $\langle\alpha,\alpha\rangle = 2$ for $\alpha\in\Delta$. Let $Q = \mathbb{Z}\Delta$ be the root lattice of $\mathfrak{g}$. There exists a bilinear map

(1.2) $\epsilon: Q \times Q \longrightarrow <\pm 1>$

(the 2-element group $<\pm 1>$ being written multiplicatively) with the property

(1.3) $\epsilon(\alpha,\alpha) = (-1)^{<\alpha,\alpha>/2}$ for $\alpha \in Q$.

The basis in (1.1) can be chosen to satisfy the bracket relations

(1.4) $$[x_\alpha, x_\beta] = \begin{cases} 0 & \text{if } <\alpha,\beta> \geqslant 0 \\ \epsilon(\alpha,\beta)x_{\alpha+\beta} & \text{if } <\alpha,\beta> = -1 \\ -\alpha & \text{if } <\alpha,\beta> = -2. \end{cases}$$

(See [6], [22].)

For many questions it is convenient to consider another basis of $\mathfrak{g}$: We set

(1.5) $k_\alpha = x_\alpha + x_{-\alpha}$, $p_\alpha = x_\alpha - x_{-\alpha}$.

Then we have

(1.6) $\mathfrak{g} = \mathfrak{k} \oplus \mathfrak{p}$,

where

(1.7) $\mathfrak{k} = \bigoplus_{\alpha\in\Delta_+} \mathbb{C}(x_\alpha + x_{-\alpha})$, $\mathfrak{p} = \mathfrak{h} \oplus \bigoplus_{\alpha\in\Delta_+} \mathbb{C}(x_\alpha - x_{-\alpha})$,

Δ_+ being a system of positive roots. This decomposition is fundamental in the theory of compact forms of semisimple Lie algebras.

2. AFFINE LIE ALGEBRAS AND VERTEX REPRESENTATIONS

We have two canonical affine Lie algebras associated with the simple Lie algebra $\mathfrak{g}$ and its decomposition (1.6):

(2.1) $\hat{\mathfrak{g}}_{\mathbf{Z}} = \mathfrak{k} \otimes \mathbb{C}[t,t^{-1}] \oplus \mathfrak{p} \otimes \mathbb{C}[t,t^{-1}] \oplus \mathbb{C}c,$

(2.2) $\hat{\mathfrak{g}}_{\mathbf{Z}+1/2} = \mathfrak{k} \otimes \mathbb{C}[t,t^{-1}] \oplus \mathfrak{p} \otimes t^{1/2}\mathbb{C}[t,t^{-1}] \oplus \mathbb{C}c$

(t being a formal variable), with the Lie bracket

(2.3) $[x\otimes t^m, y\otimes t^n] = [x,y] \otimes t^{m+n} + \langle x,y\rangle m\delta_{m+n,0}c,$

where x,y belong to $\mathfrak{k}$ or $\mathfrak{p}$ and m,n belong to $\mathbf{Z}$ or $\mathbf{Z} + \frac{1}{2}$ appropriately. We shall sometimes denote $x \otimes t^m$ by $x(m)$. We remark that the algebras $\hat{\mathfrak{g}}_{\mathbf{Z}}$ and $\hat{\mathfrak{g}}_{\mathbf{Z}+1/2}$ are isomorphic if and only if the involution defining (1.6) is inner. (We shall not need this fact.) These algebras are $\frac{1}{2}$ $\mathbf{Z}$-graded by the conditions

$$\deg x\otimes t^m = m$$
$$\deg c = 0.$$

Now we recall the construction of the vertex representations of the Lie algebras $\hat{\mathfrak{g}}_{\mathbf{Z}}$ and $\hat{\mathfrak{g}}_{\mathbf{Z}+1/2}$ (see [6], [7], [22]). Let $\hat{Q}$ be the central extension of Q by the group $\langle\pm1\rangle$ given by the bilinear cocycle ϵ. Note that ϵ is trivial on $2Q \times Q$ and $Q \times 2Q$. Define

$$\bar{Q} = Q/2Q,$$

and let $\bar{Q}^\wedge$ be the group $\hat{Q}/2Q$, or equivalently, the extension of $\bar{Q}$ by $\langle\pm1\rangle$ with cocycle given by ϵ. Let $\mathbb{C}[Q]$ be the group algebra of Q with the natural basis $\{e^\alpha \mid \alpha\in Q\}$, and let T be a faithful irreducible representation of $\bar{Q}^\wedge$. Define

(2.4) $e_\alpha: \mathbb{C}[Q] \longrightarrow \mathbb{C}[Q], \ \alpha\in Q$

(2.5) $e_\alpha \cdot e^\beta = \epsilon(\alpha,\beta)e^{\alpha+\beta}$

for $\beta\in Q$. Then

$$(2.6)\quad e_\alpha e_\beta = \epsilon(\alpha,\beta)e_{\alpha+\beta},\ \alpha,\beta\in Q.$$

One also has operators

$$(2.7)\quad e'_\alpha\colon T \longrightarrow T,\ \alpha\in Q$$

so that

$$(2.8)\quad \begin{aligned} e'_\alpha &= e'_{-\alpha},\ \alpha\in Q,\\ e'_\alpha e'_\beta &= \epsilon(\alpha,\beta)e'_{\alpha+\beta},\ \alpha,\beta\in Q.\end{aligned}$$

The vertex representations of $\hat{g}_{\mathbf{Z}}$ and $\hat{g}_{\mathbf{Z}+1/2}$ are constructed in the spaces

$$(2.9)\quad V_{\mathbf{Z}} = S(\hat{h}^-_{\mathbf{Z}}) \otimes \mathbb{C}[Q], \qquad \hat{h}^-_{\mathbf{Z}} = h \otimes t^{-1}\mathbb{C}[t^{-1}]$$

$$(2.10)\quad V_{\mathbf{Z}+1/2} = S(\hat{h}^-_{\mathbf{Z}+1/2}) \otimes T,\ \hat{h}^-_{\mathbf{Z}+1/2} = h \otimes t^{-1/2}\mathbb{C}[t^{-1}]$$

by means of the vertex operators

$$(2.11)\quad X_{\mathbf{Z}}(\alpha,z) = z^{\langle\alpha,\alpha\rangle/2} \exp\Big(\sum_{\substack{n\in\mathbf{Z}\\ n>0}} \frac{\alpha(-n)z^n}{n}\Big)\, z^{\alpha(0)} \cdot$$
$$\cdot \exp\Big(-\sum_{\substack{n\in\mathbf{Z}\\ n>0}} \frac{\alpha(n)z^{-n}}{n}\Big)e_\alpha$$

$$(2.12)\quad X_{\mathbf{Z}+1/2}(\alpha,z) = 2^{-\langle\alpha,\alpha\rangle} \exp\Big(\sum_{\substack{n\in\mathbf{Z}+1/2\\ n>0}} \frac{\alpha(-n)z^n}{n}\Big) \cdot$$
$$\cdot \exp\Big(-\sum_{\substack{n\in\mathbf{Z}+1/2\\ n>0}} \frac{\alpha(n)z^{-n}}{n}\Big)\, e'_\alpha$$

for $\alpha\in Q$. Here the Heisenberg subalgebras

$$\hat{h}_{\mathbf{Z}} = h \otimes (\bigoplus_{\substack{n\in\mathbf{Z}\\ n\neq 0}} \mathbb{C}t^n) \oplus \mathbb{C}c,$$

$$\hat{h}_{\mathbf{Z}+1/2} = h \otimes t^{1/2}\mathbb{C}[t,t^{-1}] \oplus \mathbb{C}c$$

of $\hat{g}_{\mathbf{Z}}$ and $\hat{g}_{\mathbf{Z}+1/2}$, respectively, act irreducibly on the symmetric algebras $S(\hat{h}^-_{\mathbf{Z}})$ and $S(\hat{h}^-_{\mathbf{Z}+1/2})$, respectively, in the standard way, with c acting as the identity operator. The Fourier components of these vertex operators together with the Heisenberg algebras provide representations $\pi_{\mathbf{Z}}$ and $\pi_{\mathbf{Z}+1/2}$ of the Lie algebras $\hat{g}_{\mathbf{Z}}$ and $\hat{g}_{\mathbf{Z}+1/2}$, respectively. In particular, one has for $\alpha\in\Delta$

(2.13) $$\sum_{n\in\mathbf{Z}} \pi_Z(k_\alpha(n))z^{-n} = X_Z(\alpha,z) + X_Z(-\alpha,z)$$

(2.14) $$\sum_{n\in Z} \pi_Z(p_\alpha(n))z^{-n} = X_Z(\alpha,z) - X_Z(-\alpha,z)$$

where Z denotes $\mathbf{Z}$ or $\mathbf{Z}$+1/2. Both vertex representations are irreducible.

The spaces (2.9), (2.10) are $\frac{1}{2}\mathbf{Z}$-graded by setting

$$\deg\ (S(\hat{h}^-_{\mathbf{Z}})_n \otimes e^\alpha) = n - \frac{1}{2}\langle\alpha,\alpha\rangle$$

$$\deg\ (S(\hat{h}^-_{\mathbf{Z}+1/2})_n \otimes T) = n$$

for $n\in\frac{1}{2}\mathbf{Z}$, $\alpha\in Q$; here the subscripts n denote the obvious homogeneous subspaces. For $n\in\frac{1}{2}\mathbf{Z}$, $h\in h$ and $\alpha\in Q$, the operators $h(n)$, $k_\alpha(n)$ and $p_\alpha(n)$ all have degree n on both spaces (2.9), (2.10).

We can generalize the "homogeneous" construction (2.9), (2.11) by replacing Q in (2.9) by any coset of Q in the weight lattice (the dual lattice of Q, in our A, D, E situation). We can even replace Q in (2.9) by a more general coset Q' of Q in h. However, in this case, the expansion of the vertex operator (2.11) is not given by integral powers of z. In particular, when $\langle Q',Q\rangle = \frac{1}{2}\mathbf{Z}$, the expansion is given by half-integral powers. By choosing an appropriate ϵ, one obtains various constructions of the Lie algebras

(2.1), (2.2), among others. The "twisted" construction (2.10), (2.12) can be generalized analogously.

Note that the subalgebra $\hat{\mathfrak{k}} = \mathfrak{k} \otimes \mathbb{C}[t,t^{-1}] \oplus \mathbb{C}c$ is the same in both $\hat{\mathfrak{g}}_{\mathbf{Z}}$ and $\hat{\mathfrak{g}}_{\mathbf{Z}+1/2}$. We now consider the decomposition of the vertex representations with respect to $\hat{\mathfrak{k}}$. It is easy to see that $V_{\mathbf{Z}}$ and $V_{\mathbf{Z}+1/2}$ are no longer irreducible. Let

(2.15) $\theta_Z: V_Z \longrightarrow V_Z$

be the involution such that

(2.16) $\theta_Z(h(n))\theta_Z^{-1} = -h(n),\ n\in Z,\ h\in \mathfrak{h}$

(2.17) $\theta_{\mathbf{Z}}(e^{\alpha}) = e^{-\alpha},\ \alpha\in Q$

(2.18) $\theta_{\mathbf{Z}+1/2}(t) = t,\ t\in T.$

We write

(2.19) $V_Z = V_Z^{+} \oplus V_Z^{-},$

where $V_Z^{\pm}$ are the eigenspaces of θ_Z corresponding to the eigenvalues ± 1. These $\hat{\mathfrak{k}}$-subrepresentations are decomposable in general. However, in the most important special case, when $\mathfrak{g}$ is of type E_8, they are irreducible.

3. CROSS PRODUCT ALGEBRAS AND THEIR REPRESENTATIONS

The root lattice Q of a simple Lie algebra of type A, D, E is *integral* in the sense that

$$\langle\alpha,\beta\rangle\in\mathbf{Z},\quad \alpha,\beta\in Q$$

and *even* in the sense that

$$\langle\alpha,\alpha\rangle\in 2\mathbf{Z},\quad \alpha\in Q.$$

A lattice L spanning $\mathfrak{h}$ is *unimodular* if it coincides with its $\mathbf{Z}$-dual

$$\{\alpha\in\mathfrak{h} \mid \langle\alpha,L\rangle\subset\mathbf{Z}\}.$$

Even unimodular lattices can exist only in dimensions divisible by 8. In dimension 8, there is only one such lattice, up to isometry -- the E_8-lattice. We shall denote this lattice by Γ.

We would like to note that the above vertex operator constructions remain valid for an arbitrary even (necessarily integral) lattice L spanning $\mathfrak{h}$. In particular, the spaces V_Z and the operators $X_Z(\alpha,z)$ are well defined. For $n\in\mathbf{Z}$, we set

$$L_n = \{\alpha\in L \mid \langle\alpha,\alpha\rangle=n\}.$$

If rank $\mathbf{Z}L_2$ = rank L, then V_Z decomposes into finitely many irreducible components with respect to $\hat{\mathfrak{g}}_Z$. It can happen however that L_2 is empty. The smallest example of an even unimodular lattice of this kind is the rank 24 Leech lattice Λ, which is in fact characterized by these properties (see [3], [17]). The Leech lattice gives the densest sphere packing in 24 dimensions. In the absence of an affine Lie algebra, we would still like to get some kind of vertex operator algebra from Λ.

The shortest nonzero vectors of Λ are those in Λ_4, but contrary to the root lattice case, the corresponding vertex operators generate an infinite family of vertex operators. Nevertheless, there is a way to stop the generation of new vertex operators. The resulting algebra will not be a Lie algebra but a commutative nonassociative algebra. We proceed now to its description.

Let $\mathfrak{g}$ be a commutative nonassociative algebra with a symmetric bilinear form $\langle\cdot,\cdot\rangle$, associative in the sense that

(3.1) $\langle x\times y,z\rangle = \langle x,y\times z\rangle,\quad x,y,z\in\mathfrak{g},$

where $\times$ denotes the product in $\mathfrak{g}$. We define

(3.2) $\hat{g} = \hat{g}_{\mathbb{Z}} = g \otimes \mathbb{C}[t,t^{-1}] \oplus \mathbb{C}c,$

an "affinization" of g, with the new commutative nonassociative product $\times$ determined by

$$(x\otimes t^m) \times (y\otimes t^n) = (x\times y) \otimes t^{m+n} + \langle x,y\rangle m^2 \delta_{m+n,0}c$$

(3.3)

$$c \times c = (x\otimes t^m) \times c = 0$$

where $x,y\in g$, $m,n\in\mathbb{Z}$. We shall also denote $x\otimes t^m$ by $x(m)$. We call a linear map π: $\hat{g}_{\mathbb{Z}} \longrightarrow \text{End } V$ a *representation of* $\hat{g}_{\mathbb{Z}}$ (and V a $\hat{g}_{\mathbb{Z}}$-*module*) if

(3.4) $$\pi(x(m)\times y(n)) = \\ = \frac{1}{2}\left([\pi(x(m+1)),\pi(y(n-1))] + [\pi(y(n+1)),\pi(x(m-1))]\right)$$

for $x,y\in g$, $m,n\in\mathbb{Z}$. We call the right-hand side a "cross bracket" because it is made up of two brackets that "cross". If V is of the form $\oplus_{n\in\mathbb{Z}} V_n$ and if $\pi(x(n))$ is homogeneous of degree n for all $x\in g$, $n\in\mathbb{Z}$, we say that V is a *graded* module.

To get a complete analogy with the Lie algebra case we consider g with an involution preserving the product $\times$. Let

(3.5) $g = \mathfrak{k} \oplus p$

be the decomposition with respect to the involution. We define another commutative algebra

(3.6) $\hat{g}_{\mathbb{Z}+1/2} = \mathfrak{k}\otimes\mathbb{C}[t,t^{-1}] \oplus p\otimes t^{1/2}\mathbb{C}[t,t^{-1}] \oplus \mathbb{C}c$

with the product $\times$ again given by (3.3), for $x,y\in\mathfrak{k}$ or p, $m,n\in\mathbb{Z}$ or $\mathbb{Z}+\frac{1}{2}$ appropriately. The notion of representation for $\hat{g}_{\mathbb{Z}+1/2}$ is defined as for $\hat{g}_{\mathbb{Z}}$.

It turns out that for an even lattice L, the vertex operators $X_{\mathbb{Z}}(\alpha,z)$, $\alpha\in L_4$, under the cross-bracket operation (3.4), generate commutative algebras of the types (3.2), (3.6), with a *finite-dimensional* underlying algebra $\mathfrak{g}$. In addition to the vertex operators the algebra $\hat{\mathfrak{g}}_Z$ contains operators which are the Fourier coefficients of operators of the form

(3.7) $:g(z)h(z):,\ g,h\in\mathfrak{h}$, where $g(z)=\Sigma g(n)z^{-n}$

(3.8) $z\frac{d}{dz}h(z),\ h\in\mathfrak{h}$

(3.9) $:X_Z(\alpha,z)h(z):,\ \alpha\in L_2,\ h\in\mathfrak{h}.$

Here the double dots denote the normal ordering operation, determined by the properties

$$:g(m)h(n): = \begin{cases} g(m)h(n) & \text{if } m\leqslant n \\ h(n)g(m) & \text{if } m\geqslant n, \end{cases}$$

$$:h(0)e_\alpha: = \frac{1}{2}(h(0)e_\alpha+e_\alpha h(0))$$

for $g,h\in\mathfrak{h}$, $m,n\in\frac{1}{2}\mathbf{Z}$ and $\alpha\in Q$.

Now take $L = \Lambda$. Defining operators $k_\alpha(n)$, $p_\alpha(n)$ using (2.13), (2.14) for the vertex operators with $\alpha\in\Lambda_4$, one finds a finite-dimensional commutative nonassociative algebra of the form (3.5) with

(3.10) $\mathfrak{k} = S^2(\mathfrak{h}) \oplus \sum_{\alpha\in\Lambda_4} \mathbb{C}k_\alpha$

(3.11) $\mathfrak{p} = \mathfrak{h} \oplus \sum_{\alpha\in\Lambda_4} \mathbb{C}p_\alpha.$

Here $\mathfrak{h}$ corresponds to (3.8) and $S^2(\mathfrak{h})$ to (3.7).

We can already see part of Griess's algebra at this point, represented by cross-brackets on both $V_{\mathbf{Z}}$ and $V_{\mathbf{Z}+1/2}$: The commutative nonassociative algebra $\mathfrak{k}$ is isomorphic to a subalgebra of Griess's algebra.

4. WEYL GROUPS AND THE INVOLUTION CENTRALIZER

When L is an even unimodular lattice of rank 24 (a "Niemeier lattice"), the corresponding commutative nonassociative algebra analogous to (3.5), (3.10), (3.11), has dimension 196884. Moreover, if we translate the $\mathbb{Z}$-gradation of $V_{\mathbb{Z}}$ so that deg $(1\otimes 1) = 1$, then

$$\text{(4.1)} \quad \text{ch } V_{\mathbb{Z}} = q^{-1} + \dim \mathfrak{g} + 196884q + \ldots = J(q) + \dim \mathfrak{g},$$

where $\mathfrak{g}$ is the rank 24 semisimple or abelian Lie algebra corresponding to the root system L_2 and the character of the graded module

$$\text{(4.2)} \quad V = \bigoplus_{n\geqslant -1} V_{-n}$$

is defined by

$$\text{(4.3)} \quad \text{ch } V = \sum_{n\geqslant -1} (\dim V_{-n})\, q^n.$$

(The indices n are the negatives of those used in the introduction.) Note that the character of the homogeneous vertex representation, up to an added constant, is equal to the modular function which is the central issue of the "Monstrous game". Formula (4.1) follows from the fact that

$$\text{(4.4)} \quad \text{ch } V_{\mathbb{Z}} = \frac{\theta_L(q)}{\eta(q)^{24}},$$

which in turn follows from the construction of $V_{\mathbb{Z}}$. Here

$$\text{(4.5)} \quad \theta_L(q) = \sum_{\alpha\in L} q^{\langle\alpha,\alpha\rangle/2}$$

is the θ-function of L, and

$$\text{(4.6)} \quad \eta(q) = q^{1/24} \prod_{n>0} (1-q^n)$$

is Dedekind's η-function.

Unfortunately, though, none of the 24 algebras corresponding to different Niemeier lattices is Griess's algebra, the algebra that we are looking for. To see that the Monster does not belong to Aut $\mathfrak{g}$ we need to study the subgroups of Aut $\mathfrak{g}$.

Let us look first at the case when $\mathfrak{g}$ is a simple Lie algebra of type A, D, E. Set W = Aut Q. Then W can be extended to an automorphism group $\tilde{W}$ of $\mathfrak{g}$ so that $\tilde{W}|_{\mathfrak{h}} = W$. The group $\tilde{W}$ is isomorphic to $2^{\ell}W$, 2^{ℓ} denoting the abelian group $\langle\pm 1\rangle^{\ell} \simeq \bar{Q}$ and $2^{\ell}W$ denoting an extension of W by 2^{ℓ}. For instance, for E_8, W is the Weyl group and $\tilde{W}$ is the normalizer of the 2^8-element group of elements with square 1 in a maximal torus in the corresponding Lie group. When $\mathfrak{g}$ is the above commutative algebra, W and $\tilde{W}$ can be defined analogously. For example, for the Leech lattice we obtain

$$(4.7)\qquad \tilde{W} = 2^{24}(\cdot 0)$$

where $\cdot 0$ is the Conway group, the automorphism group of the Leech lattice (see [2]).

At this point we need to inquire into finite group theory. The latter tells us that there is an important involution centralizer in the Monster which in fact characterizes the Monster among the finite simple groups. This group has the form

$$(4.8)\qquad C = \langle\pm 1\rangle 2^{24}(\cdot 0/\langle\pm 1\rangle) = 2_{+}^{1+24}(\cdot 1).$$

(The subgroup $2_{+}^{1+24} = \langle\pm 1\rangle 2^{24}$ of C is a finite Heisenberg group, called an "extraspecial group" in group theory, and the quotient group $\cdot 1 = \cdot 0/\langle\pm 1\rangle$ of C is a simple group of Conway.) This fact suggests that the Leech lattice is the most appropriate one among the 24 Niemeier lattices for our purposes. However, the discrepancy between (4.7) and (4.8) is very serious, and we need to modify our construction. We note first that though $\tilde{W} \neq C$, their actions on $\mathfrak{k}$ are identical. Moreover, $\mathfrak{k}$ constructed from the Leech lattice (see (3.10)) is isomorphic to the subalgebra of Griess's algebra on which the center of C acts trivially. Thus the problem is to extend the construction of

$\mathfrak{f}$ to the correct $\mathfrak{g}$.

We recall here that the root lattice Γ of E_8 and the Leech lattice Λ are unique in many respects. In particular, Γ is the smallest even unimodular lattice, and Λ is the smallest even unimodular lattice without vectors of square length 2. We have learned to expect many analogous phenomena from these two lattices.

5. THE MOONSHINE SPACE

In the search for the correct commutative algebra we return again to the Lie algebra case. We ask the following natural question: Is there another construction of $\hat{E}_8$ besides the straightforward constructions we considered in Section 2? Keeping in mind that $\hat{\mathfrak{f}}$ is the correct subalgebra in the commutative case, we concentrate our attention on the corresponding subalgebra of $\hat{E}_8$. It is a Lie algebra of type $\hat{D}_8$. Each of the two vertex representations of $\hat{E}_8$ decomposes into two irreducible components with respect to $\hat{D}_8$ as in (2.19). One can easily identify these as follows:

(5.1) $V_{\mathbf{Z}}^{+} \simeq V_{\text{basic}}$

(5.2) $V_{\mathbf{Z}}^{-} \simeq V_{\text{half-spin}}$

(5.3) $V_{\mathbf{Z}+1/2}^{+} \simeq V_{\text{natural}}$

(5.4) $V_{\mathbf{Z}+1/2}^{-} \simeq V_{\text{half-spin}}$

Thus we obtain all 4 level 1 standard representations of $\hat{D}_8$. The name of the representation indicates the action of the scalar subalgebra D_8 of $\hat{D}_8$ on the top graded subspace. In the case of (5.1)–(5.4), this subspace has corresponding dimensions 1, 128, 16, 128. The finite-dimensional Lie algebra D_8 has one (up to conjugacy) outer involution, and this can be lifted to $\hat{D}_8$. The induced automorphism of the spin representation of $\hat{D}_8$ interchanges the two half-spin subrepresentations. Thus the two half-spin representations of $\hat{D}_8$ are distinguishable only after we choose a basis of this Lie algebra.

Hence we can construct another Lie algebra of type $\hat{E}_8$ if we interchange the two half-spin representations. This leads us to the definition of the following modified vertex representations of $\hat{\mathfrak{f}}$:

$$(5.5) \quad V_{\mathbf{Z}}^{\natural} = V_{\mathbf{Z}}^{+} \oplus V_{\mathbf{Z}+1/2}^{-}$$

$$(5.6) \quad V_{\mathbf{Z}+1/2}^{\natural} = V_{\mathbf{Z}+1/2}^{+} \oplus V_{\mathbf{Z}}^{-}.$$

Of course, we now lose the explicit construction of the second part of the Lie algebra $\hat{E}_8$. However, what is important for us is that it exists. An immediate corollary of the construction is that the dimensions of the corresponding graded components of $V_{\mathbf{Z}}$ and $V_{\mathbf{Z}}^{\natural}$ are equal. This can be expressed as the equality of the two characters, after we suitably translate the $\frac{1}{2}\mathbf{Z}$-gradations of $V_{\mathbf{Z}}$ and $V_{\mathbf{Z}+1/2}$:

$$(5.7) \quad \mathrm{ch}\ V_{\mathbf{Z}} = \mathrm{ch}\ V_{\mathbf{Z}}^{\natural}.$$

Now let us define $V_{\mathbf{Z}}^{\natural}$ in the case of the Leech lattice and compare its character with the character of $V_{\mathbf{Z}}$. Setting deg $(1\otimes 1) = 1$ and deg $(1\otimes T) = -\frac{1}{2}$, we obtain

$$(5.8) \quad \mathrm{ch}\ V_{\mathbf{Z}}^{\natural} = q^{-1} + 0 + 196884q + \dots = J(q).$$

The constant term in (4.1) has disappeared! This discrepancy in the constant term with ch $V_{\mathbf{Z}}$ shows that the Leech lattice case has new features in comparison with the E_8 case; in particular, the algebra $\mathfrak{g}^{\natural}$ acting in $V_{\mathbf{Z}}^{\natural}$ (if such exists) need not be isomorphic to the original $\mathfrak{g}$ acting on $V_{\mathbf{Z}}$.

The space $V_{\mathbf{Z}}^{\natural}$ will be our natural module for the Monster. (This justifies our use of the symbol $\natural$ (natural).)

We can also check that the group C defined by (4.8) acts naturally on $V_{\mathbf{Z}}^{\natural}$. In fact, one can identify $\tilde{W}$ with a group of automorphisms of the group $\hat{\Lambda}$. This action preserves 2Λ and induces an action on the extraspecial group $\hat{\Lambda}/2\Lambda \simeq 2_{+}^{1+24}$. The extraspecial group acts on T (we denote this action by π), and yields a projective representation of $\tilde{W}$. We denote by $\hat{C}$ the linearization of this action. It has the following presentation:

$$\hat{C} = \{(a,a_T) \mid a \in \tilde{W},\ a_T \in GL(T),\ a = \mathrm{int}(a_T) \text{ on } \pi(\hat{\Lambda}/2\Lambda)\},$$

where $\mathrm{int}(a_T)$ denotes conjugation by a_T. The group $\hat{C}$ is a central extension of $\tilde{W}$ by the subgroup $(\mathrm{Id}, \pm \mathrm{Id}_T)$. The group $\hat{C}$ acts naturally on $V_{\mathbf{Z}} \oplus V_{\mathbf{Z}+1/2}$ as follows: Let $g = (a,a_T) \in \hat{C}$, and require that g fixes $1 \otimes 1 \in V_{\mathbf{Z}}$, $g\big|_{1\otimes T} = 1 \otimes a_T$ and

$$g\alpha(n)g^{-1} = (\bar{a}\alpha)(n)$$
$$ge_\alpha g^{-1} = a(e_\alpha).$$

for $\alpha \in \Lambda$, $n \in \frac{1}{2}\mathbf{Z}$; here $\bar{a}$ is the automorphism of Λ induced by a.

Thus we obtain a natural action of $\hat{C}$ on $V^\natural_{\mathbf{Z}} \oplus V^\natural_{\mathbf{Z}+1/2}$. But again $\hat{C}$ does not act faithfully on the first summand, as the subgroup $\pm(\mathrm{Id}, \mathrm{Id}_T)$ acts trivially there. We denote the factor group by C, which has exactly the structure given by (4.8). Thus the two quite different groups $\tilde{W}$ and C can be viewed as factor groups of one group $\hat{C}$ by two different two-element central subgroups.

6. TRIALITY FOR D_4 AND THE CONSTRUCTION OF $\hat{E}_8$

We proceed now to the construction of $g^\natural$ starting from the E_8 case. As one might expect, the construction of explicit operators from $V^+_{\mathbf{Z}}$ to $V^-_{\mathbf{Z}+1/2}$ and vice versa would be very complicated. However, if we restrict $\hat{D}_8$ to $(D_4 \times D_4)^\wedge$ the principle of triality enters the picture. The decomposition of $V^\natural_{\mathbf{Z}}$ into irreducible components becomes the following:

(6.1) $\quad V^\natural_{\mathbf{Z}} = V_0 \oplus V_1 \oplus V_2 \oplus V_3$

where

(6.2) $\quad V^+_{\mathbf{Z}} = V_0 \oplus V_1,$

(6.3) $V^-_{\mathbb{Z}+1/2} = V_2 \oplus V_3$

and V_0 contains the highest weight vector of $V^+_{\mathbb{Z}}$.

Let U_0, U_1, U_2, U_3 be the four fundamental representations of $\hat{D}_4$ of level 1, with U_0 the basic representation. One can choose bases of the two $\hat{D}_4$'s and the indexing of the U_i's in such a way that

(6.4) $V_i = U_i^1 \otimes U_i^2$, $i = 0,1,2,3$,

where the upper index refers to one of the two algebras $\hat{D}_4$. The representations U_1, U_2, U_3 have the same character. This fact is a corollary of the S_3 symmetry of the Dynkin diagram D_4

(6.5)

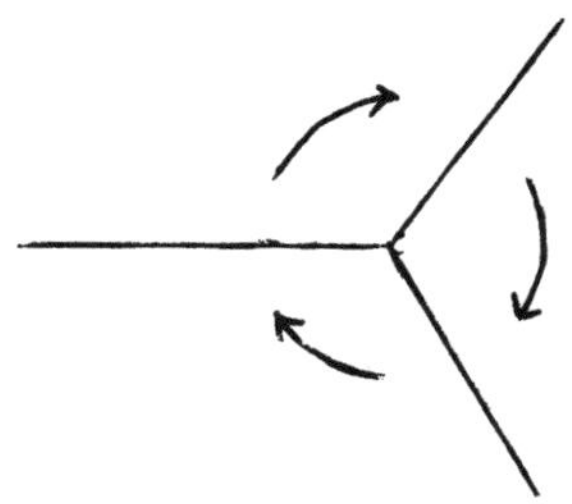

and is equivalent to the "aequatio identica satis abstrusa" of Jacobi

(6.6) $$\frac{1}{2}\left(\prod_{n=1}^{\infty}(1+q^{n-1/2})^8 - \prod_{n=1}^{\infty}(1-q^{n-1/2})^8\right) =$$

$$= 8q^{1/2}\prod_{n=1}^{\infty}(1+q^n)^8,$$

which has already arisen in supersymmetric string models [11][1]. The group S_3 permutes U_1, U_2, U_3, and correspondingly, V_1, V_2, V_3. Let

(6.7) $\mathfrak{k} = \mathfrak{k}_0 \oplus \mathfrak{p}_1$,

where $\mathfrak{k}$ has type D_8, $\mathfrak{k}_0$ has type $D_4 \times D_4$, and $\mathfrak{p}_1$ is the orthogonal complement of $\mathfrak{k}_0$ in $\mathfrak{k}$. Let

---

[1]We thank Bernard Julia for bringing this reference to our attention.

$$\bar{\mathfrak{k}} = \mathfrak{k} \otimes \mathbb{C}[t,t^{-1}], \text{ etc.,}$$

be their affinizations. We define the following operators on $V_{\mathbf{Z}} \oplus V_{\mathbf{Z}+1/2}$:

(6.8)
$$p_2 = \sigma_2 \, p_1 \, \sigma_2^{-1}, \; p_3 = \sigma_3 \, p_1 \, \sigma_3^{-1},$$
$$\bar{p}_2 = \sigma_2 \, \bar{p}_1 \, \sigma_2^{-1}, \; \bar{p}_3 = \sigma_3 \, \bar{p}_1 \, \sigma_3^{-1},$$

where σ_2, σ_3 are two involutions in S_3. Then the spaces of operators

(6.9) $\quad g^{\natural} = \mathfrak{k}_0 \oplus p_1 \oplus p_2 \oplus p_3,$

(6.10) $\quad \hat{g}^{\natural} = \bar{\mathfrak{k}}_0 \oplus \bar{p}_1 \oplus \bar{p}_2 \oplus \bar{p}_3 \oplus \mathbb{C}c$

form Lie algebras of type E_8 and $\hat{E}_8$, respectively.

Now we want to repeat this construction in the Leech lattice case. The first question we need to answer is: What is the analogue of $\mathfrak{k}_0$? To this end we shall look at $\mathfrak{k}_0$ for the E_8 case from a new point of view.

7. LATTICES FROM CODES. DEFINITION OF $\mathfrak{k}_0$

As we have mentioned, the E_8 and Leech lattices are unique in many respects. If we want to look at the more detailed structure of these lattices, we find that they can be built from no less unique binary codes, namely, the Hamming and Golay codes. The Hamming code $\mathcal{C}_\Gamma$ can be characterized as a 4-dimensional subspace of $\mathbf{F}_2^8$ whose smallest nonzero weight (number of nonzero coordinates) is 4. Similarly, the Golay code $\mathcal{C}_\Lambda$ is characterized by the property that it is a 12-dimensional subspace of $\mathbf{F}_2^{24}$ whose smallest zero weight is 8. (These properties characterize the codes up to a permutation of the different factors $\mathbf{F}_2$.) It is easy to see that there are exactly 14 code words of weight 4 in the Hamming code $\mathcal{C}_\Gamma$, and in addition this code contains the trivial code word and the code word of weight 8. Thus the weight distribution is

(7.1) $0^1\ 4^{14}\ 8^1.$

The Golay code $\mathcal{C}_\Lambda$ has the weight distribution

(7.2) $0^1\ 8^{759}\ 12^{2576}\ 16^{759}\ 24^1.$

Now let us consider the "square" lattices $\sum_{i=1}^{8} \mathbf{Z}(\frac{1}{4}\alpha_i)$, $\sum_{i=1}^{24} \mathbf{Z}(\frac{1}{4}\alpha_i)$ with the symmetric form $\langle\cdot,\cdot\rangle$ determined by the property

(7.3) $\langle\alpha_i,\alpha_j\rangle = 2\delta_{ij},$

and their sublattices Γ, Λ spanned by the vectors

(7.4) $\frac{1}{2} \sum_{i=1}^{8} \pm \epsilon_i\alpha_i,\ \epsilon \in \mathcal{C}_\Gamma,$ weight $\epsilon = 4$

(7.5) $\frac{1}{2} \sum_{i=1}^{8} (\epsilon_i - \frac{1}{2})\ \alpha_i \pm \alpha_k,\ \epsilon \in \mathcal{C}_\Gamma,\ k=1,\dots,8$

and respectively by the vectors

(7.6) $\frac{1}{2} \sum_{i=1}^{24} \pm \epsilon_i\alpha_i,\ \epsilon \in \mathcal{C}_\Lambda,$ weight $\epsilon = 8$

(7.7) $\frac{1}{2} \sum_{i=1}^{24} (\epsilon_i - \frac{1}{2})\alpha_i \pm \alpha_k,\ \epsilon \in \mathcal{C}_\Lambda,\ k=1,\dots,24,$

where the number of plus signs in (7.4), (7.6) is even, and the sign in (7.5), (7.7) is chosen in such a way that one gets $\pm \frac{3}{4}\alpha_k$. One can check easily that the lattices Γ and Λ have the characteristic properties of the E_8 and Leech lattices, respectively.

The vectors (7.4), (7.5) have square length 2 and constitute the root system of type E_8. Moreover, the division of the root system into two sets corresponds exactly to the decomposition (6.7), i.e., $\mathfrak{k}_0$ is constructed from the root subsystem (7.4), and $\mathfrak{p}_1$ from (7.5).

Similarly, the vectors (7.6), (7.7) have square length 4. Thus one can define p_1 from (7.7) for the Leech lattice algebra in exactly the same way as for E_8. However, Λ contains additional elements of square length 4:

(7.8) $\pm\alpha_i \pm \alpha_j$, $i,j = 1,\ldots,24$, $i \neq j$.

Thus (7.1) and (7.2) imply in particular that

$$|\Gamma_2| = 240,$$

$$|\Lambda_4| = 196560.$$

Moreover, in addition to the subspace constructed from the vectors (7.6) and (7.8), $\mathfrak{k}_0$ should contain $S^2(\mathfrak{h})$ (see (3.10)). With $\mathfrak{k}_0$ defined this way, we obtain a decomposition of $V_{\mathbf{Z}}^{\natural}$ into four irreducible components (6.1)–(6.3) exactly as in the E_8 case. One can show that V_1, V_2, V_3 have the same character. This suggests the existence of a group S_3 which permutes the V_i and induces a group of automorphisms of $\mathfrak{k}_0$. Precise formulas for the action of a group S_3 on $\mathfrak{k}_0$ were in fact obtained by Griess [15]. However, those calculations do not help us prove the closure of the algebra $\mathfrak{g}^{\natural}$. Thus we need to find the underlying mechanism of triality.

8. TRIALITY FOR $\hat{A}_1$

The principle of triality is very important in the construction of the exceptional Lie and Jordan algebras. It also turns out to be crucial for our construction of the commutative algebra $\mathfrak{g}^{\natural}$. The simplest manifestation of this principle is the existence of a symmetrical basis $\{y_1,y_2,y_3\}$ of $sl_2 = A_1$ satisfying

(8.1) $[y_1,y_2] = 2y_3$, $[y_2,y_3] = 2y_1$, $[y_3,y_1] = 2y_2$.

The group S_3 acts naturally on sl_2 by permuting the three indices of the basis elements. In particular, S_3 contains three involutions

(8.2) $\sigma_i\colon y_i \longleftrightarrow -y_i \qquad \sigma_i\colon y_{i+1} \longmapsto y_{i+2}$, where the indices are thought of as periodic mod 3.

We can associate to sl_2 with the fixed basis four affine Lie algebras as in Section 2:

$$\hat{\alpha}_0 = \sum_{m\in\mathbf{Z}} \mathbb{C}y_1(m) \oplus \sum_{m\in\mathbf{Z}} \mathbb{C}y_2(m) \oplus \sum_{m\in\mathbf{Z}} \mathbb{C}y_3(m) \oplus \mathbb{C}c$$

(8.3) $$\hat{\alpha}_i = \sum_{m\in\mathbf{Z}} \mathbb{C}y_i(m) \oplus \sum_{m\in\mathbf{Z}+1/2} \mathbb{C}y_{i+1}(m) \oplus$$

$$\oplus \sum_{m\in\mathbf{Z}+1/2} \mathbb{C}y_{i+2}(m) \oplus \mathbb{C}c, \quad i=1,2,3.$$

Clearly the action of the group S_3 extends naturally to $\hat{\alpha}_0$, while $\hat{\alpha}_i$ is preserved only by the involution σ_i.

Each of these algebras is of type $\hat{A}_1$ and has two fundamental representations. Both of these admit the vertex operator constructions described in Section 2.

We recall that the vertex operator construction is based on restriction to the Heisenberg subalgebra $\hat{h} = \sum_m \mathbb{C}y_1(m) \oplus \mathbb{C}c$. For $\hat{\alpha}_2$ and $\hat{\alpha}_3$ this restriction is still irreducible and therefore the representation space is isomorphic to $S(\hat{h}^-)$. For $\hat{\alpha}_0$ and $\hat{\alpha}_1$, to make it irreducible, we need to add the translation part of the Weyl group of $\hat{A}_1$. We define vertex operators by the formulas (2.11), (2.12), using a bilinear cocycle on $\mathbf{Z}(\frac{1}{4}\alpha)$ with values in $\langle \pm 1, \pm(-1)^{1/2} \rangle$ satisfying

(8.4) $$\epsilon(\alpha, \tfrac{1}{4}\alpha) = (-1)^{1/2}.$$

This nontrivial choice of ϵ will be justified when we pass to higher rank lattices. The representation spaces have the following form:

(8.5) $$\hat{\alpha}_0\colon\ S(\hat{h}_{\mathbf{Z}}^-) \otimes \mathbb{C}[\mathbf{Z}\alpha \pm \tfrac{\alpha}{4} + \tfrac{\alpha}{4}]$$

(8.6) $$\hat{\alpha}_1\colon\ S(\hat{h}_{\mathbf{Z}}^-) \otimes \mathbb{C}[\mathbf{Z}\alpha \pm \tfrac{\alpha}{4}]$$

(8.7) $\hat{\mathfrak{a}}_2$: $S(\hat{\mathfrak{h}}^-_{\mathbf{Z}+1/2}) \otimes \mathbb{C}e^{\pm\alpha/4+\alpha/4+\mathbf{Z}\alpha}$

(8.8) $\hat{\mathfrak{a}}_3$: $S(\hat{\mathfrak{h}}_{\mathbf{Z}+1/2}) \otimes \mathbb{C}e^{\pm\alpha/4+\mathbf{Z}\alpha}$

where the one-dimensional factors in (8.7), (8.8) are denoted as they are for the sake of parallel definitions of e_α and e'_α on the second tensor factors. (The notations $\mathbb{C}[\mathbf{Z}\alpha\pm\frac{\alpha}{4}]$, etc., denote the obvious subspaces of the group algebra $\mathbb{C}[\frac{1}{4}\mathbf{Z}\alpha]$.)

Although S_3 does not act on each of the $\hat{\mathfrak{a}}_i$ separately it permutes them and therefore their vertex representations. For example, there is a unique isomorphism[2]

(8.9) σ_2: $S(\hat{\mathfrak{h}}^-_{\mathbf{Z}}) \otimes \mathbb{C}[\mathbf{Z}\alpha\pm\frac{\alpha}{4}] \longrightarrow S(\hat{\mathfrak{h}}^-_{\mathbf{Z}+1/2}) \otimes \mathbb{C}e^{\pm\alpha/4+\mathbf{Z}\alpha}$

such that

$$\text{(8.10)}\quad \begin{aligned} &\sigma_2: 1 \otimes e^{\pm\alpha/4} \mapsto 1 \otimes e^{\pm\alpha/4+\mathbf{Z}\alpha} \\ &\sigma_2 \circ x(m) \circ \sigma_2^{-1} = (\sigma_2 x)(m),\ x = y_1,\ y_2 \text{ or } y_3,\ m\in\tfrac{1}{2}\mathbf{Z}. \end{aligned}$$

It is interesting to note that comparison of the characters of the spaces in (8.9) gives a standard consequence of the Jacobi triple product identity:

$$\text{(8.11)}\quad \frac{\sum_{n\in\mathbf{Z}} q^{(n+1/4)^2}}{\eta(q)} = \frac{\eta(q)}{\eta(q^{1/2})}.$$

---

[2]This isomorphism was also discovered in a different context by R. Pfister [21].

9. TRIALITY FOR HIGHER RANK

Now we have the main tool for obtaining triality in our two cases of the E_8 and Leech lattices. The basic idea is to take the product of 8 or 24 copies of the affine Lie algebras of type $\hat{A}_1$ considered in the last section. Since the two cases are completely analogous, we shall explain triality for our main example of the Leech lattice. We first note that although the Leech lattice Λ has no elements of norm square 2, its double cover

$$(9.1) \qquad \tilde{\Lambda} = \mathbb{Z}\text{-span } (\Lambda \cup \{\alpha_k\})$$

has 48 such elements, which correspond to the root system of type $(A_1)^{24}$. Let

$$(9.2) \qquad Q = \sum_{i=1}^{24} \mathbb{Z}\alpha_i$$

and let N be the unique even unimodular sublattice of $\tilde{\Lambda}$ containing Q; N is a Niemeier lattice of type A_1^{24}. Then

$$(9.3) \qquad \tilde{\Lambda} = N \cup N'$$

where

$$(9.4) \qquad N' = N + \alpha',$$

$$(9.5) \qquad \alpha' = \frac{1}{4}\sum_{i=1}^{24} \alpha_i.$$

We would like to define a bilinear cocycle ϵ on $\tilde{\Lambda}$ which has the properties (1.2), (1.3) for Λ and N. This forces us to use a cocycle with values in $<\pm 1, \pm(-1)^{1/2}>$. It will be convenient to have

$$(9.6) \qquad \epsilon(\alpha, \beta+\gamma) = \epsilon(\alpha, \beta) \text{ for } \alpha, \beta \in \tilde{\Lambda},\ \gamma \in Q$$

$$(9.7) \qquad \epsilon(\alpha_k, \alpha') = (-1)^{1/2} \text{ for } k=1,\dots,24.$$

Formula (9.6) will enable us to define explicit operators e'_α (cf. (2.7)) on the spaces $\mathbb{C}[N/Q]$ and $\mathbb{C}[N'/Q]$ below ((9.11), (9.12)). A cocycle with these properties can in fact be constructed (see [8]).

Now we consider the space

$$W = W_0 \oplus W_1 \oplus W_2 \oplus W_3 \tag{9.8}$$

where

$$W_0 = S(\hat{h}^-_{\mathbb{Z}}) \otimes \mathbb{C}[N] \tag{9.9}$$

$$W_1 = S(\hat{h}^-_{\mathbb{Z}}) \otimes \mathbb{C}[N'] \tag{9.10}$$

$$W_2 = S(\hat{h}^-_{\mathbb{Z}+1/2}) \otimes \mathbb{C}[N/Q] \tag{9.11}$$

$$W_3 = S(\hat{h}^-_{\mathbb{Z}+1/2}) \otimes \mathbb{C}[N'/Q]. \tag{9.12}$$

One can see immediately that the Fourier components of the vertex operators $X(\pm\alpha_k,z)$ acting in W_i generate a copy of the algebra $\hat{\mathfrak{a}}_i$ (8.3), i=0,1,2,3. The 24 algebras of type $\hat{A}_1$, one for each k=1,...,24, commute pairwise, so that we have algebras of type $\hat{A}_1^{24}$ acting on the four pieces. Under these algebras each piece decomposes into 2^{12} irreducible components, in the first two according to the cosets of Q in N and N', respectively, and in the last two according to the natural basis of the second factor in the tensor product.

The highest weight vectors are the vectors

$$1 \otimes e^{\alpha_C/2}, \qquad C \in \mathcal{C}_\Lambda \tag{9.13}$$

$$1 \otimes e^{\beta}, \quad \beta \text{ of the form } \frac{\pm\alpha_1 \pm \cdots \pm \alpha_{24}}{4} \tag{9.14}$$

$$1 \otimes e^{\alpha_C/2+Q}, \qquad C \in \mathcal{C}_\Lambda \tag{9.15}$$

$$1 \otimes e^{\beta+Q}, \quad \beta \text{ of the form } \frac{\pm\alpha_1 \pm \cdots \pm \alpha_{24}}{4} \tag{9.16}$$

where $\alpha_C = \sum_i \epsilon_i \alpha_i$ if $C = (\epsilon_i)_i$, $\epsilon_i \in \{0,1\}$.

Taking a tensor product of 24 copies of the spaces described in Section 8 we obtain the action of the triality group S_3 on W, which permutes the W_i, i=1,2,3. For example, there is a unique isomorphism

$$(9.17)\quad \sigma_2\colon S(\hat{h}^-_{\mathbf{Z}}) \otimes \mathbb{C}[N'] \longrightarrow S(\hat{h}^-_{\mathbf{Z}+1/2}) \otimes \mathbb{C}[N'/Q]$$

such that

$$(9.18)\quad \begin{aligned} &\sigma_2\colon 1\otimes e^{\beta} \longmapsto 1\otimes e^{\beta+Q} \text{ for } \beta \text{ of the form } \frac{\pm\alpha_1\pm\cdots\pm\alpha_{24}}{4} \\ &\sigma_2 \circ x(m) \circ \sigma_2^{-1} = (\sigma_2 x)(m) \end{aligned}$$

for $x = y_1$, y_2 or y_3 in one of the 24 $\hat{A}_1$s, and $m \in \frac{1}{2}\mathbf{Z}$.

Now let us consider products

$$(9.19)\quad y_i^{(r)}(z) y_i^{(s)}(z), \qquad i=1,2,3,$$

where r,s=1,...,24, r≠s; here each superscript designates a copy of $\hat{A}_1$. On each space W_j, each of the two factors in (9.19) has an expansion into powers z^n, $n \in Z$ ($Z=\mathbf{Z}$ or $\mathbf{Z}+\frac{1}{2}$). Thus (9.19) can always be expanded on W in integral powers of z. Each element of the triality group S_3 acting on W transforms each operator (9.19), by conjugation, into another such operator with the same r,s. When r=s, we must replace (9.19) by the corresponding normal ordered product (defined with respect to the Heisenberg algebra associated with the index i), which becomes the generating function of the Virasoro algebra associated with the corresponding copy of $\hat{A}_1$, and which is independent of i. Triality acts trivially on these operators.

For r,s = 1,...,24 and i=1,2,3, consider the expansion

$$(9.20)\quad :y_i^{(r)}(z) y_i^{(s)}(z): = \sum_{n\in\mathbf{Z}} y_i^{(r,s)}(n) z^{-n}.$$

Then all the operators $y_i^{(r,s)}(n)$, together with the identity operator, span a commutative algebra $\hat{\mathfrak{f}}_*$ under the cross-bracket operation, where

(9.21) $\mathfrak{f}_* = S^2(\mathfrak{h}) \oplus \Sigma\mathbb{C}k_\alpha,$

α ranging through the elements (7.8) (see (3.2), (3.3), (3.4)). Note that

(9.22) $S^2(\mathfrak{h}) = \sum_{r,s=1}^{24} \mathbb{C}\alpha_r\alpha_s$

breaks into the 24-dimensional space where $r=s$, corresponding to 24 Virasoro algebras, and the 276-dimensional space where $r\neq s$, corresponding to (9.20) for $i=1$. Also, the second summand in (9.21) breaks up as

(9.23) $\bigoplus_{r<s} \mathbb{C}(k_{\alpha_r+\alpha_s} + k_{\alpha_r-\alpha_s}) \oplus \bigoplus_{r<s} \mathbb{C}(k_{\alpha_r+\alpha_s} - k_{\alpha_r-\alpha_s}),$

two 276-dimensional spaces corresponding to (9.20) for $i=2,3$.

Thus we see that $\mathfrak{f}_*$ and $\hat{\mathfrak{f}}_*$ are preserved by the triality group S_3.

The structure $\mathfrak{f}_*$ does not exist in the case of the E_8-lattice Γ, and in fact will turn out to be the only difference between the algebras corresponding to Γ and Λ.

10. TRIALITY FOR VERTEX OPERATORS

Now we consider the vertex operators $X_Z(\alpha,z)$, $Z=\mathbb{Z}$ or $\mathbb{Z}+\frac{1}{2}$, for $\alpha\in N$. These operators preserve each space $W_0,\ldots,W_3$ (see (9.9)–(9.12)). The triality group S_3 acting on W transforms each of these operators by conjugation. Since there are explicit formulas for the action of S_3 on W (cf. Section 9), we can in principle compute these transforms. We are interested in particular in the vertex operators $X_Z(\alpha,z)$, α as in (7.6), because the expansion coefficients of

(10.1) $X_Z(\alpha,z) + X_Z(-\alpha,z),$

together with the operators corresponding to $\hat{\mathfrak{f}}_*$ (see (9.21)) span the subalgebra $\hat{\mathfrak{k}}_0$. It is an important fact that the family of operators $\hat{\mathfrak{k}}_0$ is preserved by triality.

Recall that

$$(10.2)\quad \sigma_2 \colon W_1 \longrightarrow W_3$$

in (9.17) induces the maps σ_2 for $s\ell_2$ (8.2) and σ_2 for $\hat{A}_1$ (8.9) (see (8.10), (9.18)). First we determine how the individual vertex operators $X_Z(\alpha,z)$, α as in (7.6), acting on W_1 (where $Z=\mathbb{Z}$) and W_3 (where $Z=\mathbb{Z}+\frac{1}{2}$), transform under σ_2. From the standard calculus of vertex operators, it follows easily that if we bracket an element of any one of the 24 $\hat{A}_1$'s with an operator $X_Z(\alpha,z)$, we obtain a linear combination of such operators. These commutation relations show that $\sigma_2\, X(\alpha,z)\sigma_2^{-1}$ is a linear combination of vertex operators of the same type, and the coefficients are uniquely determined. More precisely,

$$(10.3)\quad \sigma_2(-i)^{|S|}X(\epsilon_S\tfrac{1}{2}\alpha_C,z)\sigma_2^{-1} =$$

$$=2^{-|C|/2}\sum_{T\subset C}(-1)^{|S\cap T|}(-i)^{|T|}X(\epsilon_T\tfrac{1}{2}\alpha_C,z),$$

for $S\subset C$, $C\in\mathcal{C}_\Lambda$, where $i=(-1)^{1/2}$, α_C is as in Section 9 and ϵ_S is the automorphism of $\mathfrak{h}$ which changes the sign of those α_k for which $k\in S$; here we view the code words in $\mathcal{C}_\Lambda$ as subsets of $\{1,\ldots,24\}$.

So far, σ_2 has been defined only on W_1 and W_3. Postulating that (10.3) should hold on all of W, we are able (uniquely) to extend σ_2 to an involution σ_2 on W. This involution, however, does not preserve the subspace $V_{\mathbb{Z}}^{\natural}$ of W, and so we modify σ_2 slightly on one of the 24 coordinates to obtain an involution σ of W which does preserve $V_{\mathbb{Z}}^{\natural}$ (see [8]). Combining this with a suitable involution in C (see [8]), we obtain the triality group -- a copy of S_3 acting on $V_{\mathbb{Z}}^{\natural}$, and by conjugation, on the linear combinations of the vertex operators of the type $X(\alpha,z)$, α as in (7.6). The main content of this "triality for vertex operators" is the conjugation formula (10.3).

Now we recall that $\hat{\mathfrak{k}}_0$ is made up of $\hat{\mathfrak{k}}_*$ and the Fourier coefficients of the operators (10.1) for α as in (7.6). Combining the triality of $\hat{\mathfrak{f}}_*$ (see Section 9) with the triality for vertex operators, we obtain an action of S_3 on $\hat{\mathfrak{f}}_0$.

If we replace Λ by Γ and the Golay code $\mathcal{C}_\Lambda$ by the Hamming code $\mathcal{C}_\Gamma$, then we similarly obtain a triality for the corresponding vertex operators. The resulting triality of

$$\hat{\mathfrak{f}}_0 = (D_4 \times D_4)^\wedge$$

agrees with the usual Dynkin diagram triality discussed in Section 6.

This completes our discussion of triality. It is interesting to note that the principle of triality, which plays an important role in the construction of the exceptional Lie algebras, is applicable in a more complicated "sporadic" situation.

Using an analogue of (6.8), we obtain spaces of operators

$$(10.4)\quad \mathfrak{B} = \mathfrak{f}_0 \oplus \mathfrak{p}_1 \oplus \mathfrak{p}_2 \oplus \mathfrak{p}_3,$$

$$(10.5)\quad \hat{\mathfrak{B}} = \bar{\mathfrak{f}}_0 \oplus \bar{\mathfrak{p}}_1 \oplus \bar{\mathfrak{p}}_2 \oplus \bar{\mathfrak{p}}_3 \oplus \mathbb{C}c,$$

analogous to (6.9) and (6.10), respectively. However, in order to show that $\mathfrak{B}$ and $\hat{\mathfrak{B}}$ are algebras, we still must show

$$(10.6)\quad \bar{\mathfrak{p}}_1 \times \bar{\mathfrak{p}}_2 \subset \bar{\mathfrak{p}}_3$$

and its cyclic permutations (see (3.4)). This cannot be proved by direct calculations since the operators involved act from $V_{\mathbb{Z}}^+$ to $V_{\mathbb{Z}+1/2}^-$ and vice versa; also, triality is not helpful here. In order to complete the proof we need to extend our space further. In particular, we need to consider a quadruple cover of the Leech lattice which contains the double cover already used. This quadruple cover arises naturally in the quaternionic construction of the Leech lattice, and we proceed to its definition.

11. QUATERNIONIC CONSTRUCTIONS OF LATTICES

The quaternionic constructions of the Leech and E_8 lattices [25] are in many respects analogous to their constructions from the binary codes. Now we consider codes over the field of four elements $\mathbf{F}_4$. Again there are two remarkable self-dual codes over $\mathbf{F}_4$ in dimensions two and six, which we shall denote by $\mathcal{C}_\Gamma(\mathbf{F}_4)$ and $\mathcal{C}_\Lambda(\mathbf{F}_4)$, respectively. The first is the unique self-dual code in dimension two, and the second is characterized by the property that the shortest word has 4 nonzero coordinates. In the division ring of quaternions, we set

(11.1) $P_4 = \mathbf{Z}[i,j,k,\frac{1}{2}(1+i+j+k)]$

(11.2) $Q_4 = \{x \in \mathbf{Z}[i,j,k] \mid x\bar{x} \in 2\mathbf{Z}\}$.

Then obviously we have

(11.3) $P_4/Q_4 \cong \mathbf{F}_4$,

where we choose the following representatives of the elements of $\mathbf{F}_4$ in P_4:

(11.4) $0, 1, \alpha = \frac{1}{2}(1+i+j+k),\ \alpha^2 = \frac{1}{2}(-1+i+j+k)$.

We also denote by θ_4 an element of Q_4 such that the ideal generated by θ_4 in P_4 is Q_4, e.g., $\theta_4 = i + j$, and set $\vec{a} = (a,a,\dots,a)$ for $a \in P_4$. For $a,b \in \mathbf{F}_4$ and for $L=\Gamma$ or Λ, set

(11.5) $L^{ab} = \{\vec{a} + \theta_4 c + 2x \mid c \in \mathcal{C}_L(\mathbf{F}_4), x \in P_4^\ell, \sum_{i=1}^{\ell} x_i \equiv b \pmod{Q_4}\}$.

In particular, when $L = \Lambda$, $\ell = 6$ and we have

(11.6) $N(D_4^6) = \Lambda^{00} \cup \Lambda^{01} \cup \Lambda^{0\alpha} \cup \Lambda^{0\alpha^2}$

(11.7) $\Lambda = \Lambda^{00} \cup \Lambda^{1\alpha} \cup \Lambda^{\alpha\alpha^2} \cup \Lambda^{\alpha^2 1}$

(11.8) $\bar{\Lambda} = \Lambda^{00} \cup \Lambda^{1\alpha^2} \cup \Lambda^{\alpha 1} \cup \Lambda^{\alpha^2\alpha}$, a Niemeier lattice of type D_4^6, a Leech lattice, and its conjugate, respectively. We consider the following quadruple cover of the Leech lattice:

$$(11.9) \quad \tilde{\bar{\Lambda}} = \Lambda + N(D_4^6) = \bigcup_{a,b\in \mathbb{F}_4} \Lambda^{ab}.$$

12. $L_3(2)$ SYMMETRY

The construction of the Leech lattice and its double cover from the binary Golay code allowed us to introduce the action of 24 copies of $(sl_2)^\wedge$ in the extended vertex representation. Similarly the quaternionic construction of the Leech lattice and its quadruple cover give us an action of 6 copies of $\hat{D}_4$. The analogue of triality for sl_2 is now "$L_3(2)$-symmetry" of D_4. By $L_3(2)$ we mean the group of invertible 3×3 matrices over the 2-element field $\mathbb{F}_2$.

The simplest way to define this symmetry is to build it from two trialities of D_4: one of these trialities is induced from four sl_2 trialities (see Section 8) inside D_4, and the second follows from the symmetry of the Dynkin diagram (see Section 6). In more detail, fixing a Cartan subalgebra $\mathfrak{h}$ of D_4, we have a natural decomposition of the root system,

$$(12.1) \quad \Delta(D_4) = \Delta^{(1)} \cup \Delta^{(2)} \cup \Delta^{(3)},$$

where $\Delta^{(i)}$, i=1,2,3, is a copy of the root system of type $(A_1)^4$. Thus the root system is given by $\pm\alpha_j^{(i)}$, i=1,2,3, j=1,2,3,4. Fixing i we get the decomposition

$$(12.2) \quad D_4 = (sl_2)^4 \oplus (\mathbb{C}^2)^{\otimes 4},$$

where $\mathbb{C}^2$ is the natural representation of sl_2. Then extending the triality of sl_2 described in Section 8 to its natural representation $\mathbb{C}^2$, and taking it four times we obtain the first triality of D_4. The second is obvious from (12.1). The two groups S_3 generate a group of automorphisms of D_4 which is an extension of $L_3(2)$ by $(\mathbb{Z}/2)^3$ and which we denote by $2^3L_3(2)$. It permutes the following seven subspaces according to the way that $L_3(2)$ permutes the nonzero elements of $(\mathbb{F}_2)^3$:

(12.3) $D_4 = \bigoplus_{i=1}^{7} \mathfrak{h}_i.$

Here $\mathfrak{h}_4$ is the fixed Cartan subalgebra $\mathfrak{h}$ of D_4,

(12.4) $\mathfrak{h}_i = \sum_{j=1}^{4} \mathbb{C}(x_{\alpha_j^{(i)}} - x_{-\alpha_j^{(i)}}),\ i=1,2,3,$

(12.5) $\mathfrak{h}_{4+i} = \sum_{j=1}^{4} \mathbb{C}(x_{\alpha_j^{(i)}} + x_{-\alpha_j^{(i)}}),\ i=1,2,3.$

Each subspace in the decomposition (12.3) is a Cartan subalgebra and any two Cartan subalgebras generate a copy of $(sl_2)^4$. It is sometimes convenient to identify the index set in (12.3) with the set of nonzero elements of $(\mathbb{F}_2)^3$. This group $L_3(2)$ is clearly transitive on the 2-dimensional subspaces of $(\mathbb{F}_2)^3$. Therefore any subalgebra $(sl_2)^4$ can be conjugated to the subalgebra $\mathfrak{h}_1 \oplus \mathfrak{h}_2 \oplus \mathfrak{h}_3$. This fact and its extension to Griess's algebra are crucial in our proof of the closure of $\hat{\mathfrak{B}}$.

We associate to the Lie algebra $\mathfrak{o}_8$ of type D_4 eight affine Lie algebras similar to the four algebras associated to sl_2 in Section 8.

(12.6) $\hat{\mathfrak{a}}_0 = \mathfrak{o}_8 \otimes \mathbb{C}[t,t^{-1}] \oplus \mathbb{C}c$

(12.7) $\hat{\mathfrak{a}}_i = \mathfrak{k}_i \otimes \mathbb{C}[t,t^{-1}] \oplus \mathfrak{p}_i \otimes t^{1/2}\mathbb{C}[t,t^{-1}] \oplus \mathbb{C}c,\ i=1,\ldots,8,$

where $\mathfrak{k}_i = (sl_2)^4$, $\mathfrak{p}_i = (\mathbb{C}^2)^{\otimes 4}$. Then the group $2^3L_3(2)$ acts naturally on $\hat{\mathfrak{a}}_0$ and permutes the $\hat{\mathfrak{a}}_i$. We could then consider the basic representations of each of those algebras similar to (8.4)–(8.7) and define the action of $2^3L_3(2)$ on them.

Passing to our main picture, recall that in Section 11 we have defined the quadruple cover of the Leech lattice $\tilde{\Lambda}$, which contains the

Niemeier lattice N of type D_4^6. Let

$$(12.8)\quad \tilde{\Lambda} = N^{(0)} \cup N^{(1)} \cup N^{(2)} \cup N^{(3)}, \qquad N^{(0)} = N$$

be the coset decomposition with respect to N. We define

$$(12.9)\quad Y = \bigoplus_{i=0}^{7} Y_i$$

where

$$(12.10)\quad Y_i = S(\hat{\mathfrak{h}}_{\mathbb{Z}}^{-}) \otimes \mathbb{C}[N^{(i)}],\ i=0,1,2,3$$

$$(12.11)\quad Y_i = S(\hat{\mathfrak{h}}_{\mathbb{Z}+1/2}^{-}) \otimes \mathbb{C}[N^{(i-4)}/Q],\ i=4,5,6,7$$

(with Q the root lattice of D_4^6). The group $2^3L_3(2)$ acts on Y and on its subspace V, which has a nontrivial intersection with every subspace in (12.9), $V_i = V \cap Y_i$. The spaces of operators $\mathbb{B}$ and $\hat{\mathbb{B}}$ have natural decompositions compatible with the decomposition (12.9):

$$(12.12)\quad \mathbb{B} = \bigoplus_{i=0}^{7} \mathbb{B}_i$$

$$(12.13)\quad \hat{\mathbb{B}} = \bigoplus_{i=0}^{7} \bar{\mathbb{B}}_i \oplus \mathbb{C}c.$$

In order to prove the closure of $\hat{B}$ we need to show that

$$(12.14)\quad \bar{\mathbb{B}}_i \times \bar{\mathbb{B}}_j \subset \bar{\mathbb{B}}_{i+j} \text{ (sum in } (\mathbb{F}_2)^3).$$

Using the $L_3(2)$-symmetry we can conjugate the triple $\bar{\mathbb{B}}_i$, $\bar{\mathbb{B}}_j$, $\bar{\mathbb{B}}_{i+j}$ to $\bar{\mathbb{B}}_1$, $\bar{\mathbb{B}}_2$, $\bar{\mathbb{B}}_3$. But then we are inside $\hat{\mathfrak{f}}$, and the inclusion (12.14) is valid! This idea allows us to complete the proof that we have obtained a representation of the cross-bracket algebra $\hat{\mathbb{B}}$ in $V_{\mathbb{Z}}^{\natural}$.

13. THE MAIN THEOREM AND THE STRUCTURE OF GRIESS'S ALGEBRA

We now summarize what we have accomplished in the preceding sections:

We have constructed an infinite-dimensional graded commutative nonassociative algebra $\hat{\mathbb{B}}$ and an irreducible representation π of $\hat{\mathbb{B}}$ by cross bracket on a graded module $V = V_{\mathbb{Z}}^{\natural}$ with character $J(q)$. We have also constructed a group F_1 acting on V and as automorphisms of $\hat{\mathbb{B}}$, preserving their gradings, and such that

$$(13.1)\quad g\pi(x)g^{-1} = \pi(g\cdot x)$$

for $g\in F_1$ and $x\in\hat{\mathbb{B}}$. The algebra $\hat{\mathbb{B}}$ is the affinization of a commutative nonassociative algebra $\mathbb{B}$ of dimension 196884, equipped with a nonsingular symmetric associative bilinear form $\langle\cdot,\cdot\rangle$.

Our construction of $\hat{\mathbb{B}}$ allows us to describe $\mathbb{B}$ very explicitly. Let us identify $\mathbb{B}$ with the subspace $\mathbb{B}\otimes 1$ in $\hat{\mathbb{B}}$. Then $\mathbb{B}$ acts on

$$V = \oplus_{n\leq 1}V_n = V_1 \oplus 0 \oplus V_{-1} \oplus V_{-2} \oplus \cdots$$

preserving the grading. The nontrivial homogeneous subspace V_{-1} "just below" the trivial one-dimensional space V_1 has the same dimension as $\mathbb{B}$, in analogy with the E_8 theory. We identify $\mathbb{B}$ with V_{-1} via the linear map

$$(13.2)\quad \iota(x) = \pi(x(-2))(1\otimes 1),\ x\in\mathbb{B}.$$

Again as in the E_8 case, V_{-1} turns out to be the adjoint representation of $\mathbb{B}$, in the sense that

$$(13.3)\quad \pi(x(0))\iota(y) = \iota(x\times y),\ x,y\in\mathbb{B}.$$

This fact follows from the cross bracket identity

$$(13.4)\quad \begin{aligned}&\pi((x\times y)(m+n)) =\\ &= [\pi(x(m+\tfrac{1}{2})),\pi(y(n-\tfrac{1}{2}))] + [\pi(y(n+\tfrac{1}{2})),\pi(x(m-\tfrac{1}{2}))]\end{aligned}$$

($m,n\in\mathbb{Z}+\frac{1}{2}$, $m+n\neq 0$) for the case $m = -\frac{1}{2}$, $n = -\frac{3}{2}$, applied to $1\otimes 1$.

A similar argument shows that

$$(13.5) \quad \pi(x(2))\iota(y) = 2\langle x,y\rangle 1\otimes 1, \quad x,y\in\mathbb{B},$$

which enables us to compute the form $\langle\cdot,\cdot\rangle$.

Now recalling that the module V is made up of two different vertex representations (see (5.5)), we see that

$$(13.6) \quad \mathbb{B} \simeq S^2(\mathfrak{h}(-1))\otimes 1 \oplus \sum_{\alpha\in\Lambda_4} \mathbb{C}\cdot 1\otimes(e^{\alpha}+e^{-\alpha}) \oplus \mathfrak{h}(-\tfrac{1}{2})\otimes T.$$

One sees easily that

$$(13.7) \quad \iota(k_{\alpha}) = 1\otimes(e^{\alpha}+e^{-\alpha}).$$

Then the action of $\mathfrak{f}$ on V_{-1} can be found explicitly using standard calculations with vertex operators. Finally, using the associativity of the form $\langle\cdot,\cdot\rangle$, we find the product of any two elements in $\mathfrak{h}\otimes T$. This gives us explicitly the structure of

$$(13.8) \quad \mathbb{B} = S^2(\mathfrak{h}) \oplus \sum_{\alpha\in\Lambda_4} \mathbb{C}k_{\alpha} \oplus \mathfrak{h} \otimes T.$$

We first note that the action of F_1 decomposes V_{-1} and thus $\mathbb{B}$ into two orthogonal irreducible components, of dimensions 1 and 196883. The trivial one-dimensional module is spanned by the identity element, which we call I, of $\mathbb{B}$. The element I lies in $S^2(\mathfrak{h})$, and the operators $I(n)$ $(n\in\mathbb{Z})$ in $\hat{\mathbb{B}}$, together with the identity operator on V, span a copy of the Virasoro algebra, under Lie brackets.

The product structure on $\mathbb{B}$ is given by:

$$g^2\times h^2 = 4\langle g,h\rangle gh$$

$$g^2\times k_{\alpha} = \langle g,\alpha\rangle^2 k_{\alpha}$$

$$k_\alpha \times k_\beta = \begin{cases} 0 & \text{if } \langle\alpha,\beta\rangle = 0, \pm 1 \\ \epsilon(\alpha,\beta)k_{\alpha+\beta} & \text{if } \langle\alpha,\beta\rangle = -2 \\ \alpha^2 & \text{if } \beta = \pm\alpha \end{cases}$$

$$g^2 \times h\otimes\tau = (\langle g,h\rangle g + \frac{1}{8}\langle g,g\rangle h)\otimes\tau$$

$$k_\alpha \times h\otimes\tau = \frac{1}{8}(h - 2\langle\alpha,h\rangle\alpha)\otimes e'_\alpha\tau$$

$$h_1\otimes\tau_1 \times h_2\otimes\tau_2 = \frac{1}{2}(2h_1h_2 + \langle h_1,h_2\rangle I)\langle\tau_1,\tau_2\rangle +$$

$$+ \frac{1}{16}\sum_{\alpha\in\Lambda_4}(\langle h_1,h_2\rangle - 2\langle\alpha,h_1\rangle\langle\alpha,h_2\rangle)\langle\tau_1,e'_\alpha\tau_2\rangle k_\alpha,$$

where $g,h,h_i \in \mathfrak{h}$, $\alpha,\beta \in \Lambda_4$, $\tau,\tau_i \in T$, and $\langle\cdot,\cdot\rangle|_T$ is a nonzero $\bar{\Lambda}^\wedge$-invariant form on T (cf. Section 2).

The decomposition (13.8) is orthogonal with respect to the form $\langle\cdot,\cdot\rangle$ on $\mathcal{B}$, which is given by:

$$\langle g^2,h^2\rangle = \langle g,h\rangle^2$$

$$\langle k_\alpha,k_\beta\rangle = \delta_{\bar{\alpha}\bar{\beta}}$$

$$\langle h_1\otimes\tau_1, h_2\otimes\tau_2\rangle = \langle h_1,h_2\rangle\langle\tau_1,\tau_2\rangle,$$

where $\bar{\alpha},\bar{\beta}$ are the reductions of α,β mod 2Λ.

In a similar way we could explicitly describe the actions of C and S_3, and therefore the group F_1, on V_{-1} and $\mathcal{B}$.

Combining our constructions with certain results in [15], or the recent simplification of these results in [28] (see also [27]), we can conclude that F_1 is a finite simple group with the group C as the centralizer of an involution. Thus F_1 is the Monster. By [28],

$$F_1 = \text{Aut } \mathcal{B}.$$

Thus we have constructed a natural representation of the

Monster in the graded space V with character J(q). Recall that V is also an irreducible $\hat{\mathcal{B}}$-module and that F_1 and $\hat{\mathcal{B}}$ act compatibly on V (see (13.1)). We call the $(F_1, \hat{\mathcal{B}})$-module V the *Moonshine module*.

Finally, one can explicitly check that up to an inessential one-dimensional subspace (the span of the identity element), our algebra $\mathcal{B}$ and form $\langle \cdot, \cdot \rangle$ are isomorphic to Greiss's [15]. Another check shows that a certain involution in our triality group S_3 agrees with Griess's involution σ. Thus we obtain Griess's construction as a homogeneous "slice" of ours.

It is important to note that the infinite-dimensional vertex operator approach is more natural than the finite-dimensional approach, even though the primary object of study is a finite group. In particular, the six arbitrary parameters in Griess's construction are determined intrinsically in advance in our construction. Also, we have avoided the tedious verification that the involution σ preserves the algebra structure. Even simplification of Griess's work ([27], [28]) does not eliminate these complications. In our construction we use instead an entirely natural "calculus of vertex operators", which we consider standard, developed first for affine Lie algebras. In addition, we use a principle of triality which is based on the triality of a basis of the simplest simple Lie algebra $\mathfrak{sl}_2$.

14. THOMPSON SERIES

As another corollary of our construction, we obtain character formulas for elements of the Monster. By definition one has the Thompson series

(14.1) $$T_g(q) = \sum_{n \geq -1} (\mathrm{tr}\ g|_{V_{-n}})\, q^n, \quad g \in F_1.$$

We shall find a closed formula for the elements of C. For any $w \in W$ (see Section 4) we set

$$(14.2)\quad \eta_w(q) = \prod_{k>0} \eta(q^k)^{p_k}$$

(see (4.6)) if w has characteristic polynomial

$$(14.3)\quad \det(1-xw) = \prod_k (1-x^k)^{p_k}$$

on Λ. We also set

$$(14.4)\quad \theta_a^{\pm}(q) = \sum_{\substack{\alpha\in\Lambda \\ \bar{a}\alpha=\pm\alpha}} s(\alpha) q^{<\alpha,\alpha>/2}$$

where $a(e_\alpha) = s(\alpha)e_{\bar{a}\alpha}$, $a\in\tilde{W}$, and $\bar{a}\in W$ is the element induced by a. Then for $g\in C$ which has a preimage $(a,a_T)\in\hat{C}$ (see Section 5) we obtain

$$(14.5)\quad T_g(q) = \frac{1}{2}\left[\frac{\theta_a^+(q)}{\eta_{\bar{a}}(q)} + \frac{\theta_a^-(q)}{\eta_{-\bar{a}}(q)} + \frac{\operatorname{tr}(a_T)\eta_{\bar{a}}(q)}{\eta_{\bar{a}}(q^{1/2})} + \frac{\operatorname{tr}(-a_T)\eta_{-\bar{a}}(q)}{\eta_{-\bar{a}}(q^{1/2})}\right].$$

Note that this expression is independent of the choice of preimage (a,a_T). It is interesting to note that Conway and Norton [4] obtained another formula for the trace of an element of C:

$$(14.6)\quad T_g(q) = \frac{1}{2}\left[\frac{\theta_a^+(q)}{\eta_{\bar{a}}(q)} + \frac{\theta_a^-(q)}{\eta_{-\bar{a}}(q)} + \frac{\operatorname{tr}(a_T)\theta_a(q^2)}{\eta_{\bar{a}}(q)} + \frac{\operatorname{tr}(-a_T)\theta_{-a}(q^2)}{\eta_{-\bar{a}}(q)}\right] - c_g$$

where c_g is a constant and

$$\theta_a(q) = \sum_{\substack{\alpha\in\Lambda \\ \bar{a}\alpha=\alpha}} q^{<\alpha,\alpha>/2}.$$

Thus, provided these formulas hold for the same Moonshine module, one obtains a whole collection of identities for modular functions enumerated by the conjugacy classes of elements of C. We have seen ((4.1), (4.4), (5.8)) that in the case when $g = e$ one has

$$(14.7)\quad T_e(q) = J(q) = \frac{\theta_\Lambda(q)}{\eta(q)^{24}} - 24.$$

Similarly, if one takes $g = g_2$ (resp., $g = g_3$) to be an involution (resp., an element of order 3) in F_1 whose centralizer is the double cover of the Baby Monster (resp., the triple cover of the Fischer group F'_{24}), then using the same argument as in the case $g = e$, namely, the isomorphism of suitable subspaces of the homogeneous and twisted vertex representations, one obtains

$$(14.8)\quad T_{g_2}(q) = J_2(q) = \frac{\theta_{\Lambda_2}(q)}{[\eta(q)\eta(q^2)]^8} - 8$$

$$(14.9)\quad T_{g_3}(q) = J_3(q) = \frac{\theta_{\Lambda_3}(q)}{[\eta(q)\eta(q^3)]^6} - 6$$

where Λ_2 (resp., Λ_3) are 16-dimensional (resp., 12-dimensional) even lattices without short vectors. Those lattices are certainly not unimodular; however they are "close" to unimodular lattices in the following sense: We call L an n-*modular lattice* if the dual lattice $L^* \simeq \frac{1}{n}L$. Then Λ_n, n=2,3, is an n-modular lattice. It follows immediately from the definitions (14.8), (14.9) that $J_n(q)$, n=2,3, is invariant with respect to the Atkin-Lehner involution ω: $z \longmapsto -\frac{1}{nz}$, where $q = e^{2\pi iz}$; it is also clearly invariant with respect to $\Gamma_0(n)$. This implies the invariance of $J_n(q)$ with respect to the group $\Gamma_0(n)^+ := \langle\Gamma_0(n),\omega\rangle$, and proves that $J_n(q)$, n=2,3, is a Hauptmodul for the group $\Gamma_0(n)^+$. The above argument certainly has a more general nature.

15. TWO DUAL MODELS AND TWO SPORADIC GROUPS

It was noticed from the first construction of the basic representation of $s\ell(2)^\wedge$ (see [19]) that it can be expressed in terms of operators similar to those appearing in the dual resonance models. Later this connection became more apparent [6]. However, the central result of the dual model theories, namely, the no-ghost theorem, remained untouched. This theorem implies that the dual models in rank 24 have the best behavior in terms of symmetry and unitary structure. This result seems to be deeply related to our construction, where rank 24 again plays a special role, providing a natural setting for the representation of the largest sporadic group F_1. Trying to reduce the critical dimension of the model, physicists introduced additional fermionic degrees of freedom. In this way the critical dimension reduces to 8, and the corresponding model is known as the Neveu-Schwarz model. In this case

$$(15.1)\quad V_{\mathbf{Z}}^f = V_{\mathbf{Z}} \otimes \Lambda(\hat{h}^-_{\mathbf{Z}+1/2})$$

where $V_{\mathbf{Z}}$ is defined by (2.9). Then

$$(15.2)\quad \begin{aligned} \mathrm{ch}\, V_{\mathbf{Z}}^f &= \frac{\theta_{E_8}(q)\eta(q)^8}{\eta(q^{1/2})^8\eta(q^2)^8} = \\ &= q^{-1/2} + 8 + 276\, q^{1/2} + \dots \end{aligned}$$

One notices that this character has "moonshine" properties for the Conway group $\cdot 1$ similar to the properties of the character of $V_{\mathbf{Z}}$ with $L = \Lambda$ for the Monster. As was explained in Section 4, the natural automorphism group acting on $V_{\mathbf{Z}}^f$ is $\tilde{W}_{E_8} = 2^8 W_{E_8}$, and it is known that $\cdot 1$ has an involution centralizer C isomorphic to $2_+^{1+8}(W'_{E_8}/\langle \pm 1\rangle)$, where the prime denotes commutator subgroup. This situation is already familiar to us. In order to construct the right module we need to introduce an odd space

(15.3) $V^f_{\mathbf{Z}+1/2} = V_{\mathbf{Z}+1/2} \otimes \Lambda(\hat{h}^-_{\mathbf{Z}})$

and define $V^{f^\natural}_{\mathbf{Z}}$, $V^{f^\natural}_{\mathbf{Z}+1/2}$ as in Section 5. By analogy with the construction of F_1 one expects that

(15.4) $\mathrm{ch}\ V^{f^\natural}_{\mathbf{Z}} = \mathrm{ch}\ V^f_{\mathbf{Z}} - 8 = q^{-1/2} + 0 + 276\ q^{1/2} + \ldots$

which is in fact the case. Moreover the analogy between the two constructions exists at practically every step; this is also true in dual resonance theory. In particular one can repeat triality, $L_3(2)$ symmetry, etc. The analogue of the Conway-Norton formulas also exists, so that one can obtain a collection of identities for any conjugacy class of C. In particular for e∈C we get the identity (cf. (14.6))

$$\mathrm{ch}\ V^{f^\natural}_{\mathbf{Z}} = \frac{1}{2}\left[\frac{\theta_{E_8}(q)\eta(q)^8}{\eta(q^{1/2})^8\eta(q^2)^8} + \frac{\eta(q^{1/2})^8}{\eta(q^2)^8}\right] + \tag{15.5}$$
$$+ \frac{1}{2}2^4\left[\frac{\theta_{E_8}(q^2)\eta(q)}{\eta(q^{1/2})^8\eta(q^2)} - \frac{\eta(q^{1/2})^8}{\eta(q^2)^8}\right] - 8\cdot 2^4.$$

It is interesting to note that besides several central ingredients of dual theories like the spaces $V_{\mathbf{Z}}$, $\Lambda(\hat{h}^-_{\mathbf{Z}})$, vertex operators and the Virasoro algebra, physicists have also used many other elements which occur in our construction, such as $V_{\mathbf{Z}+1/2}$ without the factor T, $\Lambda(\hat{h}^-_{\mathbf{Z}})$, triality, mixed spaces similar to $V^\natural$ (see e.g. [13]). Even the cross bracket appears ([13, page 526])!

The remarkable relationship of the two modules for sporadic groups and the two dual resonance models suggests a deep connection between two of the most distant theories in mathematics and physics. This also gives us some hints for the continuation of our work. In the physical models, besides the "noncovariant approach" which corresponds to what we have used in the paper, there exists a "covariant" one, which starts from the vertex representation of rank

26 with signature (25,1). In this approach the symmetries of the model are displayed in a more apparent way. This fact as well as the definition of the Monster Lie algebra of rank 26 in [1] suggest that our construction can be "embedded" into a more symmetric one. One can then hope to attach to each of the 23 constructions of the Leech lattice [5] a construction of the group F_1 (cf. [1]). Thus our present construction would correspond to the construction of the Leech lattice from the binary Golay code, the next one should correspond to the complex Leech lattice construction from the ternary code, and so on.

Among the finite-dimensional simple Lie algebras, E_8 is in a sense the richest. The proper understanding of its structure yields all the exceptional Lie algebras, via the "magic square", or all the simple algebras, via root system theory. We believe that the Monster -- really a uniquely beautiful creature, -- whose full understanding will involve Lie algebras, vertex operators, modular functions and physical models, should play an analogous role in a future unified theory of the finite simple groups.

REFERENCES

[1] R. E. Borcherds, J. H. Conway, L. Queen and N. J. A. Sloane, A Monster Lie algebra?, to appear.

[2] J. H. Conway, A group of order 8,315,553,613,086,720,000, Bull. London Math. Soc. 1 (1969), 79-88.

[3] J. H. Conway, A characterization of Leech's lattice, Inventiones Math. 7 (1969), 137-142.

[4] J. H. Conway and S. P. Norton, Monstrous moonshine, Bull. London Math. Soc. 11 (1979), 308-339.

[5] J. H. Conway and N. J. A. Sloane, Twenty-three constructions for the Leech lattice, Proc. Roy. Soc. London Ser. A 381 (1982), 275-283.

[6] I. B. Frenkel and V. G. Kac, Basic representations of affine Lie algebras and dual resonance models, Inventiones Math. 62 (1980), 23-66.

[7] I. B. Frenkel, J. Lepowsky and A. Meurman, An E_8-approach to F_1, Proceedings of the 1982 Montreal Conference on Finite Group Theory, ed. by J. McKay, Springer-Verlag Lecture Notes in Mathematics (1984).

[8] I. B. Frenkel, J. Lepowsky and A. Meurman, A natural representation of the Fischer-Griess Monster with the modular function J as character, Proc. Nat. Acad. Sci. U.S.A. 81 (1984), 3256-3260.

[9] H. Garland, The arithmetic theory of loop algebras, J. Algebra 53 (1978), 480-551.

[10] H. Garland, The arithmetic theory of loop groups, Publ. Math. IHES 52 (1980), 5-136.

[11] F. Gliozzi, D. Olive and J. Scherk, Supersymmetry, supergravity theories and the dual spinor model, Nuclear Physics B122 (1977), 253-290.

[12] D. Gorenstein, Finite simple groups, Plenum Press, New York, 1982.

[13] M. B. Green and J. H. Schwarz, Supersymmetrical dual string theory, Nuclear Physics B181 (1981), 502-530.

[14] R. L. Griess, Jr., A construction of F_1 as automorphisms of a 196,883 dimensional algebra, Proc. Nat. Acad. Sci. U.S.A. 78 (1981), 689-691.

[15] R. L. Griess, Jr., The Friendly Giant, Invent. Math. 69 (1982), 1-102.

[16] V. G. Kac, An elucidation of "Infinite-dimensional...and the very strange formula" $E_8^{(1)}$ and the cube root of the modular invariant j, Advances in Math. 35 (1980), 264-273.

[17] J. Leech, Notes on sphere packings, Canadian J. Math. 19 (1967), 252-267.

[18] J. Lepowsky, Euclidean Lie algebras and the modular function j, Amer. Math. Soc. Proc. Symp. Pure Math. 37 (1980), 567-570.

[19] J. Lepowsky and R. L. Wilson, Construction of the affine Lie algebra $A_1^{(1)}$, Comm. Math. Phys. 62 (1978), 43-53.

[20] D. Mitzman, Integral bases for affine Lie algebras and their universal enveloping algebras, Ph.D. thesis, Rutgers University, 1983 and Contemporary Math., Amer. Math. Soc., to appear.

[21] R. Pfister, Spin representations of $A_1^{(1)}$, Ph.D. thesis, Rutgers University, 1984.

[22] G. Segal, Unitary representations of some infinite-dimensional groups, Comm. Math. Phys. 80 (1981), 301-342.

[23] S. Smith, On the head characters of the Monster simple group, Proceedings of the 1982 Montreal Conference on Finite Group Theory, ed. by J. McKay, Springer-Verlag Lecture Notes in Mathematics (1984).

[24] J. Thompson, Some numerology between the Fischer-Griess Monster and the elliptic modular function, Bull. London Math. Soc. 11 (1979), 352-353.

[25] J. Tits, Four presentations of Leech's lattice, in "Finite Simple Groups II", Proceedings of a London Math. Soc. Research Symposium, Durham, 1978, pp. 303-307, Academic Press, London/New York, 1980.

[26] J. Tits, Résumé de cours, Annuaire du Collège de France, 1980-81, 75-87.

[27] J. Tits, Remarks on Griess' construction of the Griess-Fischer sporadic group, I-IV, preprints, 1983.

[28] J. Tits, Le monstre, Séminaire Bourbaki, 36e année, 1983/84, no. 620 (1983).

Rutgers University
New Brunswick, NJ 08903

Mathematical Sciences Research Institute
Berkeley, CA 94720

MONODROMY, SOLITONS AND INFINITE DIMENSIONAL LIE ALGEBRAS

Michio Jimbo[†] and Tetsuji Miwa[†]

The present contribution is a brief survey of works done in Kyoto for the past 7 years. As such, it contains no new results. Rather, it is intended to cover an outline of the development that might be of some interest to the people working in vertex operators and related areas. Some more details can be found in several review articles [1], [2] as well as in the original papers [3], [4].

1. The story began with the two dimensional Ising model in statistical physics.

Consider a square lattice of size $M \times N$ on the plane. To each site (i,j) of the lattice, there is attached a random variable σ_{ij} with values $+1$ or -1, which one can think of as a "spin" pointing up or down. There are thus 2^{MN} possible configurations $\sigma = (\sigma_{ij})_{\substack{1 \leq i \leq M \\ 1 \leq j \leq N}} \in \{+1, -1\}^{MN}$. The model is defined by specifying the energy of a given configuration σ as

$$E(\sigma) = - E_1 \sum_{i=1}^{M} \sum_{j=1}^{N} \sigma_{ij}\, \sigma_{i+1,j} - E_2 \sum_{i=1}^{M} \sum_{j=1}^{N} \sigma_{ij}\, \sigma_{i,j+1}.$$

For definiteness about the boundary, one assumes here that the lattice is wound on a torus, so the index i (resp. j) is to be read mod M (resp. N).

[†] Partially supported by the National Science Foundation through the Mathematical Sciences Research Institute.

Vertex Operators in Mathematics and Physics - Proceedings of a Conference November 10-17, 1983. Publications of the Mathematical Sciences Research Institute #3, Springer-Verlag, 1984.

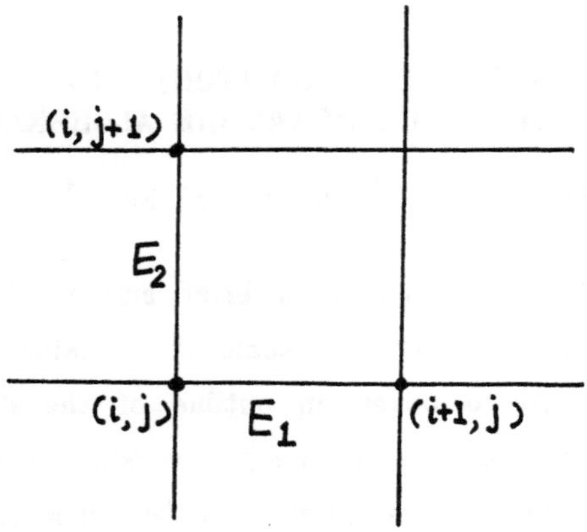

The probability of a given configuration σ to occur is then given by

$$p(\sigma) = Z_{MN}^{-1}\, e^{-E(\sigma)/kT},$$

where T = temperature, k = Boltzmann's constant, and $Z_{MN} = \sum_{\sigma} e^{-E(\sigma)/kT}$ (summed over all $\sigma_{11} = \pm 1$, $\sigma_{12} = \pm 1, \ldots, \sigma_{MN} = \pm 1$) is the normalization constant to make the total probability = 1.

One of the quantities of physical interest is the spin correlation function. Given two lattice points, say (m,n) and (0,0), the expectation value of the product $\sigma_{mn}\sigma_{00}$

$$<\sigma_{mn}\ \sigma_{00}> = \sum_{\sigma} \sigma_{mn}\ \sigma_{00}\ p(\sigma)$$

is called the 2-*point correlation function*. General n-point correlation functions are defined analogously. There is also the notion of "dual" (sometimes called also "order") correlation functions $<\mu_{mn}\mu_{00}>$, etc. These are correlation functions of the Ising model on the dual lattice, whose interaction constants K_i^* correspond in a 1:1 way to the original ones $K_i = E_i/kT$ (i=1,2). Eventually one takes the infinite lattice limit $M,N \longrightarrow \infty$ of all these quantities, so that they become functions of $(m,n) \in \mathbb{Z}^2$.

A nice structure emerges when one passes further to another limit, the continuum limit; let the lattice spacing ϵ tend to 0 while keeping $m\epsilon = x_1$, $n\epsilon = x_2$ fixed [5]. Denote by $\tau_+(x) = \lim \epsilon^{1/4}\langle\sigma_{mn}\sigma_{00}\rangle$, $\tau_-(x) = \lim \epsilon^{1/4}\langle\mu_{mn}\mu_{00}\rangle$ the scaled two-point functions, and put

$$u(x) = 2 \log \frac{\tau_-(x) + \tau_+(x)}{\tau_-(x) - \tau_+(x)}.$$

Then u(x) satisfies the following *sinh-Gordon equation*

(1) $$\Delta u = \sinh u.$$

One can show further that $\tau_\pm(x)$ depend only on the distance $t = \left(x_1^2+x_2^2\right)^{1/2}$ of the points x and 0. Putting v(t) = u(x), (1) now becomes the non-linear ordinary differential equation (NLODE)

(2) $$\left(\frac{d^2}{dt^2} + \frac{1}{t}\frac{d}{dt}\right) v = \sinh v,$$

which is an equivalent of the *Painlevé equation of the third kind*. This remarkable result (2) has been found by T.T. Wu et. al. [6]

2. Here a word is need about the Painlevé equations.

Originally they arose from the classification by Painlevé of second order NLODE whose general solution has no movable branch points [7]. (A movable branch point is one that depends on the integration constant. Example: $2yy' = 1$ has the general solution $y = \left(x-C\right)^{1/2}$, so the branch point x = C is movable.) The six new canonical types of such equations he obtained are termed after him as Painlevé equations.

Not long after Painlevé's work, it was revealed that they have another background — Riemann's monodromy problem. Consider an m × m system of linear ODE

(3) $$\frac{dY}{dx} = A(x)Y$$

with $A(x)$ rational in x, e.g. $A(x) = \sum_{\nu} \frac{A_\nu}{x-a_\nu}$. If $Y(x)$ is a fundamental matrix solution of (3), then so is its analytic continuation $Y(x)^\gamma$ along a loop γ (with a base point, say at $x_0 = \infty$). Hence one must have $Y(x)^\gamma = Y(x)M_\gamma$ with some $M_\gamma \in GL(m,\mathbb{C})$. This defines the monodromy representation

$$\rho: \pi_1(\mathbb{P}^1_{\mathbb{C}} - \{a_\nu\}; \infty) \longrightarrow GL(m,\mathbb{C}), \quad [\gamma] \longmapsto M_\gamma.$$

Riemann's problem asks the converse. Namely, given ρ, can one find a linear ODE such that its monodromy representation coincides with ρ? The answer is in fact in the affirmative. The corresponding matrix solution would then depend on the data $\{M_\gamma\}$ and $\{a_\nu\}$. Suppose we fix the monodromy $\{M_\gamma\}$ and vary the singular points $\{a_\nu\}$. Schlesinger showed that this is possible if and only if $Y(x)$ satisfies the differential equations

$$\frac{\partial Y}{\partial a_\nu} = - \frac{A_\nu}{x-a_\nu} Y. \tag{4}$$

The integrability condition of (3) and (4) gives rise to the non-linear system for the matrices A_ν

$$\frac{\partial A_\mu}{\partial a_\nu} = \begin{cases} - \dfrac{[A_\mu, A_\nu]}{a_\mu - a_\nu} & (\mu \neq \nu) \\ \displaystyle\sum_{\lambda(\neq\nu)} \frac{[A_\lambda, A_\nu]}{a_\lambda - a_\nu} & (\mu = \nu) \end{cases}$$

called the *Schlesinger equation* [8]. The fact is, all the Painlevé equations are derived as the simplest non-trivial cases of the Schlesinger equation and its suitable generalizations, obtained by allowing $A(x)$ to have multiple poles [9].

3. So it was fascinating to see that such a theory of purely mathematical origins should arise in a purely physical problem. Now the question is, what is the mechanism working behind these phenomena?

Let us go back to the Ising model on a finite lattice. There is a standard way to describe this classical statistical model in the second quantized language. View the lattice as built up by one dimensional systems (= rows) $\mathfrak{F} = M \otimes \mathbb{C}^2$, on which are acting the spin operators $\hat{\sigma}_n = I_2 \otimes I_2 \otimes \cdots \otimes \underset{\underline{n}}{\begin{pmatrix} 1 & 0 \\ 0 & -1 \end{pmatrix}} \otimes \cdots \otimes I_2$, $\hat{\mu}_n = \begin{pmatrix} 0 & 1 \\ 1 & 0 \end{pmatrix} \otimes \begin{pmatrix} 0 & 1 \\ 1 & 0 \end{pmatrix} \otimes \cdots \otimes \begin{pmatrix} 0 & 1 \\ 1 & 0 \end{pmatrix} \otimes I_2 \otimes \cdots \otimes I_2$. One may introduce an operator $V \in \mathrm{End}(\mathfrak{F})$ describing the evolution of the system in the vertical direction, in such a way that the following hold:

$$Z_{MN} = \mathrm{tr}\ V^N,$$

(5)
$$\langle \sigma_{mn}\sigma_{00} \rangle = \mathrm{tr}(\hat{\sigma}_{mn}\hat{\sigma}_{00}\ V^N)/\mathrm{tr}\ V^N$$
$$\langle \mu_{mn}\mu_{00} \rangle = \mathrm{tr}(\hat{\mu}_{mn}\hat{\mu}_{00}\ V^N)/\mathrm{tr}\ V^N$$

where $\hat{\sigma}_{mn} = V^m\hat{\sigma}_n V^{-m}$, $\hat{\mu}_{mn} = V^m\hat{\mu}_n V^{-m}$. Note that, as $N \longrightarrow \infty$, only the eigenvector $|vac\rangle$ of V corresponding to the largest eigenvalue contributes to the trace, so (5) becomes the vacuum expectation value

$$\langle \sigma_{mn}\sigma_{00} \rangle = \langle vac|\ \hat{\sigma}_{mn}\hat{\sigma}_{00}\ |vac\rangle.$$

In what follows we drop ^ and regard σ_{mn}, μ_{mn} as operators on $\mathfrak{F}$. Now define the free fermion operators p_j, q_j by

(6)
$$p_n = \mu_{n-1}\sigma_n, \quad q_n = \mu_n\sigma_n.$$

Then $p_j^2 = 1$, $q_j^2 = -1$, and all others mutually anticommute. The total algebra $A = \mathrm{End}(\mathfrak{F})$ can be regarded as the Clifford algebra generated by $W = \mathrm{span}\{p_j, q_j\}$. Onsager [10] made the crucial observation that all the relevant operators V, σ_n and μ_n in fact belong to the *Clifford group* G, i.e. the multiplicative subgroup of $A^\times$ consisting of elements g with the property

$$g\, W\, g^{-1} = W. \tag{7}$$

For example, the definition (6) implies immediately

$$\sigma_n p_j \sigma_n^{-1} = \begin{cases} -p_j & n < j \\ +p_j & n \geqslant j \end{cases}$$

$$\sigma_n q_j \sigma_n^{-1} = \begin{cases} -q_j & n \leqslant j \\ +q_j & n > j. \end{cases}$$

Passing to the continuum, one ends up with a neutral free fermion field $\psi_\pm(x)$ [11] and spin operators $\sigma(x)$, $\mu(x)$ obeying the (anti-) commutation rules

$$\sigma(a)\ \psi_\pm(x) = \begin{cases} -\psi_\pm(x)\sigma(a) & a^1 < x^1 \\ +\psi_\pm(x)\sigma(a) & a^1 > x^1, \end{cases} \tag{8}$$

where $x = (x^0, x^1)$ and $a = (a^0, a^1)$ are mutually spacelike points in the Minkowski 2-space. ($\mu(a)$ satisfies a similar relation with opposite sign.) Consider the wave function

$$w_\pm(x) = \langle vac|\ \psi_\pm(x)\ \sigma(a)\ |vac\rangle$$

$$w'_\pm(x) = \langle vac|\ \sigma(a)\ \psi_\pm(x)\ |vac\rangle.$$

A general principle in quantum field theory tells us that they are analytically continuable to one or the other side of the Euclidean space (see the figure below).

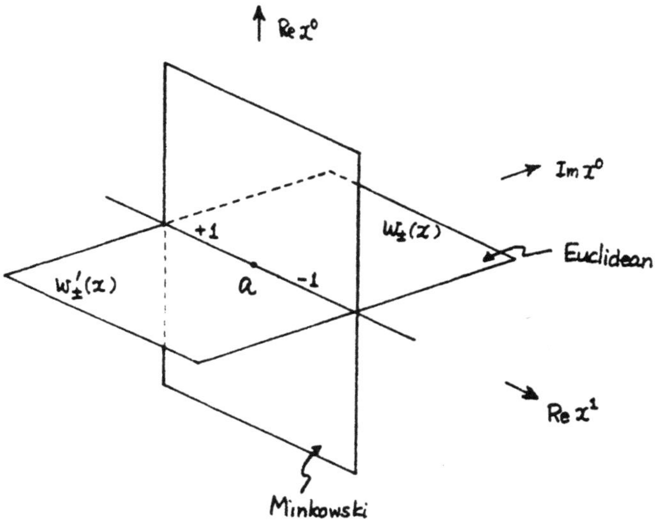

Combining this with the commutation rule (8), one concludes that $w_{\pm}(x)$ changes its sign when continued around the point x = a. Thus the characteristic property (7) of the group G entails the monodromy structure of the expectation value $w_{\pm}(x)$ [2].

The same argument applies when $\sigma(a)$ is replaced by the product $\xi_1(a_1) \cdots \xi_n(a_n)$ ($\xi_i = \sigma$ or μ). This enables one to relate n-point correlation functions to the Painlevé equations and their generalizations [3].

Aside from the Ising model, it is even possible to solve Riemann's problem in general by using quantum field operators and their expectation values [3], [13].

4. We have seen how the expectation value of the form $\langle vac | g | vac \rangle$ of a group element $g = \sigma(a), \mu(a)$ is related to special solutions of non-linear partial differential equations (sinh-Gordon and the like). We now proceed to the case where g is an arbitrary element of G.

Let us restart with the following setting.

(i) Free fermion operators ψ_n, ψ_n^* ($n \in \mathbf{Z}$) satisfying

$$(9) \quad [\psi_n, \psi_m]_+ = 0, \quad [\psi_m^*, \psi_n^*]_+ = 0, \quad [\psi_m, \psi_n^*]_+ = \delta_{mn},$$

the Clifford algebra A generated by $W = \mathrm{span}\{\psi_n, \psi_n^*\}_{n\in\mathbb{Z}}$, and

(ii) Fock representation on $\mathcal{F} = A\,|\mathrm{vac}\rangle$ and its dual space $\mathcal{F}^* = \langle\mathrm{vac}|\,A$ such that

$$(10)\qquad \begin{gathered} \psi_n|\mathrm{vac}\rangle = 0 \quad (n < 0), \quad \psi_n^*|\mathrm{vac}\rangle = 0 \quad (n \geqslant 0) \\ \langle\mathrm{vac}|\psi_n = 0 \quad (n \geqslant 0), \quad \langle\mathrm{vac}|\psi_n^* = 0 \quad (n < 0). \end{gathered}$$

Denote by $\langle a\rangle = \langle\mathrm{vac}\,|\,a\,|\mathrm{vac}\rangle$ the vacuum expectation value of $a \in A$ with regards to (9) and (10). Explicitly it is given by the rule $\langle 1\rangle = 1$, $\langle\psi_i\psi_j^*\rangle = \delta_{ij} - \langle\psi_j^*\psi_i\rangle = 1$ (if $i = j < 0$), $= 0$ (otherwise), and in general by Wick's theorem

$$\langle w_1\cdots w_r\rangle =$$

$$\begin{cases} 0 & (r \text{ odd}) \\ \displaystyle\sum_{\substack{\sigma(1)<\sigma(3)<\cdots<\sigma(r-1) \\ \sigma(1)<\sigma(2),\ldots,\sigma(r-1)<\sigma(r)}} \mathrm{sgn}\,\sigma\langle w_{\sigma(1)}w_{\sigma(2)}\rangle\cdots\langle w_{\sigma(r-1)}w_{\sigma(r)}\rangle & (r \text{ even}) \end{cases}$$

for $w_1,\ldots, w_r \in W$. In the present situation, the free fermions carry the "charge": $\deg\psi_i = +1$, $\deg\psi_i^* = -1$. This introduces a grading on $A = \oplus_{n\in\mathbb{Z}} A_n$, and on $\mathcal{F} = \oplus_{n\in\mathbb{Z}} \mathcal{F}_n$. We shall be concerned with the group that preserves this charge

$$G_0 = \{g \in A^\times \mid g\,V\,g^{-1} = V, \quad g\,V^*\,g^{-1} = V^*\}$$

where $V = \mathrm{span}\,\{\psi_i\}$, $V^* = \mathrm{span}\,\{\psi_i^*\}$. Its Lie algebra is spanned by $1 \in \mathbb{C}$ and quadratic elements $\sum a_{ij}\psi_i\psi_j^*$ (finite sum). It is possible to include here certain infinite sums as well. Introduce the normal ordering $:\psi_i\psi_j^*: = \psi_i\psi_j^* - \langle\psi_i\psi_j^*\rangle$, and put $X_A = \sum a_{ij}:\psi_i\psi_j^*:$ for $A = (a_{ij})_{i,j\in\mathbb{Z}}$. One has then

(11) $$[X_A, X_{A'}] = X_{[A,A']} + \mathrm{tr}(ACA' - A'CA)\cdot 1$$

where $C = (c_{ij})$ is given by $c_{ij} = -1/2$ $(i = j < 0)$, $= +1/2$ $(i = j \geq 0)$, $= 0$ (otherwise).

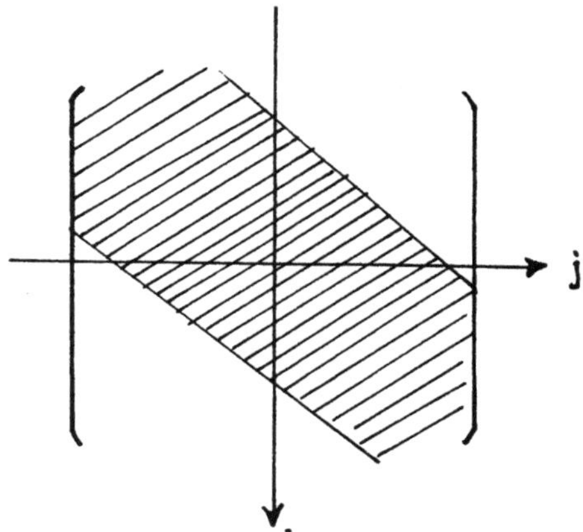

These operations (11) make sense provided that a_{ij}'s are supported on a finite band (shadowed region), i.e.

(12) $$\exists N > 0 \text{ such that } a_{ij} = 0 \text{ for } |i-j| > N.$$

In what follows we consider the Lie algebra

$$gl(\infty) = \{ X_A \mid A \text{ satisfies (12)} \} \ltimes \mathbb{C}\,1.$$

With this definition, it contains a Heisenberg subalgebra h spanned by $1 \in \mathbb{C}$ and

$$\Lambda_n = \sum_{i\in\mathbb{Z}} :\psi_i\psi^*_{i+n}:, \quad [\Lambda_n, \Lambda_m] = n\,\delta_{n+m,0}$$

with $n \neq 0$ (Λ_0 gives rise to the charge operator). It turns out that each charge n sector $\mathcal{F}_n$ is irreducible under h, and so is isomorphic to the polynomial algebra $\mathbb{C}[x] = \mathbb{C}[x_1, x_2, \dots]$ as a vector space. Let V_n denote a copy of $\mathbb{C}[x]$, and put $V = \oplus_{n\in\mathbb{Z}} V_n$. Introduce the Hamiltonian

$$H(x) = \sum_{n=1}^{\infty} x_n \Lambda_n.$$

Theorem. The following is a linear isomorphism:

$$\text{(13)} \qquad \begin{cases} \mathcal{F} = \oplus_{n\in\mathbf{Z}} \mathcal{F}_n \longrightarrow V = \oplus_{n\in\mathbf{Z}} V_n \\ a|\text{vac}\rangle \longmapsto \oplus\langle n|e^{H(x)}a|\text{vac}\rangle \end{cases}$$

where

$$\langle n| = \begin{cases} \langle \text{vac}|\ \psi_{-1}\cdots\psi_n & (n < 0) \\ \langle \text{vac}| & (n = 0) \\ \langle \text{vac}|\ \psi_0^*\cdots\psi_{n-1}^* & (n > 0). \end{cases}$$

In fact, this correspondence is lifted to operator level. Define the vertex operators

$$X(k) = \exp\Big(\sum_{n=1}^{\infty} x_n k^n\Big) \exp\Big(-\sum_{n=1}^{\infty} \frac{1}{n}\frac{\partial}{\partial x_n} k^{-n}\Big) = \sum_{j\in\mathbf{Z}} X_j k^j$$

$$X^*(k) = \exp\Big(-\sum_{n=1}^{\infty} x_n k^n\Big) \exp\Big(\sum_{n=1}^{\infty} \frac{1}{n}\frac{\partial}{\partial x_n} k^{-n}\Big) = \sum_{j\in\mathbf{Z}} X_j^* k^{-j},$$

and let the linear operators $\hat{X}_j$, $\hat{X}_j^* \in \mathrm{End}(V)$ be given by

$$\hat{X}_j f = X_{j-\ell} f \in V_{\ell+1} \text{ for } f \in V_\ell,$$

$$\hat{X}_j^* f = X_{j-\ell+1}^* f \in V_{\ell-1} \text{ for } f \in V_\ell.$$

Theorem. (13) is an A-module isomorphism via the correspondence

$$\psi_j \leftrightarrow \hat{X}_j, \qquad \psi_j^* \leftrightarrow \hat{X}_j^*.$$

In other words, the fermion Fock space $\mathcal{F}$ can also be realized as a boson Fock space V. (This is an example of the well-known Bose-Fermi correspondence in one space dimension.) In particular, an element $\psi_i\psi_j^*$ of $gl(\infty)$ has a realization as a vertex operator Z_{ij} given by

$$Z(p,q) = \frac{q}{p-q}\{\exp(\sum_{n=1}^{\infty}(p^n-q^n)x_n)\exp(-\sum_{n=1}^{\infty}\frac{1}{n}(p^{-n}-q^{-n})\frac{\partial}{\partial x_n})-1\}$$

$$= \sum_{i,j\in\mathbf{Z}} Z_{ij}\, p^i\, q^{-j}.$$

One can give a simple combinatorial rule for the action of Z_{ij} in terms of Young diagrams [2].

5. Now we return to the vacuum expectation value

$$\tau(x) = \langle vac \mid e^{H(x)}g \mid vac\rangle$$

of a group element g. (Here the group G for $gl(\infty)$ means the one generated by e^X with $X \in gl(\infty)$ locally nilpotent.) By the correspondence (13), this amounts to considering the group orbit $\mathcal{O} = G \mid vac\rangle$ of the highest weight vector $\mid vac\rangle \in \mathcal{F}$. One can give the following characterization of $\mathcal{O}$:

Theorem. For $g \in A$, $g \mid vac\rangle$ belongs to group orbit $\mathcal{O}$ if and only if

(14) $$\operatorname{Res}_{k=\infty} (X(k)\tau)(x)\ (X^*(k)\tau)(x')\ dk = 0 \text{ for any } x, x'.$$

Picking up the coefficient of k^{-1} in $(X(k)\tau)(x)(X^*(k)\tau)(x')$, one obtains an infinite set of non-linear differential equations for $\tau(x)$:

(15) $$(D_1^4 + 3D_2^2 - 4D_1D_3)\tau\cdot\tau = 0, \cdots$$

where Hirota's bilinear operator $P(D)f\cdot g$ is defined by

$$P(D_1,D_2,\ldots)f\cdot g = P(\frac{\partial}{\partial y_1}, \frac{\partial}{\partial y_2},\ldots)(f(x+y)g(x-y)) \mid_{y_1=y_2=\cdots=0}.$$

The equations (15) are typical examples of "soliton equations", known as the *Kadomtsev-Petviaschvili* (KP) *hierarchy*. It can be shown that the polynomial solutions to the KP hierarchy are exhausted by the τ-functions $\langle vac| e^{H(x)}g |vac\rangle$, $g \in G$. Looking backwards at the constructions thus far, one can state as follows: The space of solutions to the KP hierarchy is identified with $\mathcal{O}$. There is a transitive action of the group G on it [14].

6. A more conventional way to introduce the KP hierarchy is the linearization of Lax-Zakharov-Shabat. Here the non-linear equations are derived as the integrability condition of the linear system

$$\begin{cases} \dfrac{\partial}{\partial x_2} w = \left(\dfrac{\partial^2}{\partial x_1^2} + u\right) w \\ \dfrac{\partial}{\partial x_3} w = \left(\dfrac{\partial^3}{\partial x_1^3} + \frac{3}{2} u \dfrac{\partial}{\partial x_1} + v\right) w, \end{cases}$$

where $u = 2 \dfrac{\partial^2}{\partial x_1^2} \log \tau$, $v = 3 \dfrac{\partial^2}{\partial x_1 \partial x_2} \log \tau, \ldots$. The language of free fermion provides a way to unify the "linearization" in the sense above and the "bilinearization" of Hirota. Define

$$w(x,k) = X(k)\tau/\tau =$$

$$\tau(x_1 - \frac{1}{k}, x_2 - \frac{1}{2k^2},\ldots)\exp(kx_1 + k^2x_2 + \cdots)/\tau,$$

(16)

$$w^*(x,k) = X^*(k)\tau/\tau =$$

$$\tau(x_1 + \frac{1}{k}, x_2 + \frac{1}{2k^2},\ldots)\exp(-x_1k - x_2k^2,\ldots)/\tau.$$

(To put it another way, $w(x,k) = \langle 1 \mid e^{H(x)}\psi(k)\, g \mid vac\rangle$, $\psi(k) = \sum_{n\in\mathbb{Z}} \psi_n k^n$, etc.) They are formally expanded into a series at $k = \infty$

$$w(x,k) = (1 + w_1(x)k^{-1} + w_2(x)k^{-2} + \cdots) \exp(kx_1 + k^2x_2 + \cdots),$$

(17)

$$w^*(x,k) = (1 + w_1^*(x)k^{-1} + w_2^*(x)k^{-2} + \cdots)\exp(-x_1k - x_2k^2 - \cdots).$$

Theorem. Suppose two formal series $w(x,k)$, $w^*(x,k)$ of the form (17) are given. Assume that they satisfy the bilinear identity

$$\mathrm{Res}_{k=\infty}\, w(x,k)\, w^*(x',k)\, dk = 0 \qquad \text{for any } x,\ x'.$$

Then:

1) There exists a function $\tau(x)$ such that (16) holds,

2) $\tau(x)$ satisfies the KP hierarchy,

3) For each $n = 1,2,\ldots$ there exists a differential operator in $\frac{\partial}{\partial x_1}$ of the form

$$B_n(x, \frac{\partial}{\partial x_1}) = (\frac{\partial}{\partial x_1})^n + \sum_{k=0}^{n-2} b_{nk}(x)(\frac{\partial}{\partial x_1})^k$$

such that

$$\frac{\partial}{\partial x_n} w = B_n(x, \frac{\partial}{\partial x_1})w, \qquad -\frac{\partial}{\partial x_n} w^* = B_n^*(x, \frac{\partial}{\partial x_1})\, w^*,$$

where

$$B_n^*(x, \frac{\partial}{\partial x_1}) = (-\frac{\partial}{\partial x_1})^n + \sum_{k=0}^{n-2} (-\frac{\partial}{\partial x_1})^k\, b_{nk}(x)$$

denotes the formal adjoint operator.

7. Finally let us comment on the reduction of Lie algebras. By folding up the Dynkin diagram of $gl(\infty)$ (an ∞-chain $\cdots$ —o—o—o— $\cdots$), one gets various affine Lie algebras contained in $gl(\infty)$. Restricting the "symmetry algebra" $gl(\infty)$ to these subalgebras, there appear reduced hierarchies of non-linear differential equations. Many of the soliton equations (KdV, Sinh-Gordon,...) are known to fall within this category. The Painlevé type equations are obtained by one more step of reduction, whose remaining symmetry is discrete [15].

$\vdots$		$\vdots$
$\cup$		$\cup$
$gl(\infty)$	$\longleftarrow\ \longrightarrow$	KP, . . .
$\cup$		$\cup$
$A_1^{(1)}$, . . .	$\longleftarrow\ \longrightarrow$	KdV, . . .
$\cup$		$\cup$
affine Weyl group	$\longleftarrow\ \longrightarrow$	Painlevé, . . .

Hopefully, beyond the scheme described above one might be able to find a whole series of non-linear differential equations and their corresponding symmetry Lie algebras [16].

COMMENTS AND BIBLIOGRAPHY

The references given below are by no means complete.

[1] M. Jimbo, T. Miwa, M. Sato and Y. Môri, Springer Lecture Notes in Physics 116 (1980) 119.
M. Sato, T. Miwa and M. Jimbo, ibid. 126 (1980), 429.

[2] M. Jimbo and T. Miwa, Publ. RIMS, Kyoto Univ. 19 (1983) 943.

[3] M. Sato, T. Miwa and M. Jimbo, Publ. RIMS, Kyoto Univ. 14 (1978) 223, 15 (1979) 201, 577, 871, 16 (1980) 531.
M. Jimbo, T. Miwa, Y. Môri and M. Sato, Physica 1 D (1980) 80.

[4] M. Kashiwara and T. Miwa, Proc. Japan Acad. 57 A (1981) 342.
E. Date, M. Kashiwara and T. Miwa, ibid. 57A (1981) 387.
E. Date, M. Jimbo, M. Kashiwara and T. Miwa, J. Phys. Soc. Jap. 50 (1981) 3806, 3813; Physica 4 D (1982) 343; Publ. RIMS, Kyoto Univ. 18 (1982) 1077, 1111.

[5] To be more precise, one lets also T tend to the "critical temperature" Tc. In the sequel, this limit is taken from above Tc. For simplicity of presentation, we assume also $E_1=E_2$.

[6] T.T. Wu, B.M. McCoy, C.A. Tracy and E. Barouch, Phys. Rev. B 13 (1976) 316.

[7] A description of Painlevé's work can be found in the book: E.L. Ince, Ordinary differential equations, Dover, New York, 1956.

[8] L. Schlesinger, J. für Math. 141 (1972) 96.

[9] M. Jimbo, T. Miwa and K. Ueno, Physica 2D (1981) 306.

[10] L. Onsager, Phys. Rev. 65 (1944) 117.

[11] $\psi_\pm(x)$ are obtained as the limit of $p_{ij} = V^i p_j V^{-i}$, $q_{ij} = V^i q_j V^{-i}$. They satisfy the 2-dimensional Dirac equation

$$\left(\pm \frac{\partial}{\partial x^0} + \frac{\partial}{\partial x^1}\right) \psi_\pm(x) = m\, \psi_\mp(x)$$

where $m > 0$ is the mass parameter.

[12] Note that $w_\pm(x)$ solve the Dirac equation, which is elliptic in the Euclidean region. We are thus dealing with an analogue of Riemann's monodromy problem for solutions of the Euclidean Dirac equation instead of the Cauchy-Riemann equation. In the limit $m \to 0$, the former reduces to the latter.

[13] T. Miwa, Publ. RIMS, Kyoto Univ. 17 (1981) 665.

[14] It was M. and Y. Sato who first established this picture. A short account of their theory is found in: M. Sato, RIMS Kokyuroku 439 (1981) 30.

[15] K. Okamoto, Transformation groups for Painlevé equations, in preparation (Univ. of Tokyo).

[16] For the self-dual Yang-Mills equation, the transformation group of solutions is clarified by: K. Ueno and Y. Nakamura, Phys. Lett. 109 B (1982) 273, 117 B (1982) 208. See also K. Takasaki, RIMS preprint 459 (1983).

R.I.M.S.
Kyoto University
Kyoto 606, Japan

THE RIEMANN-HILBERT DECOMPOSITION AND THE KP HIERARCHY

Kimio Ueno[†]

§0.

By the Riemann-Hilbert (RH) problem, we mean the following problem. "For a given analytic curve C in $\mathbb{P}^1$, and for a given matrix function $H(\zeta)$ ($\zeta \in \mathbb{P}^1$) which is holomorphic and invertible on C, find the decomposition of $H(\zeta)$ into a product of matrix functions, $X_{\pm}(\zeta)$

(1) $$H(\zeta) = X_{-}(\zeta)^{-1} X_{+}(\zeta),$$

where $X_{+}(\zeta)$ (resp. $X_{-}(\zeta)$) is holomorphic and invertible in the inner (resp. outer) domain of C."

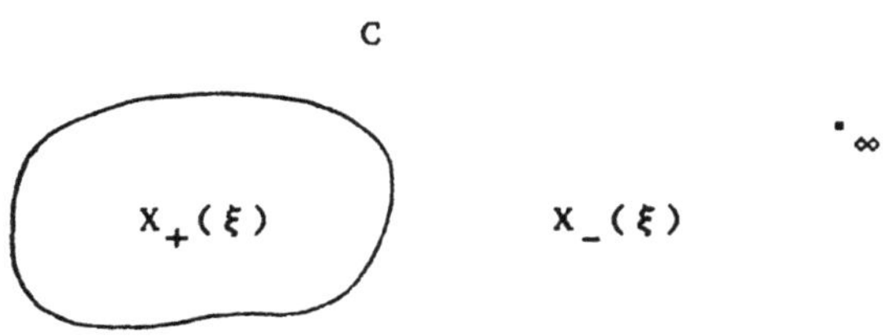

The decomposition (1) is the so-called Birkhoff decomposition of Euclidean Lie groups in the terminology of Kac-Peterson [7].

As is well-known, the RH problem, together with the inverse scattering method, has been a very useful tool to solve soliton equations [3], [4], [2]. However, to apply this viewpoint to the Kadomtsev-Petviashvili (KP) hierarchy or the Toda lattice (TL) hierarchy, we have to generalize the RH problem. In the following sections, we shall discuss this generalization.

[†] Partially supported by the National Science Foundation through the Mathematical Sciences Research Institute.

Vertex Operators in Mathematics and Physics - Proceedings of a Conference November 10-17, 1983. Publications of the Mathematical Sciences Research Institute #3, Springer-Verlag, 1984.

§1.

To begin with, we prepare some notations. Let $g\ell(\infty)$ be an algebra of all $\mathbb{Z} \times \mathbb{Z}$ matrices, that is

$$g\ell(\infty) = \{ X = \sum_{i,j\in\mathbb{Z}} x_{ij} E_{ij};\ x_{ij} \in \mathbb{C} \},$$

where $E_{ij} = (\delta_{i\mu}\delta_{j\nu})_{\mu,\nu\in\mathbb{Z}}$ is the (i,j) matrix unit. Let Λ^n be the n-th shift matrix,

$$\Lambda^n = (\delta_{\mu+n,\nu})_{\mu,\nu\in\mathbb{Z}}$$

$=$ 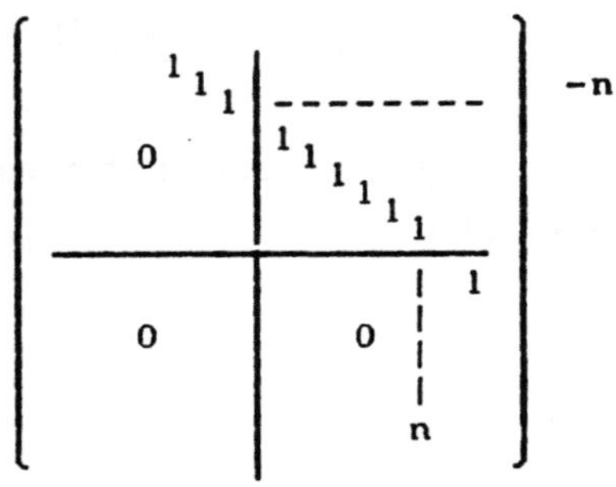

Then we should notice that any matrix $X \in g\ell(\infty)$ is expressed as

$$X = \sum_{n\in\mathbb{Z}} \operatorname{diag}(x_n(s))\ \Lambda^n.$$

Here $\operatorname{diag}(x_n(s))$ denotes a diagonal matrix $\operatorname{diag}(\cdots x_n(-1), x_n(0), x_n(1), \cdots)$, and $\operatorname{diag}(x_n(s))\ \Lambda^n$ indicates the product of $\operatorname{diag}(x_n(s))$ and Λ^n in $g\ell(\infty)$. We define the ($\pm$) part of X by

$$(X)_+ = \sum_{n\geq 0} \operatorname{diag}(x_n(s))\ \Lambda^n$$

$$(X)_- = \sum_{n<0} \operatorname{diag}(x_n(s))\ \Lambda^n.$$

Let $A \in GL(\infty)$ be a constant matrix and set

(2) $$H(x) = \exp \xi(x; \Lambda) A \exp \xi(-x; \Lambda),$$

where

$$\xi(x; \Lambda) = \sum_{n=1}^{\infty} x_n \Lambda^n,$$

and $x = (x_1, x_2, \ldots)$ is a series of an infinite number of variables.

Our first problem is stated as follows:

"Find invertible matrices of the form

$$\hat{W}^{(\infty)}(x) = \sum_{n=0}^{\infty} \mathrm{diag}(\hat{w}_n^{(\infty)}(s;\ x))\ \Lambda^{-n},$$

$$\text{with } \hat{w}_0^{(\infty)}(s;\ x) = 1 \quad \text{for all } s,$$

$$\hat{W}^{(0)}(x) = \sum_{n=0}^{\infty} \mathrm{diag}(\hat{w}_n^{(0)}(s;x))\ \Lambda^n$$

$$\text{with } \hat{w}_n^{(0)}(s;x) \neq 0 \text{ for all } s,$$

and satisfying

(3) $$\hat{W}^{(0)}(x) = \hat{W}^{(\infty)}(x)\ H(x)."$$

Obviously, this problem is a generalization of the RH problem to the group $GL(\infty)$; $H(x)$, (2) corresponds to $H(\zeta)$, (1), and $\hat{W}^{\binom{0}{\infty}}(x)$ to $X_{\pm}(\zeta)$, respectively. We shall call this problem the RH decomposition. We suppose that this problem is solved. Then one can easily verify that the solution is uniquely determined.

Now we try to show how the RH decomposition relates to a nonlinear problem. We rewrite equation (3) into

(4) $$W^{(0)}(x) = W^{(\infty)}\ A$$

where $W^{\binom{0}{\infty}}(x) = \hat{W}^{\binom{0}{\infty}}(x) \exp \xi(x;\ \Lambda)$. Set

$$B_n(x) = \partial_{x_n} W^{(\infty)}(x) \cdot W^{(\infty)}(x)^{-1}.$$

Then we have

$$(5) \qquad B_n(x) = \partial_{x_n} \hat{W}^{(\infty)}(x)\ \hat{W}^{(\infty)}(x)^{-1} + \hat{W}^{(\infty)}(x)\ \Lambda^n\ \hat{W}^{(\infty)}(x)^{-1}$$

$$= \partial_{x_n} W^{(0)}(x) \cdot W^{(0)}(x)^{-1}.$$

Since $\hat{W}^{(\infty)}(x)$ is a lower triangular matrix with the unit diagonal entries, $\partial_{x_n} \hat{W}^{(\infty)}(x) \cdot \hat{W}^{(\infty)}(x)^{-1}$ is strictly lower triangular, while $\partial_{x_n} W^{(0)}(x) \cdot W^{(0)}(x)^{-1}$ is upper triangular. Hence taking the (+) part of (5) and (6) leads to

$$B_n(x) = (\hat{W}^{(\infty)}(x)\Lambda^n\ \hat{W}^{(\infty)}(x)^{-1})_+$$

$$= (W^{(\infty)}(x)\ \Lambda^n W^{(\infty)}(x)^{-1})_+.$$

We obtain the following theorem.

<u>**Theorem**</u> **1.** We set

$$L(x) = W^{(\infty)}(x)\ \Lambda\ W^{(\infty)}(x)^{-1},$$

$$B_n(x) = (L(x)^n)_+.$$

Then the solution to the RH decomposition satisfies

$$(7) \qquad L(x)\ W^{(\infty)}(x) = W^{(\infty)}(x)\ \Lambda,$$

$$(8) \qquad \partial_{x_n} W^{(\infty)}(x) = B_n(x)\ W^{(\infty)}(x), \qquad n = 1,2,\ldots.$$

$L(x)$ and $B_n(x)$ take the forms

$$L(x) = \sum_{-\infty<n\leqslant 1} \operatorname{diag}(b_n(s;x))\Lambda^n,$$

$$B_n(x) = \sum_{m=0}^{n} \operatorname{diag}(b_{n,m}(s;x))\Lambda^{n-m}.$$

The compatibility condition for (7), (8) yields two systems of nonlinear differential equations,

(9) $\partial_{x_n} L = [B_n, L], \qquad n = 1,2,... \qquad$ (Lax equations),

(10) $\partial_{x_n} B_m - \partial_{x_n} B_n + [B_m, B_n] = 0,$

$m,n = 1,2,...,$ (Zakharov-Shabat equations).

It is not difficult to show that these systems are equivalent to each other [1].

By the way, the Lax equations (9) and the Zakharov-Shabat equations (10) are very similar to the KP hierarchy [5], [6]. This is not coincidence. In fact we can explain the link between the RH decomposition and the KP hierarchy as follows.

First of all, we should notice that from Eq. (4) we have

$$(\partial_x)^\alpha\, W^{(\infty)}(x) \cdot W^{(\infty)}(x)^{-1} = (\partial_x)^\alpha\, W^{(0)}(x) \cdot W^{(0)}(x)^{-1}$$

for any multi-index α. They are encapsulated into a single bilinear relation

(11) $W^{(\infty)}(x) \cdot W^{(\infty)}(x')^{-1} = W^{(0)}(x) \cdot W^{(0)}(x')^{-1} \qquad$ for any x, x'.

Conversely equation (4) follows from this bilinear relation. Furthermore, if we set

$$\hat{W}^{(\infty)}(x)^{-1} = \sum_{n=0}^{\infty} \operatorname{diag}(\hat{w}_n^{(\infty)*}(s+1;x))\Lambda^{-n},$$

and define

$$w^{(\infty)}(s;x;\lambda) = \left(\sum_{n=0}^{\infty} \hat{w}_n^{(\infty)}(s;x)\lambda^{-n} \right) \lambda^s \exp \xi(x;\lambda),$$

$$w^{(\infty)*}(s;x;\lambda) = \left(\sum_{n=0}^{\infty} \hat{w}_n^{(\infty)*}(s;x)\lambda^{-n} \right) \lambda^{-s} \exp \xi(-x;\lambda),$$

then we can easily show that the following residue formulas ([6]) holds;

$$\oint_{\lambda=\infty} w^{(\infty)}(s;x;\lambda) w^{(\infty)*}(s';x';\lambda) \frac{d\lambda}{2\pi i} = 0, \text{ for any } s \geqslant s' \text{ and } x,\ x'.$$

From the bilinear relation (11) (or the above formulas), we can deduce the existence of τ functions.

Theorem 2. [1] There exist τ functions, $\tau(s;x)$ which are characterized by

$$\hat{w}_n^{(\infty)}(s;x) = p_n(-\tilde{\partial}_x)\ \tau(s;x)/\tau(s;x),$$

(12)

$$\hat{w}_n^{(\infty)*}(s;x) = p_n(\tilde{\partial}_x)\ \tau(s;x)/\tau(s;x), \qquad n = 1,2,\ldots.$$

Here $p_n(x)$'s are polynomials in x defined by $\exp \xi(x;\lambda) = \sum_{n=0}^{\infty} p_n(x)\lambda^n$, and $\tilde{\partial}_x = (\partial_{x_1}, \frac{1}{2}\partial_{x_2}, \frac{1}{3}\partial_{x_3}, \ldots)$. By this condition, the τ functions are uniquely determined up to constant factors.

Substituting (12) into (11), we obtain Hirota's bilinear equations for the τ functions.

Theorem 3. [1] We have

(13) $$\sum_{j=0}^{\infty} p_{m+j}(-2a) p_j(\tilde{D}_x) e^{\langle a,D_x \rangle}\ \tau(s+m+1;x)\cdot\tau(s;x) = 0, \text{ for } m \geqslant -1.$$

Here $\tilde{D}_x = (D_{x_1}, \frac{1}{2}D_{x_2}, \ldots)$ are Hirota's bilinear operators, and $a = (a_1, a_2, \ldots)$ are indeterminate variables, and $\langle a,D_x \rangle = \sum_{n=1}^{\infty} a_n D_{x_n}$.

Equations (13) imply that each τ function $\tau(s;x)$ is a τ function for the KP hierarchy.

In this way we can understand the KP hierarchy by means of the RH decomposition, the generalization of the classical Riemann-Hilbert problem. If the matrix A in the RH decomposition belongs to the group $O(\infty)$ or $Sp(\infty)$, we have the BKP or CKP hierarchy [1], [6]. Taking the matrix A to be

$$A = I + \sum_{j=1}^{r} a_j E_{m_j n_j},$$

$$A = I + \sum_{j=1}^{N} a_j X_{p_j q_j}, \quad (X_{pq} = \sum_{\mu,\nu\in\mathbb{Z}} E_{\mu\nu} p^{\mu} q^{-\nu}),$$

we obtain rational solutions, or N-soliton solutions to the KP hierarchy [1]. X_{pq} is the so-called vertex operator [6].

§2.

In this section we shall answer the following question: "How does the RH decomposition reduce to the RH problem."

First we introduce a subalgebra $g\ell(\infty)_\ell$ defined by $g\ell(\infty)_\ell = \{X \in g\ell(\infty);\ [\Lambda^\ell, X] = 0\}$. This subalgebra is isomorphic to the loop algebra $g\ell(\ell, \mathbb{C}[[\zeta,\zeta^{-1}]])$ under the following correspondence,

(14)
$$g\ell(\ell, \mathbb{C}[[,\zeta,\zeta^{-1}]]) \xrightarrow{\sim} g\ell(\infty)_\ell$$

$$X(\zeta) = \sum_{n\in\mathbb{Z}} \operatorname{diag}(x_n(0),\dots,x_n(\ell-1))\Lambda_\ell(\zeta)^n \longmapsto X$$

$$= \sum_{n\in\mathbb{Z}} \operatorname{diag}(x_n(s))\Lambda^n.$$

In the left side, $\Lambda_\ell(\zeta)$ is defined by

$$\Lambda_\ell(\zeta) = \begin{bmatrix} 0 & 1 & & \\ & \ddots & \ddots & \\ & & \ddots & 1 \\ \zeta & & & 0 \end{bmatrix},$$

and in the right side, the entries $x_n(s)$ are ℓ-periodic, that is, $x_n(s)$

$= x_n(s+\ell)$. This correspondence is a Lie algebra isomorphism, and induces a group isomorphism.

In the RH decomposition, let A belong to $g\ell(\infty)_\ell$, and let $A(\zeta) \in GL(\ell, \mathbb{C}[[\zeta,\zeta^{-1}]])$ be the inverse image of A under the isomorphism (14). Then we find that our RH decomposition reduces to

$$\hat{W}^{(0)}(x;\zeta) = \hat{W}^{(\infty)}(x;\zeta)\ H(x;\zeta) \tag{15}$$

Here $\hat{W}^{(0)}(x;\zeta)$, $\hat{W}^{(\infty)}(x,\zeta)$ take the form

$$\hat{W}^{\binom{0}{\infty}}(x,\zeta) = \sum_{n=0}^{\infty} \operatorname{diag}(\hat{w}^{\binom{0}{\infty}}(0;x),\ldots, \hat{w}^{\binom{0}{\infty}}(\ell-1;x)\ \Lambda_\ell(\zeta)^{\pm n}$$

and $H(x;\zeta)$ is defined as

$$H(x;\zeta) = \exp\ \xi(x;\Lambda_\ell(\zeta))\ A(\zeta)\ \exp\ \xi(-x;\Lambda_\ell(\zeta)).$$

Equation (15) is nothing but the RH problem, and yields the ℓ-reduction of the KP hierarchy [6]. In fact, we obtain the following theorem.

<u>Theorem</u> 4. [1] For the solution of (15), we can choose τ functions satisfying the following two conditions;

$$\tau(s+\ell;x) = \tau(s;x) \quad \text{for any } s,$$

$$\partial_{x_n} \tau(s;x) = 0, \quad \text{for any } n \equiv 0 \bmod \ell.$$

§3.

Finally we shall discuss the Toda lattice (TL) hierarchy. The Toda lattice equation has been intensely studied by many physicists and mathematicians. The Toda lattice equation is a difference-differential equation given by

$$\partial_t^2 u(s;t) = \exp(u(s;t) - u(s-1;t)) - \exp(u(s+1;t) - u(s;t)), \ (s \in \mathbf{Z}),$$

and together with the celebrated KdV equation, it is one of the most important equations in soliton theory. The two dimensional Toda lattice equation is a natural generalization of this equation, and is defined by

$$(16) \qquad \partial_{x_1}\partial_{y_1}u(s) = \exp(u(s) - u(s-1)) - \exp(u(s+1) - u(s)),$$

$$u(s) = u(s;x_1,y_1), \qquad s \in \mathbf{Z}.$$

We should notice that equation (16) is transformed to the Zakharov-Shabat equation

$$(17) \qquad \partial_{y_1}B_1 - \partial_{x_1}C_1 + [B_1,C_1] = 0,$$

where B_1 and C_1 are given by

$$B_1 = \mathrm{diag}(\partial_{x_1}u(s)) + \Lambda, \qquad C_1 = \mathrm{diag}(e^{u(s)-u(s-1)})\ \Lambda^{-1}.$$

In order to introduce the TL hierarchy, we prepare two series of an infinite number of variables, $x = (x_1,x_2,\dots)$, $y = (y_1,y_2,\dots)$, and set

$$L(x,y) = \sum_{-\infty<n\leq 1} \mathrm{diag}(b_n(s;x,y))\ \Lambda^n$$

$$\text{with } b_1(s;x,y) = 1 \quad \text{for all } s,$$

$$M(x,y) = \sum_{-1\leq n<+\infty} \mathrm{diag}(c_n(s;x,y))\ \Lambda^n$$

$$\text{with } c_{-1}(s;x,y) \neq 0 \qquad \text{for all } s,$$

$$B_n(x,y) = (L(x,y)^n)_+, \qquad C_n(x,y) = (M(x,y)^n)_-.$$

The TL hierarchy is, by definition, a system of the Lax equations,

$$\partial_{x_n} L = [B_n, L], \qquad \partial_{y_n} L = [C_n, L],$$

(18)

$$\partial_{x_n} M = [B_n, M], \qquad \partial_{y_n} L = [C_n, L], \qquad n = 1,2,\ldots.$$

We can verify that the TL hierarchy is equivalent to a system of the Zakharov-Shabat equations,

$$\partial_{x_n} B_m - \partial_{x_m} B_n + [B_m, B_n] = 0, \; \partial_{y_n} C_m - \partial_{y_n} C_n + [C_m, C_n] = 0$$

(19)

$$\partial_{y_n} B_m - \partial_{x_m} C_n + [B_m, C_n] = 0, \; m, \; n = 1,2,\ldots.$$

We should notice that equation (17) is contained here (m = n = 1).

The linearization of the hierarchy is achieved as

$$L(x,y)\, W^{(\infty)}(x,y) = W^{(\infty)}(x,y)\Lambda, \qquad M(x,y)W^{(0)}(x,y) = W^{(0)}(x,y)\Lambda^{-1},$$

$$\partial_{x_n} W(x,y) = B_n(x,y)\, W(x,y), \; \partial_{y_n} W(x,y) = C_n(x,y)W(x,y), \; n = 1,2,\ldots,$$

$$(W(x,y) = W^{(\infty)}(x,y), \; W^{(0)}(x,y)).$$

Under the condition that L(x,y) and M(x,y) are holomorphic at x = y = 0, we may take $W^{\binom{0}{\infty}}(x,y)$ to be of the form

$$W^{(\infty)}(x,y) = \hat{W}^{(\infty)}(x,y) \exp \xi(x,\Lambda)$$

(20)

$$W^{(0)}(x,y) = \hat{W}^{(0)}(x,y) \exp \xi(y;\Lambda^{-1}),$$

where

$$\hat{W}^{\binom{0}{\infty}}(x,y) = \sum_{n=0}^{\infty} \mathrm{diag}(\hat{w}_n^{\binom{0}{\infty}}(s;x,y))\, \Lambda^{\pm n},$$

with $\hat{w}_0^{(\infty)}(s;x,y) = 1$, $\hat{w}_0^{(\infty)}(s;x,y) \neq 0$ for any s.

Instead of taking these forms, we may set

$$(21) \qquad W^{\binom{0}{\infty}}(x,y) = \hat{V}^{\binom{0}{\infty}}(x,y)\ \exp(\xi(x;\Lambda) + \xi(y;\Lambda^{-1})).$$

We shall call $W^{\binom{0}{\infty}}(x,y)$ wave matrices.

As in the KP hierarchy, we can show that there exist τ functions for the TL hierarchy, and that the matrix coefficients of the wave matrices are expressed as the τ-quotient (cf. §1, Theorem 2). Furthermore, we obtain the wave matrices if we solve the following RH decomposition,

(for (20)) $$\hat{W}^{(0)}(x,y) = \hat{W}^{(\infty)}(x,y)\ H(x,y),$$

where

$$H(x,y) = \exp\ \xi(x,\Lambda)\ A\ \exp\ \xi(-y;\ \Lambda^{-1}),$$

(for (21)) $$\hat{V}^{(0)}(x,y) = \hat{V}^{(\infty)}(x,y)\ H'(x,y),$$

where

$$H'(x,y) = \exp(\xi(x;\Lambda) + \xi(y;\Lambda^{-1}))\ A\ \exp(\xi(-x;\Lambda) + \xi(-y;\Lambda^{-1})).$$

For the details of the discussion in this note, the readers should refer to "Toda Lattice Hierarchy" by K. Ueno and K. Takasaki (RIMS preprint 425, to appear in Advanced Study in Pure Math.).

REFERENCES

[1] K. Ueno and K. Takasaki: Toda Lattice Hierarchy I. Proc. Japan Aca., 59A 167-170; ditto. II., ibid 215-218; Toda Lattice Hierarchy. RIMS preprint 425 (1983) (to appear in Advanced Study in Pure Math.)

[2] K. Ueno and Y. Nakamure; Transformation Theory for Anti-Self-Dual Equation. Publ. RIMS, Kyoto Univ. 19, No. 2 (1983).

[3] V.E. Zakharov and A.V. Mikhailov: Sov. Phys. JETP, 47, 1017 (1978).

[4] I. Hauser and F.J. Ernst: J. Math. Phys., 21, 1126 (1980).

[5] M. Sato: Kokyoroku RIMS, Kyoto Univ., No. 439, 30-46 (1981).

[6] E. Date, M. Jimbo, M. Kashiwara and T. Miwa: Transformation groups for soliton equations II. Proc. Japan Acad., 57A, 387-392 (1981); ditto. III. J Phys. Soc. Japan, 40, 3806-3812 (1981); ditto. V. Physica 4 D, 343 (1982); ditto VI. J. Phys. Soc. Japan 50, 3813-3818 (1981).

[7] V.G. Kac and D.H. Peterson: Regular Functions on Certain Infinite-dimensional Groups. To appear in "Arithmetic and Geometry" edited by M. Artin and J. Tate Birkäuser, Boston, 1983.

Department of Mathematics
Yokohama City University
22-2 Seto, Kanazawa-ku Yokohama 236, Japan

SUPERSYMMETRIC YANG-MILLS FIELDS AS AN INTEGRABLE SYSTEM AND CONNECTIONS WITH OTHER NON-LINEAR SYSTEMS

Ling-Lie Chau[†]

ABSTRACT

Integrable properties, i.e. existence of linear systems, infinite number of conservation laws, Riemann-Hilbert transforms, affine Lie algebra of Kac-Moody, and Bianchi-Bäcklund transformations, are discussed for the constraint equations of the supersymmetric Yang-Mills fields. For N$\geqslant$3 these constraint equations give equations of motion of the fields, thus integrating the constraint equations is equivalent to integrate the equation of motion. Further, these equations of motion reduce to the ordinary Yang-Mills equations as the spinor and scalar fields are eliminated. Therefore, these understandings also provide a possible method to solve the full

CONTENTS

[†] The submitted manuscript has been authored under Contract No. De-Ac02-76Ch00016 with the U.S. Department of Energy. Accordingly, the U.S. Government retains a nonexclusive, royalty-free license to publish or reproduce the published form of this contribution, or allow others to do so, for U.S. Government purposes.

Vertex Operators in Mathematics and Physics - Proceedings of a Conference November 10-17, 1983. Publications of the Mathematical Sciences Research Institute #3, Springer-Verlag, 1984.

Yang-Mills equations. Connections with other non-linear systems are also discussed.

INTRODUCTION

Beginning with Maxwell's equations and quantum electrodynamics, and the recent spectacular success of the electroweak theory of quantum flavor dynamics [1,2] gauge theories have become an essential part of physics. It has become increasingly clear that, besides its mathematical beauty, the Yang-Mills theory [1,3-5] may provide the key to our understanding of strong interactions. With the recent experimental observation of gluon jets [6], the ideas of Yang-Mills gauge theory for strong interactions is brought one step closer to reality. Despite many interesting theoretical and phenomenological observations, like confinement [7], asymptotic freedom and QCD perturbative studies [8], the Yang-Mills gauge equations are far from being solved.

In the past fifteen years or so, powerful mathematical tools have been developed in completely solving many two-dimensional nonlinear differential equations [9-11]. The characteristics of these integrable systems are: existence of the linear systems, Bianchi-Bäcklund transformation (BT), infinite number of conservation laws, soliton solutions, and even the construction of S-matrix [12]. It is our intention to see whether these powerful techniques can be applied to solve the Yang-Mills theory.

For the past few years, we have found that amazingly the self-dual Yang-Mills (SDYM) fields, in the J-formulation, possess many characteristics of those integrable systems [13-22]. Recently the affine algebra of Kac-Moody has been found in the SDYM system [23-25]. It thus provides a beautiful mathematical system in four dimensions [26-28].

Due to the extreme similarity in appearance between the SDYM equations and the two dimensional chiral fields [28-35], and the Ernst equations [36], i.e. the stationary axially symmetric equations of Einstein equations, many fruitful results have come about by investigating both systems hand in hand.

Figure 1. Some Generic Structures of integrable Non-linear systems

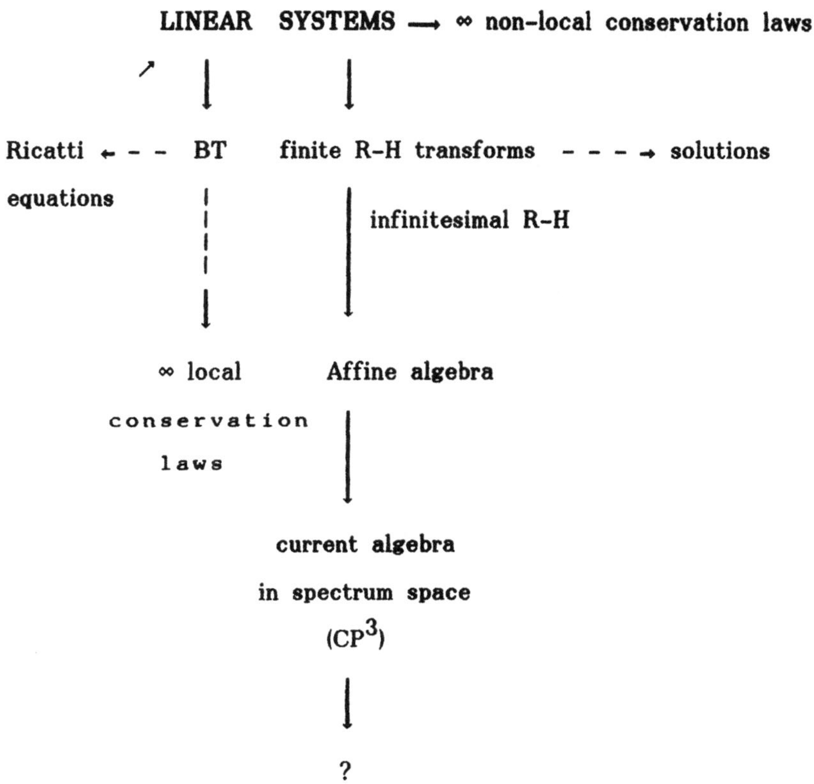

The solid lines indicate developments made for the four dimensional self-dual Yang-Mills equations, for the supersymmetric Yang-Mills constraint equations, and for the two dimensional chiral equations, and the static axially symmetric Einstein equations (i.e. the Ernst equations). The dashed lines indicate developments made only for the latter two dimensional systems.

After many year's work by many people, we now even have some systematic understandings. In Fig. 1 we give a flow chart of some logical links between various structures of these systems. At the center are the corresponding linear systems for the original non-linear equations. Their existence implies that the original non-linear equations are results of the integrability of the set of linear systems, or *the original non-linear equations are equivalent to some generalized curvatureless conditions.* Once the linear systems are obtained, many developments indicated in Fig. 1 can be made. The developments indicated by solid lines have been made both for the SDYM equations in four dimensions, as well as for the chiral equations in two dimensions and the Ernst equations. The dashed lines indicate developments made only for the latter two dimensional equations. The linear systems are immediately related to the existence of an infinite number of conservation laws. Also from these linear systems, parametric Bianchi-Bäcklund transformations (BT) can be obtained, which are good for generating new solutions (global properties have to be imposed in addition). These BT were obtained purely based upon guess work. Now we can show that they are generated by the so called Darboux transformations, which is a special form of Riemann-Hilbert transforms. From the parameters in the BT, infinite number of local conservation laws can be obtained [30,31]. For the chiral model in two dimensions, we can further show that the linear systems can be derived from the BT via the Riccati equations, [28]. This indicates that the parametric BT given in [30] for the chiral field is probably the most general one, and it is as fundamental as the linear system, and thus as the original non-linear system. So far, such development has not been done for the BT [16] of the four dimensional Yang-Mills fields.

To generate new solutions from the linear systems, for the original non-linear systems, is first to generate new solutions to the linear systems, and then via the inverse scattering method to obtain the corresponding new solutions to the original non-linear system. The matrix Riemann-Hilbert (R-H) transforms [37] provide such a method. It is also a way in which the original problem of solving differential equations in the coordinate space being transformed into solving integral equations in the spectrum space. Such methods have been very useful in obtaining many new solutions to the stationary, axially symmetric Einstein equation, i.e. the Ernst equation [36]. Actually it was shown that all solutions with proper boundary conditions can be obtained via the R-H transform, which was called the Gerosch conjecture [38]. From the finite R-H transform, the infinitesimal R-H transform can be derived. We then can show that the algebra structure satisfied by these infinitesmal R-H transforms are of the affine type, [23-25,27,28]. This has been the main topic of this workshop. Very interestingly, Dr. Friedan at this workshop has pointed out the link between the values of the central extensions to various two dimensional models in statistical mechanics. However, so far no central extensions of the loop algebra have been found in many of the nonlinear systems, e.g. Sine-Gordon, chiral, and self-dual Yang-Mills. The physical implications of the Kac-Moody algebra for these systems are yet to be understood.

Now we shall motivate how to solve differential equations on curvatures $F_{\mu\nu}$ via solving algebraic equations on them. For example, the self-dual, and anti-self-dual Yang-Mills equations are a set of algebraic constraint relations on the curvatures $F_{\mu\nu}$, i.e.,

(1) $$F_{\mu\nu} = \pm\frac{1}{2}\epsilon_{\mu\nu\gamma\delta}F^{\gamma\delta} \equiv \pm\, {}^*F_{\mu\nu}.$$

Under these constraints, the Bianchi identities

(2) $$\mathcal{D}^{\mu}\, {}^*F_{\mu\nu} = 0,$$

become the equations of motion

(3) $$\mathcal{D}^{\mu}F_{\mu\nu} = 0 \, .$$

Therefore, solving the self-dual and anti-self-dual equations, Eq. (1), can be viewed as a way of finding some special solutions of the Yang-Mills equations, Eq. (3). It was suggested by Witten [39], and by Isenberg, Yasskin and Green [40] that such concept and procedure can be generalized to more general solutions (even hopefully to all solutions) of the full Yang-Mills equations. It was demonstrated that, formulated in complex $d = 8$ dimensions, or real $d = 4$ plus $n \geqslant 3$ super dimensions, certain algebraic constraint equations do lead to Yang-Mills equations via the Bianchi identities. Recently, it has been shown that these constraint equation possess many of the integrable properties. Here we shall concentrate on the discussions of supersymmetric Yang-Mills equations.

The super symmetric Yang-Mills fields [42] formulated in the coordinate $x \equiv (x^{\alpha\dot{\beta}}, \theta^{\alpha}_{s}, \bar{\theta}^{\dot{\beta}t})$ where α, $\beta = 1,2$: $s,t = 1,2, \cdots N$, have in general six types of curvatures $F_{\mu\nu}$, $F^{st}_{\alpha\beta}$, $F_{\dot{\alpha}s,\dot{\beta}t}$, $F^{s}_{\mu,\alpha}$, $F_{\mu,\dot{\beta}t}$, $F^{s}_{\alpha,\dot{\beta}t}$. To eliminate the extra curvatures due to the introduction of super coordinates, constraints [39,42-44] were introduced onto those curvatures involving only the super coordinates, i.e., $F^{st}_{\alpha\beta}$, $F_{\dot{\alpha}s,\dot{\beta}t}$, $F^{s}_{\alpha,\dot{\beta}t}$. Interestingly for $N \geqslant 3$ supersymmetric Yang-Mills fields, these constraints can put the theory on shell [39,44] i.e., these constraints lead to equations of motion. Recently it has been shown by Volovich [45] that to these constraint equations linear systems can be introduced. And it has been demonstrated by Devchand [46] that many of the formulations [27,28] developed for the self-dual Yang-Mills system as an integrable system can also be made for these systems, e.g., infinite number of conservation laws, the existence of affine Lie algebra of Kac-Moody.

With Ge and Popowicz [47], we have developed further the integrability properties of supersymmetric Yang-Mills equations. We have constructed the Riemann-Hilbert transforms, the infinitesimal forms of which precisely give the affine Lie Algebra of Kac-Moody. From the linear systems, a two-parameter Bianchi-Bäcklund (BT) transformation can also be constructed. In the special (anti) self-dual case, the BT transformations involving only the ordinary coordinate becomes that obtained in Ref. 16.

These new developments give the hope that the full Yang-Mills equations can be solved via the supersymmetric way.

I. FORMULATION OF SUPER-SYMMETRIC YANG-MILLS FIELDS AND THE CONSTRAINT EQUATIONS

The superspace coordinates are given by $x = (x^{\mu}, \theta^{\alpha}_{s}, \bar{\theta}^{\dot{\beta}t}) = (x^{\alpha\dot{\beta}}, \theta^{\alpha}_{s}, \bar{\theta}^{\dot{\beta}t})$, where x^{μ}, $\mu = 0,1,2,3$, is the ordinary space coordinate which can also be described as $x^{\alpha\dot{\beta}} = x^{\mu}\sigma^{\alpha\dot{\beta}}_{\mu}$. The θ^{α}_{s}, $\bar{\theta}^{\dot{\beta}t}$ are the super coordinates, with $s,t = 1,2,\cdots N$, characterizing the dimensions of supersymmetry. Corresponding to these coordinates there are three types of derivatives:

$$\partial_{\mu} = \partial/\partial x^{\mu} \text{ or } \partial_{\alpha\dot{\beta}} = \partial/\partial x^{\alpha\dot{\beta}};$$

$$D^{s}_{\alpha} = \partial/\partial\theta^{\alpha}_{s} + i\bar{\theta}^{\dot{\beta}s}\partial_{\alpha\dot{\beta}},\quad \bar{D}_{\dot{\beta}t} = -\partial/\partial\bar{\theta}^{\dot{\beta}t} - i\theta^{\alpha}_{t}\partial_{\alpha\dot{\beta}}$$

with their algebraic rules

$$\left[\partial_{\alpha\dot{\beta}},\partial_{\gamma\dot{\delta}}\right] = 0,\quad \left\{D^{s}_{\alpha},D^{t}_{\beta}\right\} = 0$$

$$\left\{\bar{D}_{\dot{\beta}s},\bar{D}_{\dot{\alpha}t}\right\} = 0,\quad \left\{D^{s}_{\alpha},\bar{D}_{\dot{\beta}t}\right\} = -2i\delta^{s}_{t}\partial_{\alpha\dot{\beta}}.$$

The covariant derivatives are

$$\nabla_{\mu} \equiv \partial_{\mu} + A_{\mu},\quad \text{or } \nabla_{\alpha\dot{\beta}} \equiv \partial_{\alpha\dot{\beta}} + A_{\alpha\dot{\beta}};$$

$$\nabla^{s}_{\alpha} = D^{s}_{\alpha} + A^{s}_{\alpha};\quad \bar{\nabla}_{\dot{\beta}t} = \bar{D}_{\dot{\beta}t} + A_{\dot{\beta}t}.$$

The curvatures are

$$\left[\nabla_{\mu},\nabla_{\nu}\right] = F_{\mu\nu};\quad \left\{\nabla^{s}_{\alpha},\nabla^{t}_{\beta}\right\} = F^{st}_{\alpha\beta};\quad \left\{\bar{\nabla}_{\dot{\alpha}s},\bar{\nabla}_{\dot{\beta}t}\right\} = F_{\dot{\alpha}s,\dot{\beta}t};$$

$$\left[\nabla_{\mu},\nabla^{s}_{\alpha}\right] = F^{s}_{\mu\alpha};\quad \left[\nabla_{\mu},\bar{\nabla}_{\dot{\beta}t}\right] = F_{\mu,\dot{\beta}t};\tag{1.1}$$

$$\left\{\nabla^{s}_{\alpha},\bar{\nabla}_{\dot{\beta}t}\right\} = F^{s}_{\alpha,\dot{\beta}t} - 2i\delta^{s}_{t}\nabla_{\alpha\dot{\beta}};$$

Note that $F^{s}_{\alpha,\dot\beta t} = D^{s}_{\alpha}A_{\dot\beta t} + \bar{D}_{\dot\beta t}A^{s}_{\alpha} + \{A^{s}_{\alpha},A_{\dot\beta t}\} + 2i\delta^{s}_{t}A_{\alpha\dot\beta}$. Usually to eliminate the super components of the curvature, constraints are put on the three types of curvatures that involves only the super components:

$$(1.2)\qquad F^{st}_{\alpha\beta} + F^{st}_{\beta\alpha} = 0; \quad F_{\dot\alpha s,\dot\beta t} + F_{\dot\beta s,\dot\alpha t} = 0; \quad F^{s}_{\alpha,\dot\beta t} = 0.$$

It was noted by Volovich [45] and Devchand [46] that very similar to the self-dual Yang-Mills (SDYM) case, some of these curvatureless equations,

$$(1.3)\qquad F^{st}_{11} = 0 = F^{st}_{22};$$

$$F_{\dot1 s,\dot1 t} = 0 = F_{\dot2 s,\dot2 t}; \; F^{s}_{1,\dot1 t} = 0 = F^{s}_{2,\dot2 t};$$

can be integrated by the following form of the potentials

$$(1.4)\qquad \begin{aligned} A^{s}_{1} &= g^{-1}D^{s}_{1}g; \quad A_{\dot1 t} = g^{-1}\bar{D}_{\dot1 t}g; \\ A^{s}_{2} &= h^{-1}D^{s}_{2}h; \quad A_{\dot2 t} = h^{-1}\bar{D}_{\dot2 t}h. \end{aligned}$$

Similarly, by defining a J matrix,

$$(1.5)\qquad J \equiv hg^{-1},$$

in terms of which the curvatures become:

$$(1.6a)\qquad F^{st}_{12} = g^{-1}\{D^{s}_{1}(J^{-1}D^{t}_{2}J)\}g\Big| \,,$$

$$(1.6b)\qquad F_{\dot1 s,\dot2 t} = g^{-1}\{\bar{D}_{\dot1 s}(J^{-1}\bar{D}_{\dot2 t}J)\}g\Big| \,,$$

$$(1.6c)\qquad F^{s}_{1,\dot2 t} = g^{-1}\{D^{s}_{1}(J^{-1}\bar{D}_{\dot2 t}J) + 2i\delta^{s}_{t}\nabla_{1\dot2}\}g\Big| \,,$$

$$(1.6d)\qquad F^{s}_{2,\dot1 t} = g^{-1}\{D_{\dot1 t}(J^{-1}D^{s}_{2}J) + 2i\delta^{s}_{t}\nabla_{2\dot1}\}g\Big| \,.$$

Then the unintegrated constraints on curvatures,

$$F^{st}_{12} + F^{ts}_{12} = 0;$$

(1.7)

$$F_{\dot{1}s,\dot{2}t} + F_{\dot{1}t,\dot{2}s} = 0;\ F^{s}_{1,\dot{2}t} = 0 = F^{s}_{2,\dot{1}t}$$

give equations on J

(1.8a) $$D^s_1(J^{-1}D^t_2J) + D^t_1(J^{-1}D^s_2J) = 0,$$

(1.8b) $$\bar{D}_{\dot{1}s}(J^{-1}\bar{D}_{\dot{2}t}J) + \bar{D}_{\dot{1}t}(J^{-1}\bar{D}_{\dot{2}s}J) = 0,$$

(1.8c) $$D^s_1(J^{-1}\bar{D}_{\dot{2}t}J) + \delta^s_t 2ig\nabla_{1\dot{2}}g^{-1} = 0,$$

(1.8d) $$D_{\dot{1}s}(J^{-1}D^t_2J) + \delta^t_s 2ig\nabla_{2\dot{1}}g^{-1} = 0.$$

In these equations J^{-1} appears on the left of DJ. We call them the left-formulation. Note that Eqs. (1.8 c,d) can be viewed as defining $g\nabla_{\alpha\dot{\beta}}g^{-1}$ for $s = t$ and for $s \neq t$ they automatically follow from Eqs. (1.8 a,b).

Another formulation is

(1.9a) $$F^{st}_{12} = h^{-1}[D^s_2(JD^t_1J^{-1})]h,$$

(1.9b) $$F_{\dot{1}s,\dot{2}t} = h^{-1}[\bar{D}_{\dot{2}t}(J\bar{D}_{\dot{1}s}J^{-1})]h,$$

(1.9c) $$F^s_{1,\dot{2}t} = h^{-1}[\bar{D}_{\dot{2}t}(JD^s_1J^{-1}) + 2i\delta^s_t\nabla_{1\dot{2}}]h,$$

(1.9d) $$F^s_{2,\dot{1}t} = h^{-1}[D^s_2(J^{-1}D_{\dot{1}t}J) + 2i\delta^s_t\nabla_{2\dot{1}}]h.$$

The unintegrated constraints, Eq. (1.7), on these curvatures, give the right-equation on J:

(1.10a) $$D^s_2(JD^t_1J^{-1}) + D^t_2(JD^s_1J^{-1}) = 0,$$

(1.10b) $$\bar{D}_{\dot{2}s}(J\bar{D}_{\dot{1}t}J^{-1}) + \bar{D}_{\dot{2}t}(J\bar{D}_{\dot{1}s}J^{-1}) = 0,$$

(1.10c) $$\bar{D}_{\dot{2}s}(JD_1^tJ^{-1}) + \delta_s^t 2ih\nabla_{1\dot{2}}h^{-1} = 0,$$

(1.10d) $$D_2^s(J\bar{D}_{\dot{1}t}J^{-1}) + \delta_t^s 2ih\nabla_{2\dot{1}}h^{-1} = 0.$$

This is what we call the right-formulation.

II. LINEAR SYSTEM TO THE CONSTRAINT EQUATIONS ↔ INFINITE NUMBER OF CONSERVATIONS EQUATIONS

To these equations on J, Eqs. (1.8), a set of linear systems can be constructed [45,46]

(2.1a) $$L^s(\lambda)\Psi(\lambda) \equiv (D_1^s + \lambda D_2^s + \lambda J^{-1}D_2^sJ)\Psi(\lambda) = 0$$

(2.1b) $$M_t(\lambda)\Psi(\lambda) \equiv (\bar{D}_{\dot{2}t} + J^{-1}\bar{D}_{\dot{2}t}J + \lambda^{-2}\bar{D}_{\dot{1}t})\Psi(\lambda) = 0,$$

(2.1c) $$N(\lambda)\Psi(\lambda) \equiv [(\partial_{1\dot{2}} + g\nabla_{1\dot{2}}g^{-1}) + \lambda(\partial_{2\dot{2}} + J^{-1}\partial_{2\dot{2}}\cdot J) + \lambda^{-1}(\partial_{2\dot{1}} + g\nabla_{2\dot{1}}g^{-1}) + \lambda^{-2}\partial_{1\dot{1}}]\Psi(\lambda) = 0.$$

The integrability conditions of these linear equations give equations on J, Eqs. (1.8): $[L^s,L^t] = 0$, implies Eq. (1.8a); $[M_s,M_t] = 0$ gives Eq. (1.8b); and $[L_s,M_t] = -2i\delta_t^s N$ follows, as a result of Eqs. (1.8 c,d).

The right-formulation for the linear equation can be obtained in the followng way: multiplying Eqs. (2.1) with J on the left hand side and replacing Ψ by $J^{-1}J\Psi$, after simple manipulations we obtain

(2.2a) $$\hat{L}^s(\lambda^{-1})\hat{\Psi}(\lambda^{-1}) \equiv (D_2^s + \lambda^{-1}D_1^s + \lambda^{-1}JD_1^sJ^{-1})\hat{\Psi}(\lambda^{-1}) = 0,$$

(2.2b) $$\hat{M}_t(\lambda^{-1})\hat{\Psi}(\lambda^{-1}) \equiv (\lambda^2\bar{D}_{\dot{2}t} + \bar{D}_{\dot{1}t} + J\bar{D}_{\dot{1}t}J^{-1})\hat{\Psi}(\lambda^{-1}) = 0,$$

$$\hat{N}(\lambda^{-1})\hat{\Psi}(\lambda^{-1}) \equiv [\lambda(\partial_{1\dot{2}} + h\nabla_{1\dot{2}}h^{-1}) + \lambda^2\partial_{2\dot{2}}$$

$$+ \lambda^{-1}(\partial_{1\dot{1}} + J\partial_{1\dot{1}}J^{-1})$$

(2.2c)

$$+ (\partial_{2\dot{1}} + h\nabla_{2\dot{1}}h^{-1})]\hat{\Psi}(\lambda^{-1}) = 0.$$

One can show that the equations (1.10 a,b) are integrability consequences of Eqs. (2.2), i.e. $[\hat{L}^s, \hat{L}^t] = 0$, $[\hat{M}_s,\hat{M}_t] = 0$; and $[\hat{L}^s,\hat{M}_t] = -2i\delta^s_t\hat{N}$ follows from Eqs. (1.10 c,d).

We now construct infinite numbers of conservation laws. Note that Eqs. (1.8 a,b) can be solved in terms of two functions $X^{(1)}$, $X^{(2)}$, defined by

(2.3a) $$J^{-1}D^s_2J = D^s_1X^{(1)},$$

(2.3b) $$J^{-1}\bar{D}_{\dot{2}t}J = \bar{D}_{\dot{1}t}X^{(2)}.$$

Then an infinite number of such X's can be generated iteratively in the following way: define $X^{(n)}$ by

(2.4a) $$D^s_1X^{(n)} = [D^s_2 + J^{-1}D^s_2J]X^{(n-1)},$$

(2.4b) $$\bar{D}_{\dot{1}s}X^{(n)} = [\bar{D}_{\dot{2}s} + J^{-1}\bar{D}_{\dot{2}s}J]X^{(n-2)}.$$

Note that we define $X^{(0)} = 1$. It is easy to show that such defined $D^s_1X^{(n)}$, $\bar{D}_{\dot{1}s}X^{(n)}$ satisfy the continuity-like equation, Eqs. (1.8 a.b).

Following the procedure given in Ref. [18], i.e., multiplying Eq. (2.4 a,b) by λ^n, λ being an arbitrary complex number, summing n = 0 to ∞, and defining

(2.5) $$\Psi \equiv \sum_{n=0}^{\infty} \lambda^n X^{(n)},$$

we obtain the linear equations (2.1 a,b). The third linear equation Eq. (2.1 c) follows from Eqs. (2.1 a,b) and Eqs. (1.8 c,d).

Similarly for the right-formulation we have

(2.6a) $$JD^t_1J^{-1} = D^t_2\hat{X}^{(1)},$$

(2.6b) $$J\bar{D}_{\dot{1}t}J^{-1} = D_{\dot{2}t}\hat{X}^{(2)},$$

and

(2.7a) $$D_2^s\hat{X}^{(n)} \equiv (D_1^s + JD_1^sJ^{-1})\hat{X}^{(n-1)}$$

(2.7b) $$\bar{D}_{\dot{2}s}\hat{X}^{(n)} = (D_{\dot{1}s} + J\bar{D}_{\dot{1}s}J^{-1})\hat{X}^{(n-2)},$$

multiplying Eqs. (2.7) by λ^n and summing, and defining

(2.8) $$\hat{\Psi} \equiv \sum_{n=0}^{\infty} \lambda^n\hat{X}^{(n)},$$

we obtain the linear systems of Eqs. (2.2 a,b). As mentioned before, the third linear equation Eq. (2.2 c) follows from Eqs. (2.2 a,b) and Eqs. (1.10 c,d).

Note that these linear equations given by Devchand [46] are related to those first obtained by Volovich [45] by rewriting the linear equation for a gauge tranformed wave function $\psi = g\Psi$. This situation is very similar to that in the self-dual Yang-Mills field [21,22].

III. FINITE RIEMANN-HILBERT TRANSFORM

Now we construct the Riemann-Hilbert transform (R-H) for these linear equations, which can all be abstractly written in the following form

(3.1) $$\mathcal{D}\psi + \mathcal{B}\psi = 0,$$

where $\mathcal{D}$ denotes the differentiations in Eqs. (2.1): $D_1^s + \lambda D_2^s$, $\lambda^{-2}\bar{D}_{\dot{1}t} + \bar{D}_{\dot{2}t}$ or $\partial_{1\dot{2}} + \lambda\partial_{2\dot{2}} + \lambda^{-2}\partial_{1\dot{1}} + \lambda^{-1}\partial_{2\dot{1}}$; $\mathcal{B}$ can denote respectively their corresponding potential terms in Eqs. (2.1), $\lambda J^{-1}D_2^sJ$, $J^{-1}\bar{D}_{\dot{2}t}J$ or $g\nabla_{1\dot{2}}g^{-1} + \lambda J^{-1}\partial_{2\dot{2}}J + \lambda^{-1}g\nabla_{2\dot{1}}g^{-1}$. First note that if a new solution Ψ' can be constructed from a given solution Ψ by a Darboux-type transformation [48]

$$\Psi'(\lambda) = R(\lambda)\Psi(\lambda), \tag{3.2}$$

a new potential is generated through

$$\mathcal{B}' = R\mathcal{B}R^{-1} - (\mathcal{D}R)R^{-1}. \tag{3.3}$$

Now consider a contour C in the complex λ-plane dividing the plane into two regions C_+ and C_-. Let $R_\pm$ be analytic in the $C \cup C_\pm$ regions respectively. Then Eq. (3.3) becomes $\mathcal{B}'_\pm = R_\pm \mathcal{B} R_\pm{}^{-1} - (\mathcal{D}R_\pm)R_\pm{}^{-1}$. Requiring $\mathcal{B}'_+ = \mathcal{B}'_-$ on C, we can show, after some calculation, using the linear equations,

$$\mathcal{D}(\Psi^{-1}R_+^{-1}R_-\Psi) = 0, \text{ which implies} \tag{3.4}$$

$$\Psi^{-1}R_+^{-1}R_-\Psi = U(\lambda, \cdots), \tag{3.5}$$

where U is a group element of the theory and independent of the three differentiations. In general U is still a function of linear combinations of $x_{\alpha\dot{\beta}}$, θ^s_α, $\bar{\theta}^{\dot{\beta}t}$ i.e.,

$$\lambda(\theta^1_s - i\bar{\theta}_{\dot{\beta}s}x^{1\dot{\beta}}) - (\theta^2_s - \bar{\theta}_{\dot{\beta}s}x^{2\dot{\beta}});$$

$$\lambda^{-2}(\bar{\theta}^{\dot{2}t} - i\theta^t_\alpha x^{\alpha\dot{2}}) - (\bar{\theta}^{\dot{1}t} - i\theta^t_\alpha x^{\alpha\dot{1}});$$

$$x^{1\dot{2}} - \lambda^{-1}x^{2\dot{2}} + \lambda^2 x^{1\dot{1}} - \lambda x^{2\dot{1}};$$

$$x^{1\dot{2}} + \lambda^{-1}x^{2\dot{2}} - \lambda^2 x^{1\dot{1}} - \lambda x^{2\dot{1}}; \text{ and} \tag{3.6}$$

$$x^{1\dot{2}} - \lambda^{-1}x^{2\dot{2}} - \lambda^2 x^{1\dot{1}} + \lambda x^{2\dot{1}}.$$

From Eq. (3.5) we obtain $R_- = R_+\Psi U\Psi^{-1}$ or alternative $R_+ - R_- = R_+\Psi(1-U)\Psi^{-1}$, from which a dispersion relation can be written.

$$R(\lambda) =$$

(3.7a)

$$1 - \frac{\lambda}{2\pi i}\int_C \frac{d\lambda'}{\lambda'(\lambda'-\lambda)} R(\lambda')\Psi(\lambda')[U(\lambda')-1]\Psi^{-1}(\lambda'),$$

where we assume the boundary condition $R(0) = 1$, and that one subtraction is needed in the dispersion integral. Using Eq. (3.2) we obtain from Eq. (3.7a) the integral equation for Ψ',

(3.7b)
$$\Psi'(\lambda) =$$

$$\Psi(\lambda) - \frac{\lambda}{2\pi i}\left\{\left[\int_C \frac{d\lambda'}{\lambda'(\lambda'-\lambda)} \Psi'(\lambda')[U(\lambda'-1]\Psi^{-1}(\lambda')\right\}\Psi(\lambda).$$

These are the R-H transforms. *It is well known that* R-H *transforms have been very useful in generating new solutions in many non-linear systems* [36-38]. *This is a very fascinating prospect for the supersymmetric Yang-Mills fields.*

Following the same procedure as for the left-formulation linear equations, we can formulate R-H transform for the right formulation linear equations,

(3.8a)
$$\hat{R}(\lambda^{-1}) = 1 - \frac{1}{2\pi i}(\lambda^{-1}) \int_{C'} \frac{d(\lambda'^{-1})}{(\lambda'^{-1})(\lambda'^{-1}-\lambda^{-1})}$$

$$\hat{R}(\lambda'^{-1})\hat{\Psi}(\lambda'^{-1})[\hat{U}(\lambda'^{-1}) - 1]\hat{\Psi}^{-1}(\lambda'^{-1}),$$

and

$$\hat{\Psi}'(\lambda^{-1}) = \hat{\Psi}(\lambda^{-1}) - \frac{1}{2\pi i}(\lambda^{-1})[\int_{C'} \frac{d(\lambda'^{-1})}{(\lambda'^{-1})(\lambda'^{-1}-\lambda^{-1})}$$

(3.8b)
$$\hat{\Psi}'(\lambda'^{-1})[\hat{U}(\lambda'^{-1})-1]\ \hat{\Psi}^{-1}(\lambda'^{-1})]\hat{\Psi}(\lambda^{-1}),$$

the boundary conditions here is $\hat{R}(0) = 1$, $\hat{\Psi}'(0) = \hat{\Psi}(0)$ at

$\lambda \longrightarrow \infty$.

Comparing the left-formulation and the right formulation linear equations, Eqs. [2.1, 2.2], we find the $\hat{\Psi}$ and Ψ are related by the following relation [49]:

$$\hat{\Psi}(\lambda^{-1}) = J\Psi(\lambda)\rho(\lambda, \cdots), \tag{3.9}$$

where ρ is an arbitrary function and "$\cdots$" denotes those variables of Eq. (3.6) which are independent of the differentiations in Eqs. (2.2).

IV. INFINITESIMAL RIEMANN-HILBERT TRANSFORM AND THE AFFINE ALGEBRA OF KAC-MOODY

Now let us discuss the infinitesimal R-H transform, i.e., $U(\lambda' \cdots) - 1 \equiv v(\lambda', \cdots)$ with $v(\lambda', \cdots)$ belonging to the algebra. Keeping only the first order in variation, Eq. (3.7b) becomes

$$[\delta_v \Psi(\lambda)]\Psi^{-1}(\lambda) = -\frac{\lambda}{2\pi i}\int_C \frac{d\lambda'}{\lambda'(\lambda'-\lambda)}\Psi(\lambda')v(\lambda', \cdots)\Psi^{-1}(\lambda'). \tag{4.1a}$$

Similarly for Eq. (3.8b),

$$[\delta_{\hat{v}}\hat{\Psi}(\lambda^{-1})]\hat{\Psi}^{-1}(\lambda^{-1}) = -\frac{\lambda}{2\pi i}\int_{C'} \frac{d\lambda'^{-1}}{\lambda'^{-1}(\lambda'^{-1}-\lambda^{-1})}\Psi(\lambda'^{-1})\hat{v}(\lambda'^{-1}, \cdots)\hat{\Psi}^{-1}(\lambda'^{-1}). \tag{4.1b}$$

Take different form of v, $\hat{v}$ we can obtain different transforms. Using the appropriate v, $\hat{v}$ we can obtain the affine Lie algebra of Kac-Moody [23,28], choosing

$$V(\lambda', \cdots) = V_a^m(\lambda') = \eta^m(\lambda)t_a, \tag{4.2}$$

where $[t_a, t_b] = C^c_{ab} t_c$, and $\eta(\lambda) = \prod_i \frac{\lambda - \lambda_i}{\lambda - \lambda_i^*}$, one obtains

$$\delta^m_a \Psi(\lambda) = \frac{\lambda}{2\pi i} \left\{ \int \frac{d\lambda'}{\lambda'(\lambda'-\lambda)} \eta^m(\lambda')[\Psi(\lambda')t_a \Psi(\lambda')^{-1}] \right\} \Psi(\lambda)$$

(4.3)

$$\equiv \frac{\lambda}{2\pi i} \left\{ \int \frac{d\lambda'}{\lambda'(\lambda'-\lambda)} \eta^m(\lambda') S_a(\lambda') \right\} \Psi(\lambda),$$

where $S_a(\lambda') \equiv \Psi(\lambda')t_a \Psi^{-1}(\lambda')$. To obtain the algebraic structure of such variation, we can make another infinitesimal transformation to obtain $\delta^n_b \delta^m_a \Psi(\lambda)$. After lengthy calculations and manipulations involving contour deformations [50,51], we obtain

$$(\delta^m_a \delta^b - \delta^n_b \delta^m_a)\Psi(\lambda)$$

$$= - \frac{\lambda}{2\pi i} \int \frac{d\lambda'}{\lambda'(\lambda'-\lambda)} \eta^{m+n}(\lambda')[S_a(\lambda'), S_b(\lambda')]$$

(4.4)

$$= - \frac{\lambda}{2\pi i} \int \frac{d\lambda'}{\lambda'(\lambda'-\lambda)} \eta^{m+n}(\lambda') C^c_{ab} S_c(\lambda')$$

$$= C^c_{ab} \delta^{m+n}_c \Psi(\lambda),$$

i.e. it is an affine Lie algebra of Kac and Moody.

The implications of such infinite dimensional algebras on the physical systems are yet to be understood.

V. PARAMETRIC BIANCHI-BÄCKLUND TRANSFORMATIONS

Now we derive a two-parameter Bianchi-Bäcklund transformation (BT). From the Darboux-type transformation Eq. (3.2), and specifying R in the following form [48]

$$R = \xi I + f(\lambda) J'^{-1} J, \tag{5.1}$$

where ξ is a constant parameter and $f(\lambda)$ is an arbitrary function of λ. Substituting Eq. (5.1) into Eq. (3.3) we obtain the following BT transformations

$$f(\lambda)D_1^s(J'^{-1}J) = \lambda\xi(J^{-1}D_2^sJ - J'^{-1}D_2^sJ'), \tag{5.2}$$

$$\lambda^{-2}f(\lambda)\bar{D}_{\bar{1}t}(J'^{-1}J) = \xi(J^{-1}\bar{D}_{\bar{2}t}J - J'^{-1}\bar{D}_{\bar{2}t}J'), \tag{5.3}$$

$$\begin{aligned} &f(\lambda)[\partial_{\bar{1}2}(J'^{-1}J) + \lambda^{-2}\partial_{1\bar{1}}(J'^{-1}J) + \lambda^{-1}\partial_{2\bar{1}}(J'^{-1}J) \\ &\quad = \xi\{(g\nabla_{1\bar{2}}g^{-1} - g'\nabla'_{1\bar{2}}g'^{-1}) + \lambda(J^{-1}\partial_{2\bar{2}}J - J'^{-1}\partial_{2\bar{2}}J') \\ &\qquad + \lambda^{-1}(g\nabla_{2\bar{1}}g^{-1} - g'\nabla'_{2\bar{1}}g'^{-1})\} \\ &\qquad + f(\lambda)[(J'^{-1}J)(g\nabla_{1\bar{2}}g^{-1} + \lambda^{-1}g\nabla_{2\bar{1}}g^{-1}) \\ &\qquad - (g'\nabla'_{1\bar{2}}g'^{-1} + \lambda^{-1}g'\nabla'_{2\bar{1}}g'^{-1})(J'^{-1}J)]. \end{aligned} \tag{5.4}$$

We next restrict ourselves to the self-dual case, i.e., under the additional imposed constraints [45,46], $g\nabla_{1\bar{2}}g^{-1} = 0$, $A_{1\bar{2}} = g^{-1}\partial_{1\bar{2}}g$, $A_{2\bar{1}} = h^{-1}\partial_{2\bar{1}}h$. By requiring the constraints Eq. (1.2, 1.10) to be held in the primed quantities, Eq. (5.4) can be reduced to

$$\begin{aligned} &f(\lambda)\partial_{1\bar{2}}(J'^{-1}J) + f(\lambda)\lambda^{-2}\partial_{1\bar{1}}(J'^{-1}J) \\ &= \lambda\xi(J^{-1}\partial_{2\bar{2}}J - J'^{-1}\partial_{2\bar{2}}J') + \lambda^{-1}\xi(J^{-1}\partial_{2\bar{1}}J - J'^{-1}\partial_{2\bar{1}}J'). \end{aligned} \tag{5.5}$$

Now specifying $f(\lambda) = \lambda$ and comparing the same powers in λ, we obtain from Eqs. (5.2, 5.3)

$$D_1^s(J'^{-1}J) = \xi(J^{-1}D_2^sJ - J'^{-1}D_2^sJ'), \tag{5.6a}$$

$$\bar{D}_{\bar{1}t}(J'^{-1}J) = \eta(J^{-1}\bar{D}_{\bar{2}t}J - J'^{-1}D_{\bar{2}t}J'), \tag{5.6b}$$

and Eq. (5.5) yields

$$\partial_{1\bar{2}}(J'^{-1}J) = \xi(J^{-1}\partial_{2\bar{2}}J - J'^{-1}\partial_{2\bar{2}}J'), \tag{5.6c}$$

$$\partial_{1\bar{1}}(J'^{-1}J) = \xi(J^{-1}\partial_{2\bar{1}}J - J'^{-1}\partial_{2\bar{1}}J') \tag{5.6d}$$

where $\eta = \xi\lambda^1$, another free-parameter. Similarly if we choose $g\nabla_{2\dot{1}}g^{-1} = 0$, $A_{1\dot{2}} = h^{-1}\partial_{1\dot{2}}h$ and $A_{2\dot{1}} = g^{-1}\partial_{2\dot{1}}g$ and same for the primed qualities we find that necessarily $f(\lambda) = \lambda^2$, and correspondingly Eqs. (5.6) become

$$D_1^s(J'^{-1}J) = \xi(J^{-1}D_2^sJ - J'^{-1}D_2^sJ'), \tag{5.7a}$$

$$\bar{D}_{\dot{1}t}(J'^{-1}J) = \xi(J^{-1}D_{\dot{2}t}J - J'^{-1}\bar{D}_{\dot{2}t}J'), \tag{5.7b}$$

$$\partial_{2\dot{1}}(J'^{-1}J) = \xi(J^{-1}\partial_{2\dot{2}}J - J'^{-1}\partial_{2\dot{2}}J') \tag{5.7c}$$

$$\partial_{1\dot{1}}(J'^{-1}J) = \xi(J^{-1}\partial_{1\dot{2}}J - J'^{-1}\partial_{1\dot{2}}J'), \tag{5.7d}$$

where $\zeta\eta = \xi^2$.

Note these BT transformations given by Eqs. (5.7) coincide precisely with those given in Ref. [16] for the ordinary SDYM fields.

VI. DISCUSSIONS ON THE CONSTRAINT EQUATIONS AND EQUATION OF MOTION

The main purpose of the constraints Eqs. (1.2) are to eliminate fields that are unphysical, or to make sure that fields in the super dimensions are to be determined in terms of fields in the ordinary space. Witten [39] gave a geometric description of how these constraint equations came about: let us consider light-like lines in the ordinary space

$$x^{\alpha\dot{\alpha}} = c^{\alpha\dot{\alpha}} + t\lambda^{\alpha}\bar{\lambda}^{\dot{\alpha}}, \tag{6.1}$$

where λ^{α} with $\alpha = 1,2$ is an arbitrary pair of complex numbers, $c^{\alpha\dot{\alpha}}$ and λ^{α}'s are given but t arbitrary. Translation in the light-like direction $\lambda^{\alpha}\bar{\lambda}^{\dot{\alpha}}$ are generated by

(6.2) $$D = i\, \lambda^{\alpha}\bar{\lambda}^{\dot{\alpha}}\, \partial/\partial x^{\alpha\dot{\alpha}}.$$

In super space, translations in light-like directions are square roots of D. In fact

(6.3) $$T \equiv 2^{-1/2}(\lambda^{\alpha}D^{s}_{\alpha} - \bar{\lambda}^{\dot{\alpha}}\bar{D}_{\dot{\alpha}s}),$$

gives

(6.4) $$T^{2} = D.$$

So we can view

(6.5) $$T^{s} = \lambda^{\alpha}D^{s}_{\alpha}, \qquad \bar{T}_{s} = -\bar{\lambda}^{\dot{\alpha}}\bar{D}_{\dot{\alpha}s},$$

as the accompanying light line in the super space.

Integrability on a line means that the translation operators along that line satisfy a commutation relation unmodified in the presence of gauge fields. Integrability along a line in the ordinary space is trivial. But integrability along a light-like line including the super directions puts restrictions on the curvature. Integrability along the generalized light-like line means [39],

(6.6) $$\{T^{s},T^{t}\} = 0 = \{\bar{T}_{s},\bar{T}_{t}\}, \; \{T^{s},\bar{T}_{t}\} = 2\delta^{s}_{t}D,$$

which is true for arbitrary λ^{α}, $\bar{\lambda}^{\dot{\alpha}}$, thus

(6.7a) $$\{D^{s}_{\alpha}, D^{t}_{\beta}\} + \{D^{t}_{\alpha}, D^{s}_{\beta}\} = 0,$$

(6.7b) $$\{\bar{D}_{\dot{\alpha}s},\bar{D}_{\dot{\beta}t}\} + \{\bar{D}_{\dot{\alpha}t}, \bar{D}_{\dot{\beta}s}\} = 0,$$

and

(6.7c) $$\{D^{s}_{\alpha}, \bar{D}_{\dot{\beta}t}\} = -2i\delta^{s}_{t}\partial/\partial x^{\alpha\dot{\beta}}.$$

In the presence of gauge fields, the unalteration of these algebras precisely gives the constraint equations Eqs. (1.2).

We can see that as the super dimension N increases, the number of constraints increases. It was pointed out by Sohnius [44] and Witten [39] that at N = 3, these constraints on curvature via Bianchi identities actually give equations of motion [52]. This feature was considered not so desirable due to the lack of Lagrangian formulation.

However, from our point of view, for the N = 3,4 super-symmetic Yang-Mills theories, this provides the interesting possibility that solving these constraint equations is equivalent to solving the full supersymmetric N = 3,4 Yang-Mills theories.

VII. OUTLOOK

Equations of motion for N = 3,4 supersymmetric Yang-Mills fields are results of integrable conditions in the generalized light-like directions in super coordinates. This gives the hope that the theory can be solved using the integration technique for non-linear systems. The important directions to pursue are: firstly, to see what kind of classical solutions the finite R-H transforms, and the BT can provide; secondly, to study the quantum inverse scattering for these systems, [53], or alternatively to construct Lagrangian for Eqs. (8) so that quantum field theory for such system can be formulated.

REFERENCES

[1.] C.N. Yang and R.L. Mills, Phys Rev. 96 (1954) 191.

[2.] For electroweak theory, see S.L. Glashow, Nucl. Phys. 22, (1961) 579, A. Salam and J.C. Ward, Phys. Lett. 13, (1964) 168, and S. Weinberg, Phys. Rev. Lett. 19,(1967) 1264. For the discovery of intermediate bosons predicted by the electroweak theory, see UA1, CERN, G. Arnison, et al., Phys. Lett. 122B (1983) 103, ibid. 126B, (1983) 398; UA2, CERN, M. Banner, et al., ibid. 122B, (1983) 476. For the theory that predicted the existence of the charm quark based upon electroweak theory, see. G. Glashow, J. Iliopoulos and L Maiani, Phys. Rev. D2,

(1970) 1285. For the discovery of charm, see J.J. Aubert, et.al., Phys. Rev. Lett. 33, (1974) 1404, Samuel C.C. Ting, Rev. Mod. Phys. 49, (1977) 235, J. E. Augustin, et al., Phys. Rev. Lett. 33, (1974) 1406, B. Richter, Rev. Mod. Phys. 49 (1977) 251. For charm particle observation in $\nu_\mu p$ reaction in the BNL 7-foot buble chamber, see E.G. Cazzoli, et al., Phys. Rev. Lett. 34, (1975) 1125; in e^+e^- rection, see G. Goldhaber, et al., Phys. Rev. Lett. 37, (1976) 255, in pp reaction, See M. Basile, et al., Nuovo Cim. 63A, (1981) 230; and for a review see A. Zichichi, Rapporteur Talk, Proc. EPS High Energy Conf., Lisbon, July 1981 and CERN preprint EP/82/30.

[3.] T.-T. Wu and C.N. Yang, Phys. Rev. D12 (1075) 3845;

C.-H. Gu and C.N. Yang, Sicentia Sinica 18 (1975) 483; C.N. Yang, Proc. of Sixth Hawaiian Topical Conf. in Particle Phys., Univ. of Hawaii Press (1975); T.-T. Wu and C.N. Yang, Nucl. Phys. B107 (1976) I; C.N. Yang, "Monopoles and Fiber Bundles", in Understanding of the Fundamental Constituents of Matter, ed. by A. Sichichi, (1976).

[4.] A. Belavin, A. Polyakov, A. Schwartz and Y. Tyupking, Phys. Lett. 59B (1975) 85;
M.F. Atiyah, V.G. Drinfield, N.J. Hitching and Yu. I. Manin, Phys. Lett. 65A (1978) 185;
M.F. Atiyah and R.S. Ward, Comm. Math. Phys. 55 (1977) 117.

[5.] A.M. Polyakov, JETP Lett. 20 (1974) 194;
G. 't Hooft, Nucl. Phys. B79 (1974) 276;
M.K. Prasad and C.M. Sommerfield, Phys. Rev. Lett. 35 (1975) 760;
E.B. Bogomolny, Sov. J. Nucl. Phys. 24 (1976) 861.

[6.] For jet productions, in e^+e^- reactions see, MARK J Collaborationn, D.P. Barber et al., Phys. Rev. Lett 43 (1979) 830;
TASSO Callaboration, R. Brandelik et al, Phys. Lett. 86B (1975) 243;
PLUTO Collaboration, C. Berger et al., Phys. Lett. 86B (1979) 418;
JADE Collaboration, W. Bartel et al., DESY preprint 79/80 (1979);
in hadronic reactions see, Spectrometer Collaboration at ISR, CERN, T. Akesson et al., Phys. Lett. 118B (1982) 185, and 193;
UA2 Collaboration at CERN $\bar{p}p$ Collider, M. Banner et al., Phys. Lett. 119B (1982) 203;

UA1 Collaboration at CERN $\bar{p}p$ Collider, G. Arnison et al, Phys. Lett. 123B (1983) 115.

[7.] K. Wilson, Phys. Rev. D10 (1974) 2445;
A.M. Polyakov, Phys. Lett. 59B (1975) 82;
G. 't Hooft, Phys. Rev. D14 (1976) 3432;
For recent work, see
M. Creutz, Phys. Rev. Lett. 43 (1979);
M. Creutz, L. Jacobs, C. Rebbi, Phys. Rev. D20 (179) 1915);

[8.] G. t'Hooft, Nucl. Phys. B33 (1971) 173, and private communication;
H.D. Politzer, Phys. Rev. Lett. 30 (1973) 1346;
D.J. Gross and F.A. Wilczek, Phys. Rev. Lett. 30 (1973) 1343.

[9.] See Bäcklund Transformations, The Inverse Scattering Method, Solitons and Their Applications, Lecture Notes in Mathematics, edited by R.M. Miura, Vanderbilt U., 1974, Springer-Verlag; and references therein.

[10.] Proceedings of "Group Theoretical Methods in Physics", Istanbul, Turkey, 1982, Lecture Notes in Physics, #180, edited by M. Serdaroglu and Inönü.

[11.] Proceedings of "Nonlinear Phenomena", Oaxtepec, Mexico, Nov. 19-Dec. 17, 1982, Lecture Notes in Physics #189, edited by K.B. Wolf.

[12.] A.B. Zamolodchikov, Comm. Math. Phys. 55 (1977) 183;
A.B.B. Zamoldchikov and A.B. Zamolodchikov, Nucl. Phys. B133 (1978) 525.

[13.] The Main feature of what I call integrability here is the existence of certain curvatureless conditions that lead to the corresponding linear systems. The relations of such defined integrability and other integrabilities defined as solving the initial-value problem or showing the existence of action-angle variables in a hamiltonian formalism is unclear, and is interesting to be clarified. See also Rev. 26.

[14.] C.N. Yang, Phys. Rev. Lett 38 (1977) 1377;
S. Ward, Phys. Lett. 61A (1977) 81;
Y. Brihaye, D.B. Fairlie, J. Nuyts, R.F. Yates, JMP 19 (1978) 2528.

[15.] A.A. Belavian and V.E. Zakharov, Phys. Lett.

73B (1978) 2528.

[16.] M.K. Prasad, A. Sinha, L.-L. Chau Wang, Phys. Rev. Lett. 43 (1979) 750.

[17.] M.K. Prasad, A. Sinha, L.-L. Chau Wang, Phys. Lett. 87B (1979) 237.

[18.] L.-L. Chau Wang, Proc. Guanzhou (Canton) Conf. on Theoretical Particle Physics, p. 1082, 1980.

[19.] L.-L. Chau, Oroc. Int'l. School of Subnuclear Physcis, "The High Energy Limit", ed. Z. Zichichi (Plenum Press), p. 249, 1983; Proc. of Int'l. Workshop on Higher Energy Physics, Protvine, Serpukhov, USSR, p. 402, 1980.

[20.] K. Pohlmeyer, Comm. Math. Phys. 72 (1980) 37.

[21.] L.-L. Chau, Lectures at the 18 Winter School of Theoretical Physics, Karpacz, Poland, 1981.

[22.] L.-L. Chau, M.K. Prasad, A. Sinha, Phys. Rev. D23 (1981) 2321; ibid D24 (1981) 1574.

[23.] For the affine Lie algebra of Kac-Moody in the SDYM field, see L.-L. Chau, M.-L. Ge and Y.-S. Wu, Phys. Rev. D25 (182) 1086; and Ref. 24-25.

[24.] K. Ueno and Y. Nakamura, Phys. Lett. 117 (1982) 208.

[25.] L.-L. Chau, Y.-S. Wu, Phys. Rev. D26 (1983) 3581;
L.-L. Chau, M.-L. Ge, A. Sinha, Y.-S. Wu, Phys. Lett. 121B (1983) 391.

[26.] Integrability of SU(2) SDYM in the sense of Painlevé, see M. Jimbo, M.D. Kruskal, and T. Miwa, Phys. Lett. 92A (1982) 59.

[27] For a summary in the development of SDYM fields, see lectures by L.-L. Chau, "Self-dual Yang-Mills as a Totally Integrable System" in Ref. 10.

[28.] For systematic discussions on the SDYM systems and the chiral system and their similarities, see lectures by L.-L. Chau, "Chiral fields, Self-dual Yang-Mill Fields, as Integrable Systems and the Role of the Kac-Moody Algebra" in the proceedings of Ref. 11.

[29.] For linear system of the chiral model, see V.E. Zakharov and A.V. Mikhailov, Sov. Phys. JETP 47 (1978) 1017.

[30.] For parametric BT, see A.T. Ogielski, M.K. Prasad, A. Sinha, L.-L. Chau Wang, Phys. Lett. 91B (1980) 387; As β = 0, it is the BT in K. Pohlmeyer, Comm. Math. Phys. 46 (1976) 207; I.V. Cheredrik, Teor. Mat. Fiz. 38 (1979) 120.

[31.] For local conservation las in chiral model, see ref. (30), I.V. Cheredrik, Teor. Mat. Fiz. 38 (1979) 120, and K.-C. Chou and X.-C. Song, Scientia Sinica, vol. XXV, No. 8 (1982) 716, ibid XXV (1982) 825m Science Exploration, Vol. 1, No. 3 (1982) 38.

[32.] For non-local conservation laws for the O(3) σ model, see M. Lüscher and K. Pohlmeyer, Nucl. Phys. B137 (1978) 44; M. Lüscher, Nucl. Phys. B135 (1978) 1. For non-local conservation laws of the chiral model, see E. Brezin, C. Itzykson, J. Zinn-Justin, and J.B. Zuber, Phys. Lett. 82B (1979) 442.

[33.] For connection of parametric BT to Riccati equation and the linear system, see L.-L. Chau (1979), unpublished, and her lecture in Ref [11]. P. Winternitz in the proceedings of Ref. 11, and J. Harnad, Y'. Saint-Aubin, and S. Shmider, Comm. Math. Phys. 93, 33-56 (1984).

[34.] For the affine algebra of Kac-Moody in the chiral model, see L. Dolan and A. Roos, Phys. Rev. D22 (1980) 20; B.Y. Hou, Comm. of Theor. Phys. (a new Journal in English published in China) vol. 1, (1981) 333; L. Dolan, Phys. Rev. Lett. 47 (1981) 1371; M.L. Ge and Y.S. Wu, Phys. Lett. 108B (1982) 411; C. Devchand and D.B. Fairlie, Nucl. Phys. B194 (1982) 232; L.-L. Chau, Y.S. Wu, B.Y. Hou and M.L. Ge, Scientia Sinica, A25, 907 (1982); K.C. Chou and X.C. Song, Comm. in Theor. Phys. (Beijing, China), 2, 1391 (1983).

[35.] For developments in supersymmetric chiral fields see P. DiVecchia and S. Ferrara, Nucl. Phys. B130 (1977) 93; E. Witten, Phys. Rev. D16 (1977) 2991; E. Cremmer and J. Sherk, Phys. Lett. 74B (1978) 341; J.F. Shonfeld, Nucl. Phys. B169 (1980) 49; Z. Popowicz and L.-L. Chau Wang, Phys. Lett. 98B (1981) 1080; L.-L. Chau, M.L. Ge, Y.-S. Wu,

Phys. Rev. D25, (1982) 1080.

[36.] F.J. Ernst, Phys. Rev. 167 (1968) 1175, ibid. 168 (1968) 1415; E. Ernst, J. Math. Phys. 20 (1978) 871; V.A. Velinskii and V.E. Zakharov, Zh. Teor. Fiz. 75 (178) 1953 [Sov. Phys. JETP 50 (1979) 1]; D. Maison, Phys. Rev. Lett. 41 (1978) 521, J. Math. Phys. 20 (1979) 871.

[37.] For Riemann-Hilbert transform for the non-linear systems, see V.E. Zakharov and A.D. Shabat, Func. Anal. and Appl. 13, 13 (1979); V.E. Zakharov and A.V. Mikhailov, Comm. Math. 74, 21 (1980): for a brief review see, e.g., A.V. Mikhailov, CERN preprint TH-3194, 1981, and references therein; in general relativity see, e.g., I. Hauser and F.J. Ernst Phys. Rev. 20, 362 (1979), 20, 1783 (1979); for SDYM see Y. Nakamura and K. Ueno, Phys. Lett. 117B, 208 (1982); see also Y.-S. Wu, comm. Math. Phys. 90, 461 (1983).

[38.] R. Geroch, J. Math. Phys, 12 918 (1971), ibid 13, 394 (1972), and I. Huaser and F.J. Ernst, J. Math. Phys. 22, 1051 (1981).

[39.] E. Witten, Phys. Lett. 77B (1978) 394.

[40.] J. Isenberg, P.B. Yasskin and P.S. Green, Phys. Lett. 73B (1978) 462.

[41.] For discussions on integrabililty of Yang-Mills field formulated in extra ordinary dimensions see. P. Forgacs, Z. Horvath. L. Palla, Phys. Lett. 115B 463 (1982).

[42.] Y.A. Gol'fand and E.P. Likhtman, JETP Lett. 13, 323 (1971); D.V. Volkov and V.P. Akulov, Phys. Lett. 46B, 109 (1973); J. Wess and B. Zumino, Nucl. Phys. B70, 39 (1974). For reviews on supersymmetry theories see P. Fayet and S. Ferrara, Phys. Rep. 32C, 249 (1977); A. Salam and J. Strathdee, Phys. Rev. D11. 1521 (1975), Fortschritte der Physik 26, 57 (1978). P. Van Niewenhuizen, Phys. Rept. 68, No. 4, 189 (1981).

[43.] R. Grimm, M. Sohnius, and J. Wess, Nucl. Phys. B133, 275 (1978).

[44.] M. Sohnius, Nucl. Phys. B136, 461 (1978)

[45.] I.V. Volovich, Phys. Lett. 129B, 429 (1983); Theo. and Math. Phys. vol. 57, No. 3, Dec. 1983.

[46.] C. Devchand, Nucl. Phys. B238, 333 (1984).

[47.] L.-L. Chau, M.-L. Ge, and Z. Popowicz, Phys. Rev. Lett. 52 1940 (1984).

[48.] This form is well known in the study of soliton equations, D. Levi, O. Ragnisco, and A. Sym, lett. Al Nuovo Cimento, 33, 401 (1982); D. Levi and O. Ragnisco, Phys. Lett. 87!, 31 (1982), we would like to thank Dr. Levi for bringing our attention to this method.

[49.] Note there is another relation between $\hat{\Psi}$ and Ψ, which is not used in this paper $\hat{\Psi}(\lambda^{-1}, \theta_1^s, \theta_2^s, \bar{\theta}_{1t}, \bar{\theta}_{2t}, x_\mu) =$

$\psi(\lambda, \theta_2^s, \theta_1^s, \bar{\theta}_{2t}, \bar{\theta}_{1t}, x_\mu)\ \eta(\lambda, \cdots)$ where $(\cdots)$ denotes variables given in Eq. (3.6).

[50.] The method used here is similar to that in H. Eihenherr, Phys. Lett. 115B (1982) 385. T. Koikawa and R. Sasaki, Phys. Lett. 124B (1983) 85. See also Chou and Song in Ref. [34].

[51.] G. Bhattacharya and L.-L. Chau. Invited Talk presented by Chau at the Colloq. of Group Theoretical Methods in Physics, Trieste, Italy, Sept. 5-10, 1983.

[52.] It is known that for $N \geq 5$, the supersymmetric Yang-Mills equations do not exist.

[53.] For recent work on the quantum inverse scattering formulation for the chiral model, see H.J. de Vega, H. Eichenherr, J.M. Maillet Comm. Math. Phys. 92, 507 (1984). A. Polyakov and P.B. Wiegman, Phys. Lett. 131B, 121 (1983).

Physics Department
Brookhaven National Laboratory
Upton, N.Y. 11973

LAX PAIRS, RIEMANN-HILBERT TRANSFORMS AND AFFINE ALGEBRAS FOR HIDDEN SYMMETRIES IN CERTAIN NONLINEAR FIELD THEORIES

Yong-Shi Wu and Mo-Lin Ge

ABSTRACT

Two recently proposed approaches, the H-transformation method and the Riemann-Hilbert transform method, to the group structure of infinite-parameter hidden symmetries in certain nonlinear field theories are briefly reviewed, and the relationship between them and their respective advantages are discussed. Once the Lax pair or linearization system in question is known, these methods provide simple and systematic procedures for both finding explicit expressions of hidden symmetry transformations and deriving the infinitely many commutation relations among them. The hidden symmetry algebras (usually of the affine type) for, e.g., 2-D chiral models, 4-D self-dual Yang-Mills equations and 2-D reduced gravity have been obtained or extended by applying these methods.

I. INTRODUCTION

The interesting field theories in particle physics are exclusively relativistic nonlinear ones, such as general relativity, Yang-Mills theories, chiral scalar models (another name being nonlinear sigma models) and their supersymmetric generalizations. The knowledge of symmetries in these theories should be quite useful for developing appropriate nonperturbative treatments which are required by the nonlinearity in question. The idea of the approach to this problem in recent years has been the following: perhaps it is better to study first some analogous but simpler systems, and then try to generalize the results to more complicated and realistic cases. This is the reason why the study of (infinite-parameter) symmetries in

Vertex Operators in Mathematics and Physics - Proceedings of a Conference November 10-17, 1983. Publications of the Mathematical Sciences Research Institute #3, Springer-Verlag, 1984.

two-dimensional nonlinear field theories and four-dimensional Yang-Mills equations with the constraint of self-duality has recently attracted a lot of attention.

At the first sight, it seems surprising that there exist infinitely many symmetries in certain nonlinear field theories. However, it becomes not surprising once one recognizes the existence of infinite number of (nonlocal) conservation laws in, e.g., 2-D chiral models and 4-D self-dual Yang-Mills systems [1]. The problems are how to find the "hidden" symmetries leading to these conservation laws and to uncover their group structure. They have been solved in the last several years for many simpler cases and form the subject of this review talk.

Here we will review two approaches, i.e., the H-transformation method [2-5] (as we call it) and the Riemann-Hilbert transform method [6-8], and discuss the connection between them. (Essentially, they are equivalent to each other.) These methods, starting with Lax pair or the linearization system appropriate to the nonlinear theory under consideration, provide us with simple and systematic procedures of finding explicit expressions for hidden symmetries and of evaluating their commutators. Both of the methods have been extensively used to deal with the group structure of hidden symmetries in various nonlinear problems. The list, which may be incomplete, includes

2-D principal chiral models [2-5;7]

2-D nonlinear sigma models on coset spaces [9,10;11]

2-D super-chiral or graded-chiral models [12,13]

2-D reduced gravity [14-16;6,7]

2-D Heisenberg (or Landau-Lifshitz) model [17;18]

4-D self-dual Yang-Mills equations [19,20;7]

4-D N=4 supersymmetric Yang-Mills theories [21;22]

(Papers using other methods are not included here.)

It is impossible to review all these applications here, so I will be more concentrated on the basic ideas and methodological aspects of the two approaches, using the simplest models for the purpose of illustration. For the most physically interesting cases of 4-D self-dual Yang-Mills and supersymmetric Yang-Mills theories, see L. L. Chau's review in these proceedings.

II. THE H-TRANSFORMATION AND THE HALF AFFINE ALGEBRA IN 2-D PRINCIPAL CHIRAL MODELS

In this section we use the 2-D principal chiral model as prototype to illustrate the H-transformation method. The Lagrangian of the model is

$$\mathscr{L} = \frac{1}{4} \operatorname{tr} (\partial_\xi g \partial_\eta g^{-1}) \tag{2.1}$$

where $\xi = t + x$, $\eta = t - x$ are light-cone coordinates and the field $g(\xi,\eta)$ takes values in matrix Lie group G. The equations of motion can be written as

$$\partial_\xi A_\eta + \partial_\eta A_\xi = 0 \tag{2.2}$$

Here we introduce the "potential" $A_\mu \in g$ (the Lie algebra of G) as

$$A_\mu = g^{-1}\partial_\mu g \qquad (\mu = \xi,\eta) \tag{2.3}$$

which automatically satisfies the "curvature-free" condition

$$\partial_\xi A_\eta - \partial_\eta A_\xi + [A_\xi, A_\eta] = 0 \tag{2.4}$$

The existence of an infinite set of nonlocal conservation laws in this model can be readily seen by the well-known BIZZ (Brezin, Itzykson, Zinn-Justin and Zuber) construction [23]. First, it is obvious that $J^{(1)}_\mu \equiv A_\mu$ is conserved. Suppose we have already the n-th conserved current $J^{(n)}_\mu$. The BIZZ construction recursively generates the (n+1)-st conserved current $J^{(n+1)}$ as follows:

$$(2.5)\quad \begin{cases} \partial_\xi X^{(n)} = J^{(n)}_\xi, \ \partial_\eta X^{(n)} = -J^{(n)}_\eta \\ J^{(n+1)}_\mu = (\partial_\mu + A_\mu) X^{(n)} \equiv D_\mu X^{(n)} (\mu = \xi, \eta) \end{cases}$$

Here the existence of the solution $X^{(n)}$ is due to the conservation of $J^{(n)}_\mu$, and the conservation of $J^{(n+1)}_\mu$ is due to the induction hypothesis and Equations (2.2) and (2.4).

Introducing (ℓ being a parameter, called the spectral parameter)

$$(2.6)\qquad U(\ell;\xi,\eta) = \sum_{n=0}^{\infty} \ell^n X^{(n)} (\xi,\eta)$$

the infinite number of equations defining $X^{(n)}$ can be simply summarized by the following linear system [24]

$$(2.7)\qquad \partial_\xi U = \ell D_\xi U, \ \partial_\eta U = -\ell D_\eta U$$

or

$$(2.8)\qquad \partial_\xi U = \frac{\ell}{1-\ell} A_\xi U, \ \partial_\eta U = \frac{-\ell}{1+\ell} A_\eta U$$

These are just the Lax-pair or Zakharov-Mikhailov linearization system for the model [25]. We can always normalize U so that

$$(2.9)\qquad U(\ell = 0; \xi,\eta) = 1$$

The key point here is that the necessary and sufficient conditions for the existence of solution U to the system (2.8) for arbitrary ℓ are just Equations (2.2) and (2.4) under consideration. From a solution to Equation (2.8), we can reconstruct

$$(2.10)\qquad A_\xi = \partial_\xi \dot{U} (\ell=0), \ A_\eta = -\partial_\eta \dot{U} (\ell=0)$$

(where a dot denotes differentiation with respect to ℓ), and then obtain $g(\xi,\eta)$ by integration. Therefore, it is not surprising that the linear system (2.8) plays important roles in the study of the original nonlinear problem (2.2)-(2.4).

The so-called H-transformation approach to the hidden symmetries (leading to the infinite set of conservation laws discussed above) also exploits the linear system (2.8). In this approach, the solution U to Equations (2.8) and (2.9) is used to construct explicitly an infinitesimal transformation of the basic field $g(\xi,\eta)$ in question, which contains the parameter ℓ via U and, when expanded in powers of ℓ, generates an infinite set of infinitesimal symmetry transformations [2]:

$$(2.11) \quad g^{-1}\delta g = -U(\ell;\xi,\eta)\, T_\alpha U(\ell;\xi,\eta)^{-1} = \sum_{n=0}^{\infty} \ell^n\, g^{-1}\delta^{(n)}g$$

with $T_\alpha = \alpha^a T_a$ (T_a being generators of the Lie algebra $\mathfrak{g}$, and α^a infinitesimal constants). Using Equations (2.8) and (2.9) one obtains

$$(2.12) \qquad \delta A_\xi = \frac{-1}{\ell}\, \partial_\xi(UT_\alpha U^{-1}), \quad \delta A_\eta = \frac{1}{\ell}\, \partial_\eta(UT_\alpha U^{-1})$$

Therefore, the equations of motion (2.2) and (2.3) is invariant under the transformation (2.11). In other words, if $g(\xi,\eta)$ is a solution to Equations (2.2) and (2.3), so is $g+\delta g$ and, therefore, so are $g+\delta^{(n)}g$ ($n\geq 0$). If the starting $g(\xi,\eta)$ is not a solution, there does not exist a U satisfying the Lax pair (2.8). In this case, however, we can use one equation in the Lax pair, e.g.,

$$(2.13) \qquad \partial_\xi U = \frac{\ell}{1-\ell}\, A_\xi U, \quad U(\ell;\xi = -\infty, \eta) = 1$$

to define the function U used in Equation (2.11). Using Equation (2.13) only, it can be shown that the change of the Lagrangian is a total divergence either under δg or under each $\delta^{(n)}g$ ($n\geq 0$). Thus, the transformation (2.1) leads, via the generalized Noether analysis [26,2], to an infinite set of conservation laws equivalent to those derived from the BIZZ procedure. Hereafter, we call a transformation of the type (2.11) as H-transformation. It represents off-shell hidden symmetries in the entire field-configuration space.

A remark on the structure of the H-transformation (2.1) is in order. It is obvious that the global transformation $g^{-1}\delta g = T_\alpha$, for constant T_α, is a symmetry of the theory. The H-transformation (2.11) is to sandwich T_α by U and U^{-1}. What is amusing here is that when $U(\ell;\xi,\eta)$ is a solution to the Lax pair, this sandwiching also leads to a symmetry which contains a parameter and implies actually infinitely many symmetries!

To identify the infinite dimensional Lie algebra of the transformations $\delta^{(n)}g$ $(n\geq 0)$, it is better to work out first the commutation relations between parametric H-transformations and then to expand them in powers of the parameters. To explicitly display the dependence on the parameters, we denote δg in Equation (2.11) by $\delta_\alpha(\ell)g$. We need to compute the commutator

$$[\delta_\alpha(\ell),\delta_\beta(\ell')]g \equiv \delta_\alpha(\ell)\{g+\delta_\beta(\ell')g\} - \delta_\alpha(\ell)g - (\alpha\rightleftarrows\beta,\ell\rightleftarrows\ell')$$

(2.14)

$$= -g[\delta'_\beta U\, U^{-1}, UT_\alpha U^{-1}] + gU'T_\beta U'^{-1}UT_\alpha U^{-1} - (\alpha\rightleftarrows\beta, U\rightleftarrows U')$$

Here for brevity we have used $U \equiv U(\ell;g)$, $U' \equiv U(\ell';g)$ and $\delta'_\beta U = U(\ell;g + \delta_\beta(\ell')g) - U(\ell;g)$, $\delta_\alpha U' = U(\ell';g + \delta_\alpha(\ell)g) - U(\ell';g)$. To express, e.g., $\delta_\alpha U'$ in terms of U and U', we can consult the differential equation and boundary condition satisfied by $\delta_\alpha U'$ which follow from Equations (2.13) and (2.12):

(2.15)
$$\begin{cases} \partial_\xi(\delta_\alpha U') = \dfrac{\ell}{1-\ell}A_\xi(\delta_\alpha U') - \dfrac{1}{1-\ell}\partial_\xi(UT_\alpha U^{-1})U' \\ \delta_\alpha U(\ell';\xi = -\infty,\eta) = 0 \end{cases}$$

It is easy to check that

$$\delta_\alpha U'\cdot U'^{-1} = \frac{\ell'}{\ell'-\ell}\{UT_\alpha U^{-1} - U'T_\alpha U'^{-1}\} \tag{2.16}$$

since both sides satisfy the same differential equation and boundary condition (2.15). Using this result it follows that [3,4]

(2.17) $[\delta_\alpha(\ell),\delta_\beta(\ell')]g = \alpha^a\beta^b C^c_{ab} \dfrac{\ell\delta_c(\ell)g - \ell'\delta_c(\ell')g}{\ell-\ell'}$

where $g^{-1}\delta_\alpha(\ell)g = g^{-1}\delta_c(\ell)g\cdot\alpha^c$. C^c_{ab} are structure constants of the Lie algebra $\mathfrak{g}$. Upon expanding in powers of ℓ and ℓ', this parametric commutation relation leads immediately to the infinite set of commutation relations among $\delta^{(n)}_a g$ $(n\geqslant 0)$:

(2.18) $$[\delta^{(n)}_a, \delta^{(m)}_b]\, g = C^c_{ab}\, \delta^{(n+m)}_c g \qquad (n,m\geqslant 0)$$

This is just the "half" affine algebra $\mathfrak{g}\otimes\mathbb{C}[t]$ first obtained in Reference [27] by a different method. The utilization of the spectral parameter to summarize an infinite set of transformations $\delta^{(n)}g$ in the H-transformation (2.11) greatly simplifies the derivation of infinite number of commutation relations (2.18).

We know that the generators of an affine Kac-Moody algebra [28] are indexed by all the integers. To make it possible to apply the powerful techniques and interesting results in the representation theory of Kac-Moody algebras in the mathematical literature (see, e.g., references [29,30] and references therein), it is necessary to extend the above symmetry transformations and their Lie algebras. This will be sketched in the next section. The fact that this extension [5] was first made by the H-transformation method again shows the power of the method.

III. THE H-TRANSFORMATON AND THE FULL AFFINE ALGEBRA IN 2-D PRINCIPAL CHIRAL MODELS

First let us make sure of the existence of another infinite set of nonlocal conserved currents which can be indexed with negative integers. For this purpose, it is interesting to observe [31,5] that the above-mentioned BIZZ procedure can be inverted to generate recursively the (-n)-th conserved current $J^{(-n)}_\mu$ from the (-n+1)-st one $J^{(-n+1)}_\mu$ $(n\geqslant 1)$:

$$(3.1)\quad \begin{cases} D_\xi \chi^{(-n)} = J_\xi^{(-n+1)}, \; D_\eta \chi^{(-n)} = J_\eta^{(-n+1)} \\ J_\xi^{(-n)} = \partial_\xi \chi^{(-n)}, \; J_\eta^{(-n)} = -\partial_\eta \chi^{(-n)} \end{cases}$$

Here the integrability of the equations defining $\chi^{(-n)}$ in terms of $J_\mu^{(-n+1)}$ can be easily verified by using Equations (2.2) and (2.4) and the induction hypothesis. By construction, $J_\mu^{(-n)}$ is conserved:

$$(3.2)\quad \partial_\eta J_\xi^{(-n)} + \partial_\xi J_\eta^{(-n)} = 0$$

Similar to Equation (2.6), we introduce

$$(3.3)\quad W(\ell;\xi,\eta) = \sum_{n=0}^{\infty} \ell^n \chi^{(-n-1)}(\xi,\eta)$$

Then from Equation (3.1) we have

$$(3.4)\quad \partial_\xi W = \frac{1}{\ell-1} A_\xi W, \; \partial_\eta W = -\frac{1}{\ell+1} A_\eta W$$

This is just the Lax pair (2.8) with the substitution

$$(3.5)\quad \ell \to 1/\ell$$

If $W(\ell;\xi,\eta)$ satisfies the same boundary condition as $U(1/\ell;\xi,\eta)$ does, then we will have the simple relation

$$(3.6)\quad W(\ell;\xi,\eta) = U(1/\ell;\xi,\eta)$$

Thus, the expansion of W around $\ell=0$ can be viewed as the expansion of U around $\ell=\infty$. Since $U(\ell;\xi,\eta)$, as the solution to Equation (2.8), is singular at $\ell=\pm 1$, the coefficients of its two expansions around $\ell=0$ and $\ell=\infty$ respectively are linearly independent of each other.

This guarantees that the infinite set of transformations $\{\tilde{\delta}_\alpha^{(n)} g, \; n \geqslant 0\}$ summarized by the following parametric transformation [5]

(3.7) $g^{-1}\tilde{\delta}_{\alpha}(\ell)g = W(\ell;\xi,\eta)T_{\alpha}W(\ell;\xi,\eta)^{-1} = \sum_{n=0}^{\infty} \ell^{n}g^{-1}\tilde{\delta}^{(n)}_{\alpha}g$

are linearly independent of the previous set $\{\delta^{(n)}_{\alpha}g,\ n\geq 0\}$. Here when $g(\xi,\eta)$ is not a solution, we require $W(\ell;\xi,\eta)$ should satisfy

$$(3.8)\qquad \partial_{\xi}W = \frac{1}{\ell-1}A_{\xi}W,\quad W(\ell,\xi=-\infty,\eta) = 1$$

As before, it is easy to check that the Lagrangian (2.1) is changed by a total divergence and the equations of motion (2.2) and (2.3) are invariant under $\tilde{\delta}^{(n)}_{\alpha}g$ and, therefore, each $\tilde{\delta}^{(n)}_{\alpha}g$. Hence, these $\tilde{\delta}^{(n)}_{\alpha}g$ ($n\geq 0$) defined by the H-transformation (3.7) are hidden symmetry transformations for the model too. It can be shown [5] that they lead to an infinite set of conservation laws equivalent to those derived from the inverse BIZZ procedure.

Now to identify the Lie-algebraic structure of the generators corresponding to $\delta^{(m)}g$ and $\tilde{\delta}^{(n)}g$ ($m,n\geq 0$), we need to compute the commutation relations $[\tilde{\delta}_{\alpha}(\ell),\tilde{\delta}_{\beta}(\ell')]g$ and $[\delta_{\alpha}(\ell),\tilde{\delta}_{\beta}(\ell')]g$. We can directly do the computation by using the same trick for evaluating $[\delta_{\alpha}(\ell),\delta_{\beta}(\ell')]g$ discussed in the last section. A simpler way is to apply the substitution (3.5) together with $\delta_{\alpha}(\ell)g\to-\tilde{\delta}_{\alpha}(\ell)g$ to the already known commutator (2.17), giving

$$(3.9)\qquad [\tilde{\delta}_{\alpha}(\ell),\tilde{\delta}_{\beta}(\ell')]g = \alpha^{a}\beta^{b}C^{c}_{ab}\,\frac{\ell'\tilde{\delta}_{c}(\ell)g - \ell\tilde{\delta}_{c}(\ell')g}{\ell-\ell'}$$

$$(3.10)\qquad [\delta_{\alpha}(\ell),\tilde{\delta}_{\beta}(\ell')]g = \alpha_{a}\beta_{b}C^{c}_{ab}\,\frac{\ell\ell'\delta_{c}(\ell)g + \tilde{\delta}_{c}(\ell')g}{1-\ell\ell'}$$

Upon expanding in powers of ℓ and ℓ', the commutators (3.9) and (3.10) lead to somewhat complicated commutation relations between $\tilde{\delta}^{(m)}_{a}g$ and $\tilde{\delta}^{(n)}_{b}g$ and between $\delta^{(m)}_{a}g$ and $\tilde{\delta}^{(n)}_{b}g$. However, if we redefine the linear combination $\delta^{(0)}_{a}g + \tilde{\delta}^{(0)}_{a}g$ as the zero-th generator and relabel the index of $\tilde{\delta}^{(n)}_{a}g$ ($n\geq 1$) by the negative integer ($-n$), then we are led to [5]

(3.11) $$[Q^{(m)}_a,\ Q^{(n)}_b] = C^c_{ab}Q^{(m+n)}_c \qquad (n,m\in\mathbf{Z})$$

where $Q^{(m)}_a$ ($m\in\mathbf{Z}$) are defined by

(3.12) $$Q^{(m)}_a = \begin{cases} \delta^{(m)}_a & \text{if } m>0 \\ \delta^{(0)}_a + \tilde{\delta}^{(0)}_a & \text{if } m=0 \\ \tilde{\delta}^{(-m)}_a & \text{if } m<0 \end{cases}$$

This is the full affine algebra without central extension $g\otimes C[t,t^{-1}]$. If we define

(3.13) $$Q_a(\theta) = \sum_{n=-\infty}^{+\infty} e^{in\theta}Q^{(n)}_a$$

then Equation (3.11) can be written as the gauge algebra (or current algebra) on a circle, or the loop algebra

(3.14) $$[Q_a(\theta),\ Q_b(\theta')] = C^c_{ab}Q_c(\theta)\delta(\theta-\theta')$$

We note that the parameter $e^{i\theta}$ in the loop algebra is essentially the spectral parameter ℓ in the Lax pair (2.8) or (3.4). Moreover, if we had not generalized the indices of the generators to negative integers, we would not have the amusing form (3.14).

Finally, we remark that as shown in reference [5], either the discrete symmetry $g\to g^{-1}$ of the model or the parallel development with $B_\mu = \partial_\mu g\cdot g^{-1}$ cannot lead to new hidden symmetry transformations independent of the above ones.

IV. SIMPLIFIED DERIVATION AND EXTENSION OF THE GEROCH GROUP IN 2-D REDUCED GRAVITY

When there are two Abelian symmetries, such as translations or uni-axial rotations, the vacuum Einstein equations in four dimensions can be reduced to a 2-D nonlinear problem [32]. The cases of physical interest are stationary axisymmetric gravity and gravitational plane waves. To my knowledge, the Geroch group [33] found in this case is the first known example of an infinite-dimensional invariance group for a certain nonlinear system. Here we sketch a new, simplified rederivation [14] and extension [15] of the Geroch group, using the H-transformation method, which avoids completely the introduction of the double infinite hierarchy of potentials in the current approach [34].

First, let us briefly review the Hauser-Ernst linearization system [35] for the problem under consideration, which is the starting point of our H-transformation approach. For, e.g., gravitational plane waves

$$(4.1) \qquad -ds^2 = f(t,z)(-dt+dz^2) + g_{ij}(t,z)dx^i dx^j \quad (i,j = 1,2)$$

the vacuum Einstein equations reduce to

$$(4.2) \qquad \partial_\xi(\alpha g^{-1}\partial_\eta g) + \partial_\eta(\alpha g^{-1}\partial_\xi g) = 0 \quad (\alpha^2 = \det g)$$

and a pair of equations which determines f in terms of g by simple integration. Here $g = (g_{ij})$ is a 2 x 2 symmetric matrix. Instead of g we can use the complex matrix Ernst potential

$$(4.3) \qquad E = g + i\psi, \ \partial_\xi\psi = \alpha^{-1}g\epsilon\partial_\xi g, \ \partial_\eta\psi = \alpha^{-1}g\epsilon\partial_\eta g$$

where $\epsilon = \begin{bmatrix} 0 & 1 \\ -1 & 0 \end{bmatrix}$. The necessary and sufficient conditions for E to be the Ernst potential for some space-time are

$$(4.4) \qquad (1) \ \ \mathrm{Re}E(\equiv g) \text{ is symmetric with } \alpha\equiv\det(\mathrm{Re}E)>0$$

(4.5) (2) $\mathrm{Im}(E+E^{+}) = 2\beta\epsilon$, where $*d\beta = d\alpha$

(4.6) (3) $2(\beta+\alpha^{*})dE = (E+E^{+})(i\epsilon)dE$

The last equation is the equation of motion for E, which is equivalent to Equation (4.2). Here + means hermitian conjugate;

$$d = \partial_{\xi}d\xi + \partial_{\eta}d\eta, \quad *d = \partial_{\xi}d\xi - \partial_{\eta}d\eta$$

The H–E linearization system (Lax pair) for Equation (4.6) is

$$dF(\tau) = \Gamma(\tau)(i\epsilon)F(\tau) \tag{4.7}$$

$$\Gamma(\tau) \overset{d}{=} [1-2\tau(\beta+\alpha^{*})]^{-1}\tau dE \tag{4.8}$$

τ is the (complex) spectral parameter. It can be shown that we have freedom to choose

$$\begin{cases} F(0)=1, \quad \dot{F}(0) = E(i\epsilon) \\ \left[(1-2\beta\tau)^{2}-(2\alpha\tau)^{2}\right]^{1/2} \det F(\tau) = 1 \\ F(\tau)^{x}(i\epsilon)[1-\tau(E+E^{+})(i\epsilon)]F(\tau) = i\epsilon \end{cases} \tag{4.9}$$

Here $F(\tau)^{x} = [F(\bar{\tau})]^{+}$, $\dot{F}(\tau) = \partial F(\tau)/\partial\tau$. These requirements fix the τ dependence of $F(\tau)$ to a large extent. We also need the continuation form of the H–E equation from τ to $1/\tau$:

$$\begin{cases} dG(\tau) = \tilde{\Gamma}(\tau)(i\epsilon)G(\tau) \\ \tilde{\Gamma}(\tau) \overset{d}{=} [\tau-2(\beta+\alpha^{*})]^{-1} dE \end{cases} \tag{4.10}$$

$$\begin{cases} dE = -2(\beta+\alpha^{*})dG(0)\cdot G(0)^{-1}(i\epsilon) \\ \left[(\tau-2\beta)^{2}-(2\alpha)^{2}\right]^{1/2} \det G(\tau) = 1 \\ G(\tau)^{x}(i\epsilon)[\tau-(E+E^{+})(i\epsilon)]G(\tau) = i\epsilon \end{cases} \tag{4.11}$$

Now we are ready to propose the H-transformations appropriate to the linearization system (4.7) or (4.10):

$$\text{(4.12)} \quad \delta_\alpha(\ell)E = \sum_{n=0}^{\infty} \ell^n \delta_\alpha^{(n)} E = \frac{-1}{\ell}\{F(\ell)T_\alpha F(\ell)^{-1} - T_\alpha\}(i\epsilon)$$

$$\text{(4.12')} \quad \tilde{\delta}_\alpha(\ell)E = \sum_{n=0}^{\infty} \ell^n \tilde{\delta}_\alpha^{(n)} E = \ell\{G(\ell)T_\alpha G(\ell)^{-1} - T_\alpha\}\,(i\epsilon)$$

where $F(\ell)$ and $G(\ell)$ are solutions to Equations (4.7–9) and (4.10,11) respectively, which are analytic around $\ell=0$; $T_\alpha = \alpha^a T_a \in sl(2,\mathbb{R})$, i.e., $\mathrm{tr}T_\alpha = 0$ (α^a being infinitesimal). It can be shown that $E+\delta_\alpha(\ell)E$ and $E+\tilde{\delta}_\alpha(\ell)E$ satisfy all Equations (4.4–6) for the Ernst potential. Thus, $\delta_\alpha(\ell)E$, $\tilde{\delta}_\alpha(\ell)E$ and, therefore, all $\delta_\alpha^{(n)}E$ and $\tilde{\delta}_\alpha^{(n)}E$ ($n\geq 0$) are hidden symmetries in the present problem. Back to the original field variables g, we have

$$\text{(4.13)} \quad \delta_\alpha(\ell)g = \sum_{n=0}^{\infty} \ell^n \delta_\alpha^{(n)} g = -\frac{1}{\ell}\mathrm{Re}\{F(\ell)T_\alpha F(\ell)^{-1}(i\epsilon)\}$$

$$\text{(4.13')} \quad \tilde{\delta}_\alpha(\ell)g = \sum_{n=0}^{\infty} \ell^n \tilde{\delta}_\alpha^{(n)} g = \ell\ \mathrm{Re}\{G(\ell)T_\alpha G(\ell)^{-1}(i\epsilon)\}$$

In addition to the equations of motion (4.2), they also preserve the determinant $\alpha \equiv \sqrt{\det\ g}$ and the property of g being symmetric.

To calculate the commutators among the transformations (4.12 and 12'), we note that we can identify $\delta_\alpha(\ell)F(\tau)$ as

$$\text{(4.14)} \quad \delta_\alpha(\ell)F(\tau) = \frac{\tau}{\tau-\ell}\,\{F(\ell)T_\alpha F(\ell)^{-1} - F(\tau)T_\alpha F(\tau)^{-1}\}F(\tau)$$

by showing that the right-hand satisfies all the defining equations (4.7–9) for $F(\tau)+\delta F(\tau)$. Using this equation one finds

$$\text{(4.15)} \quad [\delta_\alpha(\ell),\delta_\beta(\ell')]E = \alpha^a\beta^b C^c_{ab}\,\frac{\ell\delta_c(\ell)E - \ell'\delta_c(\ell')E}{\ell-\ell'}$$

where $[T_a,T_b] = C^c_{ab}T_c$. It leads to the half affine algebra

$\mathfrak{sl}(2,\mathbb{R})\otimes\mathbb{R}(t)$ for $\delta^{(m)}_{\alpha}E$ $(m\geq 0)$. Repeating the same trick as in Section III one can obtain the full affine algebra $\mathfrak{sl}(2,\mathbb{R})\otimes\mathbb{R}[t,t^{-1}]$

(4.16) $$[\delta^{(n)}_{a},\ \delta^{(m)}_{b}]E = C^{c}_{ab}\delta^{(n+m)}_{c}E \quad (m,n = 0,\ \pm 1,\ \pm 2,\ \ldots)$$

after relabelling

(4.17) $$\tilde{\delta}^{(m)}_{a}E = \delta^{(-m)}_{a}E\ (m\geq 1), \quad (\tilde{\delta}^{(0)}_{a}E = 0)$$

(Very recently, by considering the function f in Equation (4.1) which we do not consider, Julia has succeeded in finding the central extension of the algebra (4.16) [36].)

By comparing our expression (4.12) with the Riemann-Hilbert problem approach of the Geroch group, we found [14] that our $\delta^{(n)}_{\alpha}E$ $(n\geq 0)$ can be identified with those generators of the Geroch group which give rise to nontrivial transformations for the metric tensor g. In the usual approach the Geroch group is realized by means of transformations acting on $F(\tau)$, and it is well known that half the set of its generators are gauge transformations, which do not change E or g at all. However, our $\tilde{\delta}^{(n)}_{\alpha}E$ $(n\geq 1)$ is clearly nontrivial. Thus, we claim that Equation (4.12') represents an extension of the Geroch group. It would be interesting to use the extended half to generate new solutions from an old one as people have done with the unextended Geroch group.

V. INFINITESIMAL RIEMANN-HILBERT TRANSFORMS AND THE AFFINE ALGEBRA OF HIDDEN SYMMETRIES

The Riemann-Hilbert transform has been a very powerful method for obtaining solutions to many nonlinear problems [37]. Now it is also an alternative systematic approach to hidden symmetries in addition to the H-transformation method described above. This is because the infinitesimal (regular) RH transforms have been recently turned out [7,8] to be particularly appropriate to deal with hidden symmetries and their group structure. Here we will review the explicit expressions for infinitesimal RH transforms as hidden symmetries and present extremely simple derivations of their commutators. The connection to and the comparison with the H-transformation method will be discussed also in this section.

Although most discussions below are generally true in a large class of nonlinear problems, it is better to present them in a concrete model, which we choose, again, to be the 2-D principal chiral model discussed in Section II.

The RH transform method starts with the Lax pair (or the linearization system) too. The basic idea of it for solving, e.g., Equation (2.2) is as follows [37]: First we generate a new solution U' to the linearization system (2.8) (more precisely, a new U' which satisfies a linear system of the same form as Equation (2.8)) from an old solution U by means of the RH transform. Then we can construct a new solution g' to Equation (2.2) from U' using Equation (2.10). Concretely, the RH transform from U to U' consists of the following steps:

1) Select a small circle C, in the complex ℓ-plane, with the center at the origin such that $U(\ell)$ is analytic on $C_+ \cup C_-$. ($C_\pm$ denotes the inside and outside of C, respectively.)

2) Choose a matrix function of only ℓ, $u(\ell) \in G$ (called the Riemann-Hilbert data), which is analytic on C, and form the kernel

$$K(\ell) = U(\ell) \cdot u(\ell) \cdot U(\ell)^{-1}, \quad (\ell \in C) \tag{5.1}$$

3) Consider the following RH problem of finding a pair of functions $X_\pm(\ell)$ which are analytic in $C_\pm$ and continuous on C, respectively, such that

$$X_-(\ell) = X_+(\ell) K(\ell) \text{ for } \ell \in C \tag{5.2}$$

We can always normalize $X_\pm(\ell)$ so that

$$X_+(\ell=0) = 1 \tag{5.3}$$

4) Construct a new U' by

$$
(5.4)\qquad U'(\ell) = \begin{cases} X_+(\ell)U(\ell) & \text{in } C_+ \\ X_-(\ell)U(\ell)u(\ell)^{-1} & \text{in } C_- \end{cases}
$$

We will be restricted to the case in which det $X_\pm(\ell)\neq 0$ for all ℓ, the so-called regular RH transform.

It is easy to verify that

$$
(5.5)\qquad \begin{aligned} dU'\cdot U'^{-1} &= dX_+\cdot X_+^{-1}+X_+\Omega X_+^{-1} \\ &= dX_-\cdot X_-^{-1}+X_-\Omega X_-^{-1} \end{aligned}
$$

with

$$
(5.6)\qquad d = \partial_\xi d\xi+\partial_\eta d\eta,\ \Omega = \frac{\ell}{1-\ell}A_\xi d\xi - \frac{\ell}{1+\ell}A_\eta d\eta
$$

Since $X_\pm(\ell)$ is analytic in $C_\pm$ respectively, Equation (5.5) implies that $dU'\cdot U'^{-1}$ is analytic in $C_\pm$ with its singularities determined by Ω. Thus, it must have the form

$$
(5.7)\qquad dU'\cdot U'^{-1} = \frac{\ell}{1-\ell}A'_\xi d\xi - \frac{\ell}{1+\ell}A'_\eta d\eta
$$

That is, U' satisfies a linear system of the same form as Equation (2.8). Hence, we have the new A'_μ, by using Equation (2.10),

$$
(5.8)\qquad A'_\xi = A_\xi+\partial_\xi \dot{X}_+(0),\ A'_\eta = A_\eta-\partial_\eta \dot{X}_-(0)
$$

By standard procedure, the solution $X_+(\ell)$ to the RH problem can be shown to satisfy the integral equation

$$
(5.9)\qquad X_+(\ell)-1+\frac{1}{2\pi i}\int_C d\ell'\,\frac{X_+(\ell')[K(\ell')-1]}{\ell'(\ell'-\ell)} = 0\ (\ell\in C_+)
$$

Thus, the RH transform $U(\ell) \longmapsto U'(\ell)$ can be written as an integral equation too:

$$(5.10)\quad U'(\ell)+\frac{1}{2\pi i}\int_C d\ell'\frac{\ell}{\ell'(\ell'-\ell)}U'(\ell')[u(\ell')-1]U(\ell')^{-1}U(\ell) = U(\ell)$$

For the infinitesimal RH transform,

$$(5.11)\qquad u(\ell)-1 = \nu(\ell) \in \mathfrak{g}$$

We have $U(\ell) \longmapsto U(\ell)+\delta U(\ell)$ with [7]

$$(5.12)\quad \delta U(\ell) = \frac{-1}{2\pi i}\int_C d\ell'\frac{\ell}{\ell'(\ell'-\ell)}U(\ell')\nu(\ell')U(\ell')^{-1}\cdot U(\ell)$$

With no loss of generality we can assume

$$(5.13)\qquad \nu(\ell) = \alpha^a T_a \ell^{-k} \quad (k\in \mathbf{Z})$$

where α^a are infinitesimal constants, T_a generators of $\mathfrak{g}$. (A generic infinitesimal (regular) RH transform is a linear combination of such ones.) The corresponding $\delta U(\ell)$ turns out to be [8]

$$\delta_\alpha^{(k)} U(\ell) = U(\ell)(-\alpha^a T_a \ell^{|k|}) \quad \text{for } k<0$$

$$(5.14)\qquad = -\alpha^a [U(\ell),T_a] \quad \text{for } k=0$$

$$= -\alpha^a \sum_{n=0}^{\infty} \ell^{n+1} T_a^{(n+k+1)}(\xi,\eta)\cdot U(\ell) \quad \text{for } k>0$$

where the functions $T_a^{(m)}(\xi,\eta)$ $(m\geq 0)$ are defined by the expansion

$$(5.15)\qquad U(\ell;\xi,\eta)T_a U(\ell;\xi,\eta)^{-1} = \sum_{m=0}^{\infty} \ell^m T_a^{(m)}(\xi,\eta)$$

We emphasize that for $k<0$, $\delta_\alpha^{(k)}U(\ell)$ is only a right multiplication with a matrix independent of (ξ,η) so that it does not change the original basic field $g(\xi,\eta)$. In this sense, all the infinitesimal (regular) RH transforms with $k\leq -1$ are trivial, though formally they give global [(ξ,η)-independent] right-multiplication transforms for $U(\ell)$.

Since now we have the explicit expressions, Equation (5.14), for the action of infinitesimal RH transforms on $U(\ell)$, we can use them to check directly that these actions on $U(\ell)$ provide a representation of the affine algebra $\mathfrak{g}\otimes\mathbb{C}[\ell,\ell^{-1}]$:

$$[\delta^{(j)}_{a},\delta^{(k)}_{b}]U = C^{c}_{ab}\delta^{(j+k)}_{c}U \quad (j,k\in\mathbb{Z}) \tag{5.16}$$

In fact, for both j and k<0 the proof is trivial. The case of j<0 and k>0 is easy to check by using

$$\delta^{(j)}_{a}T^{(n)}_{b}(\xi,\eta) = -C^{c}_{ab}T^{(n+j)}_{c}(\xi,\eta) \text{ for } j<0 \tag{5.17}$$

which can be derived from the action of $\delta^{(j)}_{a}$ (j<0) on Equation (5.15). The case of both j,k>0 can so be easily checked as follows: First consider

$$\delta_{\alpha}(\ell')U(\ell) = \sum_{k=0}^{\infty}\ell'^{k}\delta^{(k)}_{\alpha}U(\ell) \tag{5.18}$$

It is very simple to verify that

$$\begin{aligned}\delta_{\alpha}(\ell')U(\ell)\cdot U(\ell)^{-1} &= -\alpha^{a}\sum_{k=0}^{\infty}\sum_{n=k+1}^{\infty}\ell'^{k}\ell^{n-k}T^{(k)}_{a}(\xi,\eta)\\ &= \frac{\ell}{\ell-\ell'}\{U(\ell')T_{\alpha}U(\ell')^{-1}-U(\ell)T_{\alpha}U(\ell)^{-1}\}\end{aligned} \tag{5.19}$$

Using this formula the computation of $[\delta_{\alpha}(\ell'),\delta_{\beta}(\ell'')]U$ becomes quite simple too. Finally, just expand the commutator in powers of ℓ' and ℓ'', as we did in Section II.

Here I would like to point out an alternative derivation of the group structure of infinitesimal RH transforms, which is extremely simple in that it needs nearly none of real computation. It is based on the observation that the composition of two RH transforms corresponding to the RH data $U_1(\ell)$ and $U_2(\ell)$, respectively, is an RH transform corresponding to the product $U_2(\ell)U_1(\ell)$ as the RH data [35]. This can be easily seen as follows: From Equations (5.1), (5.2) and (5.4) we have

$$U_1(\ell) = X_+^{(1)}(\ell)U_0(\ell), \quad U_2(\ell) = X_+^{(2)}(\ell)U_1(\ell) \tag{5.20}$$

with $X_+^{(i)}(\ell)$(i=1,2) being the solution to the following RH problems

$$\begin{cases} X_-^{(1)}(\ell) = X_+^{(1)}(\ell)U_0(\ell)u_1(\ell)U_0(\ell)^{-1} \\ X_-^{(2)}(\ell) = X_+^{(2)}(\ell)U_1(\ell)u_2(\ell)U_1(\ell)^{-1} \quad (\ell \in C) \end{cases} \tag{5.21}$$

It is straightforward to see that

$$U_2(\ell) = X_+^{(2)}(\ell)X_+^{(1)}(\ell)U_0(\ell) \equiv X_+(\ell)U_0(\ell) \tag{5.22}$$

with $X_\pm(\ell) \equiv X_\pm^{(2)}(\ell)X_\pm^{(1)}(\ell)$ satisfying the RH problem

$$X_-(\ell) = X_+(\ell)U_0(\ell)u_2(\ell)u_1(\ell)U_0(\ell)^{-1} \tag{5.23}$$

which obviously has, as the RH data,

$$u(\ell) = u_2(\ell)\cdot u_1(\ell) \tag{5.24}$$

In particular, Equation (5.24) is true when $u_1(\ell)$, $u_2(\ell)$ are infinitesimal. Writing $u_i(\ell) = 1+v_i(\ell)$ (i=1,2) and denoting the corresponding infinitesimal RH transforms as $\delta_{v_i}U(\ell)$, we have immediately from Equation (5.24)

$$[\delta_{v_1}, \delta_{v_2}]U = \delta_{[v_1,v_2]}U \tag{5.25}$$

The commutation relations (5.16) follow directly when each v_i takes the form of Equation (5.13). In other words, the composition rule (5.24) provides a homomorphism between the set of RH transforms and the multiplicative group of RH data; and the latter is clearly of the affine type. In our opinion, this is the reason why infinitesimal RH transforms are particularly appropriate to deal with the group structure of hidden symmetries.

It is interesting to notice that Equation (5.19) is identical to

Equation (2.16) in the H transformation approach. Actually, if we consider Equation (5.18) as a parametric transformation and use Equation (2.10) and the formula

(5.26) $$\delta A_\mu = \delta(g^{-1}\partial_\mu g) = \partial_\mu(g^{-1}\delta g)+[A_\mu, g^{-1}\delta g]$$

to extract the corresponding transformation on g, we will obtain

(5.27) $$\delta_\alpha(\ell')g = -gU(\ell')T_\alpha U(\ell')^{-1}$$

Thus, we can identify the infinitesimal RH transforms with $k \geq 0$ with our H transformations (2.11). However, as emphasized before, another half set of RH transforms do not lead always to nontrivial symmetry transformations in the space of basic fields. This is not surprising since RH transforms act on the auxiliary quantities $U(\ell)$ in the theory, in contrast to the H transformations which always act on the basic field variables.

To conclude this section, we point out that a complication may occur in the attempts of applying the RH transform method to deal with hidden symmetries in more complicated cases. Normally, the RH data $u(\ell)$ has to satisfy certain constraints in order that the RH transform does not go out of the space of field configurations we are interested in. (In our previous example, it is the requirement $u(\ell) \in G$ that guarantees $\delta g(\xi,\eta)$ to take values in G.) However, in some cases these constraints on $u(\ell)$ are very hard to formulate. For example, in 4-D self-dual Yang-Mills systems the constraints on $u(\ell)$ for maintaining the reality and hermitian properties of the potentials have never been worked out yet [7]. Similar problems also obstruct the application of the RH transform method to hidden symmetries in nonlinear σ-models on coset spaces.

In this aspect, the H-transformation method has the advantage that it is relatively easy to formulate the proper conditions keeping H transformations in the same configuration space. For instance, the H transformations in 4-D self-dual Yang-Mills equations which maintain the reality and hermitian properties of the potentials have been found for some time. (For details, see [20] or L. L. Chau in these proceedings.) This is the reason why we do not think the RH transform method can supercede the H transformation method at all.

VI. RICHER GROUP STRUCTURE OF RH TRANSFORMS

From the last section one still does not know what in the RH transform approach corresponds to the second infinite set of H transformations (3.7). Answering this question has led to the discovery of richer group structure of RH transforms [8]. Here we only state the results for 2-D principal chiral models and refer the readers who are interested in the details to the original paper.

Just like the case of H transformations (3.7), we start with the linearization system (3.4) and consider the RH transforms for it in a similar way to the last section (with minor technical modifications). Equivalently, by the correspondence (3.6), we can also consider the original linearization system (2.8), but choose a large circle $\tilde{C}$ such that $U(\ell)$ is analytic outside of it. (Note that the homotopy of C and $\tilde{C}$ must be different. This is because the singularities of the coefficients in Equation (2.8) are at $\ell = \pm 1$. So these two points must be outside of C, but inside of $\tilde{C}$.) Either way we have the second set of RH transforms, indexed by all integers too, $\tilde{\delta}^{(n)}_{\alpha} U(\ell)$ or $\tilde{\delta}^{(n)}_{\alpha} W(\ell)$ $(n \in \mathbb{Z})$. Similar to the situation in the last section, half the set (say, $n \leqslant 0$) are trivial in that they do not lead to any change in the basic field $g(\xi,\eta)$; and another half (say, $n \geqslant 1$) can be identified with the second set of H transformations (3.7). Also, the algebra for the second set of RH transforms turns out to be an affine algebra $g \otimes \mathbb{C}[\ell,\ell^{-1}]$.

It is interesting to consider both the sets of RH transforms combined together forming a larger set of transformations in the space of $U(\ell;\xi,\eta)$. What is the algebra for the enlarged set of RH transforms? It contains at least two subalgebras, formed respectively by $\delta^{(n)}_{\alpha}$ and $\tilde{\delta}^{(n)}_{\alpha}$ $(n \in \mathbb{Z})$, each of them isomorphic to $g \otimes \mathbb{C}[\ell,\ell^{-1}]$. All the commutators have been worked out [8]. The two affine subalgebras turned out not to commute with each other. And it is amusing to notice that there are two more affine subalgebras. One is formed by those transformations which lead to nontrivial H transformations (2.11) and (3.7), the other one leading to the trivial transformation in $g(\xi,\eta)$. They do not commute with each other either.

ACKNOWLEDGEMENTS

The authors are grateful to Professer S. S. Chern, Professor J. Lepowsky and Professor S. Mandelstam for their kind invitation and warm hospitality. They are indebted to L. L. Chau and Hou Bo-Yu for pleasant collaboration, and to I. Frenkel, B. Julia, A. Sinha, H. C. Tze, and E. Witten for helpful discussions. In particular, the authors thank Professor C. N. Yang for the encouragement and enlightening discussions he gave when the authors began study in this field. The first author also thanks Professor S. L. Adler and the Institute for Advanced Study for the warm hospitality he received when most of this work was done. This work was supported in part by the U.S. Department of Energy.

REFERENCES

[1] For a review, see, e.g., L.L. Chau in Proc. 1980 Guangzhou Conf. on Theoretical Particle Physics, Science Press, (1981); and talk at Internat. School of Subnuclear Physics, Erice, Italy, July 1980.

[2] B.Y. Hou, Yale preprint 80-29, 1980 (unpublished); B.Y. Hou, M.L. Ge and Y.S. Wu, Phys. Rev. D24 (1981) 2238.

[3] M.L. Ge and Y.S. Wu, Phys. Lett. 108B (1982) 411.

[4] C. Devchand and D.B. Fairlie, Nucl. Phys. B194 (1982) 232.

[5] Y.S. Wu, Nucl. Phys. B211 (1983) 160.

[6] I. Hauser and F.J. Ernst, Phys. Rev. D20 (1979) 362, 978.

[7] K. Ueno, RIMS preprint 374, 1981; K. Ueno and Y. Nakamura, Phys. Lett. 109B (1982) 273.

[8] Y.S. Wu, Commun. Math. Phys. 90 (1983) 461.

[9] Y.S. Wu and M.L. Ge, Phys. Energ. Fort. et Phys. Nucl. (in Chinese) 7 (1983) 427.

[10] L.L. Chau and B.Y. Hou, BNL preprint 33866, 1983.

[11] L.L. Chau and T. Koikawa, Phys. Lett. 123B (1983) 47.

[12] L.L. Chau, M.L. Ge and Y.S. Wu, Phys. Rev. D25 (1982) 1080.

[13] L.L. Chau, Z. Popowicz and J. Lukierski, BNL 32458, 1982.

[14] Y.S. Wu and M.L. Ge, J. Math. Phys. 24 (1983) 1187.

[15] Y.S. Wu, Phys. Lett. 96A (1983) 179.

[16] B.Y. Hou, M.L. Ge and Y.S. Wu, in Proc. of the Third Grossmann Meeting on Gravitation, Shanghai, Sept. 1982.

[17] M.L. Ge and Y.S. Wu, Lanzhou preprint, 1983.

[18] H. Eichenherr, Phys. Lett. 115B (1983) 385.

[19] L.L. Chau, M.L. Ge and Y.S. Wu, Phys. Rev. D25 (1982) 1086.

[20] L.L. Chau and Y.S. Wu, Phys. Rev. D26 (1982) 3581; L.L. Chau, M.L. Ge, A. Sinha and Y.S. Wu, Phys. Lett. 121B (1983) 391.

[21] I.V. Volovich, Phys. Lett. 123B (1983) 329, Teop. Mat. Φu3. 57 (1983) 163; C. Devchand, Durham preprint, 1983.

[22] L.L. Chau and M.L. Ge, BNL preprint, 1983.

[23] E. Brezin, C. Itzykson, J. Zinn-Justin and J.B. Zuber, Phys. Lett. 82B (1979) 442.

[24] A.T. Ogielski, M.K. Prasad, A. Sinha and L.L. Chau Wang, Phys. Lett. 91B (1980) 387.

[25] V.E. Zakharov and A.V. Mikhailov, JETP (Sov. Phys.) 47 (1978) 1017.

[26] L. Dolan and A. Roos, Phys. Rev. D22 (1982) 2018.

[27] L. Dolan, Phys. Rev. Lett. 47 (1981) 1371. See also T.L. Cartright and C.K. Zachos, Phys. Rev. D24 (1981) 2661.

[28] V.G. Kac, Math. USSR Izv. 2 (1968) 1311; R.V. Moody, J. Algebra, 10 (1968) 211.

[29] See the articles in these proceedings; A.J. Feingold and I.B. Frenkel, IAS preprint, 1983.

[30] J. Lepowsky and R.L. Wilson, Commun. Math. Phys. 62 (1978) 43; I.B. Frenkel and V.G. Kac, Invent. Math. 62 (1980) 23; G. Segal, Comm. Math. Phys. 80 (1981) 301-342; I.B. Frenkel, J. Fun. Anal. 44 (1981) 259.

[31] B.L. Markowsky and R.P. Zaikov, Dubna preprint E2-80-654, 1980.

[32] For a review, see, e.g., B. Julia, Invited talk at the John Hopkins Workshop on Particle Theory, May 1981 (LEPTENS 82-14); Lectures given at International School of Cosmology and Gravitation, Erice, May 1982 (LEPTENS 8-18).

[33] R. Geroch, J. Math. Phys. 12 (1971) 918; 13 (1972) 394.

[34] W. Kinnersley and D. Chitre, J. Math. Phys. 18 (1977) 1538; 19 (1978) 1926.

[35] I. Hauser and F.J. Ernst, J. Math. Phys. 21 (1980) 1126.

[36] B. Julia, preprint LPTENS 82/18 (1982) and in these proceedings.

[37] V.E. Zakharov and A.B. Shabat, Funct. Anal. Appl. 13 (1979) 13; For a brief review, see, e.g., A.V. Mikhailov, CERN preprint TH-3194, 1981, and references therein.

Yong-Shi Wu
Department of Physics PM-15
University of Washington
Seattle, WA 98195

Mo-Lin Ge
Department of Physics
Lanzhou University
Lanzhou, China

MASSIVE KALUZA-KLEIN THEORIES AND BOUND STATES IN YANG-MILLS

L. Dolan[†]

As we have seen in the last few years, the role of affine symmetry algebras in particle theories has become more far reaching and more clear. Probably the most fundamental use of symmetry transformations in quantum field theory is that the particle spectrum, the physical states of the system, usually form linear multiplets, i.e. carry a representation of the global invariance group.

It sometimes happens, as has been calculated in two and three-dimensional sigma models, that whereas the physical bound states carry unitary representations, the fundamental fields in the Lagrangian density transform non-linearly under the symmetry group. The role of affine symmetries might thus be quite naturally understood in that while the gauge fields transform non-linearly (as evidenced by the self-dual expressions), the physical spectrum, the Yang-Mills bound states, composites of the fundamental gluon fields, will form linear representations. Since representations of infinite parameter groups are infinite dimensional, the bound states must come in an infinite tower.

The basic unitary representations associated with certain affine Kac-Moody algebras are constructed in terms of the vertex operators of the dual string model. (String theories share the property of having an infinite number of states.) This now provides a way to establish a connection between non-perturbative Yang-Mills theory and the string, i.e. the physical hadronic spectrum.

There has been a lot of work, in various theories, on realizations of affine symmetries. Most recently, in massless models, they have been found in the non-compact extended supergravity models, where it was suggested that they serve as a guide in

---

[†]Partially supported by the National Science Foundation through the Mathematical Sciences Research Institute

---

Vertex Operators in Mathematics and Physics - Procedures of a Conference November 10-17, 1983. Publications of the Mathematical Sciences Research Institute #3, Springer-Verlag, 1984.

specifying the representation of the non-compact groups appropriate to the physical particle content.[1] If the affine symmetry also exists in pure Yang-Mills, presumably this invariance of supergravity is a shadow of the invariance of the simpler gauge theory contained in it. We note that in supergravity, the affine transformations are realized on the gauge field strengths as duality rotations.[2] Thus the gauge potential itself is transforming non-locally, as is the case for the known affine expressions on the self-dual set.

In the first part[3] of this talk, I will describe the simplest case of a different class of theories in which an affine Kac-Moody algebra (more precisely a certain loop algebra) also exists, but where most of it is spontaneously broken. It is a theory with an infinite number of fields both massless and massive, and is obtained via Kaluza-Klein from pure gravity in five dimensions with ground state $M^4 \times S^1$. Relatively little is known about the massive sector of Kaluza-Klein theories, and the affine symmetry provides an exact analysis of the mixing between the massless and massive states. It also predicts that the zero-mode scalar particle is not massless.

Although all of this is illustrated in an abelian model, these infinite parameter symmetries will also appear in higher dimensional non-abelian and supergravity Kaluza-Klein theories, the latter resulting in affine Lie superalgebras.[4]

In the second part of this talk, it is shown how the massive spin-2 particles in the theory form an infinite dimensional representation of the infinite parameter symmetry, which generalizes the Salam-Strathdee[5] SO(1,2). In this sense, it is a spectrum generating algebra; and we can make contact with the action-angle variables of the quantum inverse scattering method. The spectral properties of the representation of the infinite-dimensional algebra give the masses of the spin-2 states, even though the symmetry is spontaneously broken to U(1) × Poincaré and the members of the multiplet are not degenerate in mass.

In the last part of the talk, inspired by the analysis of composite fields of scalars valued on non-compact coset spaces,[6,1] I suggest, in the confining compact non-abelian gauge theory, where there is a mass gap, how one might construct the bound states, and the relationship of this to the linear representations of the affine symmetry.

5-DIMENSIONAL KALUZA-KLEIN[3]

Symmetries of the four-dimensional Lagrangian, including the massive fields, obtained from a five-dimensional Kaluza-Klein theory are described by a non-compact infinite parameter algebra involving (i) the affine algebra of the Poincaré algebra, (ii) a Virasoro algebra of internal symmetries for which the Salam-Strathdee SO(1,2) is the finite dimensional subalgebra, (iii) a mixing of internal and external symmetries. All symmetries are spontaneously broken except for Poincaré x U(1). The vector and scalar Goldstone bosons provide masses for the spin-2 tower.

Define the coordinate in five dimensions to be

$$z^M = \begin{cases} x^\mu \\ x^5 = \frac{\theta}{m}, \quad 0 \leq \theta \leq 2\pi \end{cases}$$

where m^{-1} is the radius of the internal space S^1, $m^{-1} \sim m^{-1}_{planck}$. The metric tensor is γ_{MN}. The action is

$$I = \frac{1}{2\pi K^2} \int d\theta \int d^4x \, (|\gamma|)^{1/2} R(\gamma). \tag{1}$$

It is invariant under the infinitesimal general coordinate transformation $z^M \longrightarrow z^M - \xi^M(z)$. The ground state is <u>assumed</u> to be $M^4 \times S^1$. This means that $\gamma_{MN}(x,x^5)$ is periodic in x^5.

$$\gamma_{MN}(x,x^5) = \sum_{n=-\infty}^{\infty} \gamma_{MN_n}(x) \, e^{in\theta}. \tag{2}$$

To make contact with a familiar theory, a parametrization of γ_{MN} is chosen which gives rise to the standard gauge expression in four dimensions:

$$(3)\qquad \gamma_{MN}(x,x^5) = \phi^{-1/3} \begin{bmatrix} g_{\mu\nu} + K^2 A_\mu A_\nu & K\phi A_\mu \\ K\phi A_\nu & \phi \end{bmatrix}.$$

Then

$$g_{\mu\nu}(x,x^5) = \sum_{n=-\infty}^{\infty} g_{\mu\nu n}(x)\, e^{in\theta}$$

$$(4)\qquad A_\mu(x,x^5) = \sum_{n=-\infty}^{\infty} A_{\mu n}(x)\, e^{in\theta}$$

$$\phi(x,x^5) = \sum_{n=-\infty}^{\infty} \phi_n(x)\, e^{in\theta}.$$

Equation (4) describes n tensor, n vector and n scalar four-dimensional fields, $\phi_n^* = \phi_{-n}$ etc. With (4), the four-dimensional Lagrangian density becomes, from (1),

$$(5)\qquad \mathcal{L}(x) = \frac{1}{2\pi K^2} \int d\theta\, (|\gamma|)^{1/2} R(\gamma)$$

$$= (|g_0|)^{1/2} \{\frac{R(g_0)}{K^2} - \frac{1}{4}\phi_0 F_{\mu\nu 0} F_0^{\mu\nu} - \frac{1}{6K^2} \frac{\partial_\mu \phi_0 \partial^\mu \phi_0}{\phi_0^2}\}$$

+ infinite tower of massive charged spin-2 states.

n=0 terms

$g_{\mu\nu 0}(x)$, $A_{\mu 0}(x)$ are the usual massless spin 2 and spin 1 gauge fields. ϕ_0 has zero canonical dimension. Define $\phi_0 = e^{K\tilde{\phi}}$. Then $\partial_\mu \ln \phi_0 = K\partial_\mu \tilde{\phi}$ and $-\dfrac{1}{6K^2} \dfrac{\partial_\mu \phi_0 \partial^\mu \phi_0}{\phi_0^2} = -\dfrac{1}{6} \partial_\mu \phi \partial^\mu \phi$, the usual sigma model form. Indices are raised and lowered by $g_{\mu\nu 0}$. $\partial^\mu = g^{\mu\nu}\, \partial_\nu$ etc.

n≠0 terms

These describe an infinite tower of charged massive, purely

spin 2 particles with charges $q_n = n\ K\ m$ and masses $m_n = |n|m$. In addition to the quadratic mass terms there are also interaction terms. The affine algebra to be discussed is a symmetry of (5).

First look at the n=0 terms. The n=0 action is invariant under

1) four-dimensional general coordinate transformations:

$$\Delta g_{\mu\nu 0} = \partial_\mu \zeta_0^\rho\, g_{\rho\nu 0} + \partial_\nu \zeta_0^\rho\, g_{\rho\mu 0} + \zeta_0^\rho\, \partial_\rho g_{\mu\nu 0}$$

(6)
$$\Delta A_{\mu 0} = \partial_\mu \zeta_0^\rho\, A_{\rho 0} + \zeta_0^\rho \partial_\rho A_{\mu 0}$$

$$\Delta \phi_0 = \zeta_0^\rho \partial_\rho \phi_0.$$

2) local U(1)

(7)
$$\Delta A_{\mu 0} = \frac{1}{K}\, \partial_\mu \zeta_0^5$$

3) global scale transformations

$$\Delta A_{\mu 0} = \lambda A_{\mu 0}$$

(8)

$$\Delta \phi_0 = -2\ \lambda \phi_0.$$

Symmetries of the vacuum are determined by the vacuum expectation values

$$< 0 \mid g_{\mu\nu 0}(x) \mid 0 > = \eta_{\mu\nu} = \begin{bmatrix} -1 & & & \\ & 1 & & \\ & & 1 & \\ & & & 1 \end{bmatrix}$$

(9)
$$< 0 \mid A_{\mu 0}(x) \mid 0 > = 0$$

$$< 0 \mid \phi_0(x) \mid 0 > = 1.$$

The choice (9) is equivalent to the assumption that the vacuum of the five-dimensional theory is $M^4 \times S^1$. In order to express (5) in terms of the physical fields $h_{MN}(x)$ of the theory we define

(10) $$\gamma_{MN}(x,x^5) = \gamma_{MN_{classical}}(x,x^5) + h_{MN}(x,x^5),$$

$$\text{i.e. } g_{\mu\nu} = \eta_{\mu\nu} + \bar{g}_{\mu\nu}, \ A_\mu = \bar{A}_\mu, \ \phi = 1 + \bar{\phi},$$

where $\gamma_{MN_{classical}}$ is a solution to the five-dimensional Einstein equations. $M^4 \times S^1$ means

(11) $$\gamma_{MN_{classical}} = \begin{bmatrix} \eta_{\mu\nu} & 0 \\ 0 & 1 \end{bmatrix}.$$

(11) is consistent with (9) as well as with

(12)
$$< 0 \mid g_{\mu\nu n}(x) \mid 0 > = 0$$
$$< 0 \mid A_{\mu\nu}(x) \mid 0 > = 0$$
$$< 0 \mid \phi_n(x) \mid 0 > = 0$$

for $n \neq 0$.

The choice (9) determines which symmetries are spontaneously broken.

Whenever there is an infinitesimal transformation, there exists an operator $U = e^{iQ}$ where

(13) $$\phi' = U\phi U^{-1} \text{ so } \Delta\phi = [Q,\phi].$$

The states transform as $\mid 0 >' = U \mid 0 >$. Spontaneous symmetry breaking is the statement that $\mid 0 >' \neq \mid 0 >$ or equivalently $Q \mid 0 > \neq 0$. If $< 0 \mid \Delta\phi \mid 0 > \equiv < 0 \mid [Q,\phi] \mid 0 > \neq 0$, then $Q \mid 0 > \neq 0$, so whether or not $< 0 \mid \Delta\phi \mid 0 >$ vanishes is a test for spontaneous breakdown.

Consider the symmetry in (6) when $\zeta_0^\mu(x) = a^\mu$ (translations). Then from (6),

$$\Delta_\mu \phi_0 = \partial_\mu \phi_0 \equiv [P_\mu, \phi_0]$$

and

(14) $$< 0 \mid [P_\mu, \phi_0] \mid 0 > = < 0 \mid \partial_\mu \phi_0 \mid 0 >$$

$$= \partial_\mu < 0 \mid \phi_0 \mid 0 >$$

$$= \partial_\mu 1$$

$$= 0.$$

Therefore $P_\mu \mid 0 > = 0$ is consistent with $< 0 \mid \phi_0 \mid 0 > = 1$, and translations are not spontaneously broken in this vacuum. In the same way,

(15) $$M_{\mu\nu} \mid 0 > = 0 \qquad \text{and} \qquad Q \mid 0 > = 0$$

where $M_{\mu\nu}$ and Q generate Lorentz transformations and the global scale transformation respectively. All other transformations in (6), (7), (8) have non-vanishing vacuum expectation values. Therefore the symmetry of the vacuum is Poincare $\times$ $\mathbb{R}$.

In particular, the global scale symmetry (8) is spontaneously broken, since

(16) $$< 0 \mid -2\lambda\phi_0 \mid 0 > = - 2\lambda \neq 0.$$

So the apparent masslessness of ϕ_0 is due to ϕ_0 being a Goldstone boson associated with the breakdown of the global scale invariance. The masslessness of $A_{\mu 0}$ and $g_{\mu\nu 0}$ is due to local gauge invariance and general covariance; they will not get a mass in the quantum theory.

Now look at the symmetry of the full four-dimensional $\mathcal{L}$ with $n \neq 0$ modes. The crucial observation is that the assumed topology of the ground state $M^4 \times S^1$ restricts us to general coordinate transformations which are periodic in θ:

(17a) $$\zeta^{\rho} = \sum_{k=-\infty}^{\infty} \zeta_k^{\rho}(x)\, e^{ik\theta}$$

(17b) $$\zeta^{5} = \sum_{k=-\infty}^{\infty} \zeta_k^{5}(x)\, e^{ik\theta}.$$

Under general coordinate transformations,

(18) $$\Delta\gamma_{MN} = \partial_M \zeta^{\rho}\gamma_{\rho N} + \partial_N \zeta^{\rho}\gamma_{\rho M} + \zeta^{\rho}\partial_{\rho}\gamma_{MN}.$$

Then, for example,

(19) $$\Delta\phi = \zeta^{\rho}\partial_{\rho}\phi + \zeta^{5}\partial_5\phi + 3\phi\partial_5\zeta^5 + 3K\phi A_{\rho}\partial_5\zeta^{\rho}$$

$$= \sum_{n=-\infty}^{\infty} \Delta\phi_n(x)\, e^{in\theta}.$$

Use $\frac{1}{2\pi}\int_0^{2\pi} d\theta\, e^{i(k+\ell)\theta} = \delta_{k,-\ell}$ to identify the kth transformation on the nth field corresponding to (17) and (18):

(20) $$\frac{1}{2\pi}\int_0^{2\pi} d\theta\, e^{-in\theta}\,\Delta\phi = \Delta\phi_n \equiv \sum_{k=-\infty}^{\infty} \Delta_k\phi_n$$

etc. Then we find

(21a) $$\Delta_k\phi_n =$$

$$\zeta_k^{\rho}\partial_{\rho}\phi_{n-k} + im(n+2k)\zeta_k^5\phi_{n-k} + K(3imk)\zeta_k^{\rho} \sum_{\ell=-\infty}^{\infty} \phi_{n-k-\ell}\, A_{\rho\ell}$$

(21b) $$\Delta_k A_{\mu n} = \frac{1}{K}\left(\delta_{nk}\partial_{\mu}\zeta_k^5 + im\; k\; \zeta_k^{\rho} \sum_{\ell=-\infty}^{\infty} \phi^{-1}_{n-\ell-k}\, g_{\mu\rho\ell}\right)$$

$$+ \partial_{\mu}\zeta_k^{\rho}A_{\rho(n-k)} + \zeta_k^{\rho}\,\partial_{\rho}A_{\mu(n-k)} + im(n-2k)\,\zeta_k^5 A_{\mu(n-k)}$$

$$- K(im\; k)\zeta_k^{\rho} \sum_{\ell=-\infty}^{\infty} A_{\mu(n-k-\ell)}A_{\rho\ell}.$$

(21c) $$\Delta_k g_{\mu\nu n} = \partial_{\mu}\zeta_k^{\rho}g_{\rho\nu(n-k)} + \partial_{\nu}\zeta_k^{\rho}g_{\rho\mu(n-k)} + \zeta_k^{\rho}\partial_{\rho}g_{\mu\nu(n-k)}$$

$$+ imn\;\zeta_k^5\, g_{\mu\nu(n-k)}$$

$$+ K\ (imk)\zeta^{\rho}_{k} \sum_{\lambda=-\infty}^{\infty} \{g_{\mu\nu(n-k-\ell)}A_{\rho\ell}$$

$$+ g_{\mu\rho(n-k-\ell)}A_{\nu\ell} - g_{\nu\rho(n-k-\ell)}A_{\mu\ell}\}.$$

In (21), the zero mode (k=0) transformations do not mix fields of different spin or mass. The $k \neq 0$ modes do mix fields of different mass, and of different spin.

The generators of this symmetry can be defined on the fields or on the coordinates. For simplicity, choose their realization on the coordinates,

(22) $$M \equiv \Delta x^{\nu}\partial_{\nu} + \Delta x^{5}\ \partial_{5}$$

where

(23) $$\Delta x^{\rho} = \zeta^{\rho} = \sum_{k=-\infty}^{\infty} \zeta^{\rho}_{k}(x)\ e^{ik\theta}$$

and

$$\Delta x^{5} = \zeta^{5} = \sum_{k=-\infty}^{\infty} \zeta^{5}_{k}(x)\ e^{ik\theta}.$$

To get a Lie algebra, we must have global transformations. Choose

(24) $$\zeta^{\rho}_{k}(x) = a^{\rho}_{k} + \omega^{\rho}_{\nu k}x^{\nu}$$

$$\zeta^{5}_{k}(x) = c_{k}.$$

Define

(25) $$P^{n}_{\mu} \equiv e^{in\theta}\partial_{\mu}$$

$$M^{n}_{\mu\nu} = e^{in\theta}(x_{\nu}\partial_{\mu} - x_{\mu}\partial_{\nu})$$

$$Q^{n} = ie^{in\theta}\partial_{\theta}.$$

The algebra is

$$[P^n_\mu, P^m_\nu] = 0$$

$$[M^n_{\mu\nu}, P^m_\sigma] = \eta_{\mu\sigma} P^{n+m}_\nu - \eta_{\nu\sigma} P^{n+m}_\mu$$

$$[M^n_{\mu\nu}, M^m_{\rho\sigma}] = \eta_{\mu\rho} M^{n+m}_{\nu\sigma} + \eta_{\nu\sigma} M^{n+m}_{\mu\rho} - \eta_{\mu\sigma} M^{m+n}_{\nu\rho} - \eta_{\nu\rho} M^{m+n}_{\mu\sigma}$$

(26)

$$[Q^n, Q^m] = (n-m) Q^{n+m}$$

$$[Q^n, P^m_\mu] = - m P^{n+m}_\mu$$

$$[Q^n, M^m_{\mu\nu}] = - m M^{n+m}_{\mu\nu}.$$

This algebra contains Poincaré $\otimes$ $\mathbb{C}[t,t^{-1}]$ (with no central extension), Virasoro, and a mixing of internal and space-time symmetries which is evidenced by the mixing of different spins in (21).

Although (26) is a symmetry algebra of the four-dimensional $\mathcal{L}(x)$ (Eq. (5)), the symmetry of the vacuum is Poincaré $\times$ U(1).

The identification of the Goldstone bosons from the broken symmetries is as follows. Normally, the Goldstone fields are those which transform non-linearly (i.e. with an inhomogeneous constant piece) under the broken generators. For example, in the O(2) Higgs model:

$$\Delta\phi_a = \epsilon_{ab}\phi_b.$$

If we choose the vacuum expectation value (VEV) such that $\langle\phi_1\rangle = \hat{\phi}$ and $\langle\phi_2\rangle = 0$ then $\phi_1(x) = \bar{\phi}_1(x) + \hat{\phi}$, $\phi_2(x) = \bar{\phi}_2(x)$, so $\Delta\bar{\phi}_1 = \bar{\phi}_2$ but $\Delta\bar{\phi}_2 = - \hat{\phi} - \bar{\phi}_1$, i.e. $\langle \Delta\bar{\phi}_2 \rangle = - \hat{\phi} \neq 0$ is a way to identify that $\bar{\phi}_2$ is (at least part of) the Goldstone boson of the broken O(2) symmetry. This criterion, applied to the local symmetries $\Delta^5_k\phi_n$ and $\Delta^\rho_k A_{\mu n}$ in (21) identifies for $n \neq 0$, that ϕ_n and $A_{\mu n}$ are the corresponding Goldstone fields. The fields $g_{\mu\nu n}$, with two degrees of freedom for each n, then acquires a mass by absorbing the two degrees of freedom of each vector Goldstone boson $A_{\mu n}$ and the one degree of freedom of each scalar Goldstone boson ϕ_n to yield a pure spin-2 massive particle with five degrees of freedom.

For $n = 0$, the gauge parameters $\zeta_0^{\rho}(x)$ and $\zeta_0^{5}(x)$ represent general covariance and U(1). The global scale transformation (8) corresponds to a Weyl rescaling $\Delta\gamma_{MN_0} = \frac{2\lambda}{3}\gamma_{MN_0}$ together with a general coordinate transformation

(27) $$\zeta^5 = -\frac{\lambda\theta}{m}.$$

Since (27) is forbidden by the periodicity requirement, the global scale symmetry will not be an invariace of the full Lagrangian; so ϕ_0 is a pseudo Goldstone boson and its masslessness an artifact of the tree approximation.

The global scale symmetry, however, is an example of an invariance of the zero mode sector which can be derived in the Kaluza-Klein framework by considering new (non-periodic) coordinate transformations. This is the approach used to look for the affine algebra in the full Yang-Mills theory as a special choice of coordinate transformation in a 4+K dimensional pure gravity theory.[7]

The infinite parameter symmetries discussed here will also appear in $d > 5$ Kaluza-Klein theories. The scalar harmonics on S^1 will be replaced by the scalar, vector, and tensor harmonics of the internal manifold. For example, for $S^2 = \frac{SO(3)}{SO(2)}$, (23) is replaced by

(28a) $$\zeta^{\mu}(x,\theta) = \sum_{\ell} \zeta_{\ell}^{\mu}(x)\, Y_{\ell m}(y)$$

(28b) $$\zeta^{\alpha}(x,\theta) = \sum_{\ell} \zeta_{\ell}(x)\, V_{\ell}^{\alpha}(y)$$

where $Y_{\ell m}$ and V_{ℓ}^{α} are the spherical and vector harmonics on S^2. It is straightforward to see that the infinite parameter algebra will involve (from (28a)) products of spherical harmonics, so that the structure constants of the algebra will be given by Clebsch Gordan coefficients.[3,8]

Finally, if we Fourier expand local supersymmetry parameters, in higher dimensions, we will generate supersymmetric generalized affine algebras. For example, in extended supergravity theories that can be derived via Kaluza-Klein compactification, these new symmetries will relate the massless and massive supermultiplets.[4] In the particular case of extended N=8 supergravity compactified on S^7 from 11 dimensions, the generators of the infinite parameter symmetry algebra will be maps from S^7 to the finite parameter symmetry of the vacuum, combined with diffeomorphisms of S^7. Algebras defined as maps from odd-dimensional compact spaces to certain finite dimensional groups can admit a non-trivial extension.[8] Thus in the N=8 theory, it will be interesting to see if this c-number term appears in the quantum theory.

SPECTRUM GENERATING ALGEBRAS

The finite parameter subalgebra P^0_μ, $M^0_{\mu\nu}$, and Q^{-1}, Q^0, Q^1 of (26) also closes. This is Poincaré × SO(2,1). Salam and Strathdee have observed that the massive tower of spin-2 states form an irreducible representation of SO(1,2). The mass operator is $P^2 = P_\mu P^\mu$. Therefore

$$[P^2, g_{\mu\nu n}(x)] = \Box\, g_{\mu\nu n}. \tag{29}$$

Also, from (21c), for $\zeta^\mu_k = 0$ and $\zeta^5_k = \delta_{k,0}c_k$ we derive $\Delta_0 g_{\mu\nu n}$ to be

$$\Delta_0 g_{\mu\nu n} = [Q^0, g_{\mu\nu n}(x)] = imn\, g_{\mu\nu n}(x). \tag{30}$$

From the equations of motion which are to lowest order $(\Box + (mn)^2)g_{\mu\nu n} = 0$, then $[(Q^0)^2, g_{\mu\nu n}] \sim [P^2, g_{\mu\nu n}]$ on shell, so $(Q^0)^2$ can also be called the mass operator. Specifically,

$$[(Q^0)^2, g_{\mu\nu n}(x)] = -(mn)^2 g_{\mu\nu n}(x) + 2imn\, g_{\mu\nu n}(x)\, Q_0. \tag{31}$$

Since $Q^0 \mid 0 > = 0$, then

$$\begin{aligned} [(Q^0)^2, g_{\mu\nu n}(x)] \mid 0 > &= -(mn)^2\, g_{\mu\nu n}(x) \mid 0 > \\ &= (Q^0)^2\, g_{\mu\nu n}(x) \mid 0 > \\ &= (Q^0)^2 \mid g_{\mu\nu n} > \\ &= -(mn)^2 \mid g_{\mu\nu n} >. \end{aligned}$$

Therefore the spectrum of $(Q^0)^2$ is $-m^2n^2$, i.e. the spectrum of masses.

In this sense, $[Q^0, g_{\mu\nu n}(x)] = imn\, g_{\mu\nu n}(x)$ is like

$$[H, b(k,t)] = ik\, b(k,t), \tag{32}$$

where $Q^0 \leftrightarrow H$, $g_{\mu\nu n}(x) \leftrightarrow b(k,t)$, $n \leftrightarrow k$. Equation (32) is familiar from the quantum inverse scattering method. The operator b(k,t) and its canonical conjugate a(k,t) are the scattering data. $[H, a(k,t)] = 0$ so a(k,t) generates conservation laws $M_\ell(t)$:

$a(k,t) \sim \sum_k k^{-\ell} M_\ell(t)$, $\dot{a}(k,t) = 0$ implies $\dot{M}_\ell(t) = 0$. Also, since $H \mid 0 > = 0$,

$$\begin{aligned}[H, b(k,t)] \mid 0 > &= H\, b(k,t) \mid 0 > \\ &= ik\, b(k,t) \mid 0 > \\ &= ik \mid b(k,t) >. \end{aligned} \tag{33}$$

b(k,t) creates eigenstates $\mid b(k,t) >$ of H.

In general, for any symmetry transformation $Q \mid 0 > = 0$, which has

$$[Q, \phi_n(x)] = f(n)\, \phi_n(x) \tag{34}$$

where n=1,2,...,N, then eigenvalues of Q in an $N \times N$ matrix representation are f(n).

For the compact generator of SO(1,2), we have

$$[Q^0, g_{\mu\nu n}(x)] = (T^0)_{nk}\, g_{\mu\nu n}(x) = imn\, g_{\mu\nu n}(x) \tag{35}$$

where $(T^0)_{nk}$ is an infinite dimensional matrix representation of Q_0 in SO(1,2). Therefore, knowing the spectral properties of $(T^0)_{nk}$, one can find the masses to be mn; i.e. there exists a discrete infinite-dimensional representation T^0 of Q^0 in SO(1,2) which has eigenvalues which increase in steps of unity (m) with no upper bound.[9]

Clearly, this T^0 is also a representation of the Q^0 in the full Virasoro algebra, and $g_{\mu\nu n}$ is the infinite dimensional multiplet of the infinite parameter symmetry.

BOUND STATES IN YANG-MILLS[10]

In the above discussion of five-dimensional Kaluza-Klein, the fundamental fields in $\mathcal{L}(x)$ transform linearly under the Virasoro generators. The perturbative spectrum is thus given by a representation of this algebra. Since the masses are of order m_{planck}, there will be non-perturbative corrections to the physical spectrum.[11] In a confining theory, however, we are only interested in the non-perturbative spectrum.

Recently there has been interest to compute the affine symmetry transformations in the full (non-self-dual) Yang-Mills theory. We assume, for the moment, that such a symmetry will be found; and from that assumption I now argue how it can be used to describe the physical states of the theory. The physical states are bound states in this confining model and thus inherently non-perturbative. The Yang-Mills theory is ultraviolet free and should have a mass gap.

The affine symmetry will be realized nonlinearly on the fundamental fields of the Lagrangian (this we know from the self-dual expressions[7]), and I speculate that the physical particles will be composites (of gluons and eventually of quarks) which transform linearly in the infinite dimensional representation of the unbroken generators. Since this is an infinite parameter algebra, the representation will be infinite dimensional and there will be an infinite number of bound states; i.e. hadrons. Since the observed hadronic spectrum is not degenerate, the affine symmetry is an exact symmetry of the Lagrangian and is spontaneously broken down to SU(3) $\times$ Poincaré. That is to say, we are assuming that an infinite set of infinitesimal symmetry transformations $\Delta_n^a A_\mu(x)$ exist for the SU(3) Yang-Mills Lagrangian. Here $A_\mu = A_\mu^a(x) T^a$ where T^a are the generators of SU(3), $a=1,\ldots,8$ and $n = -\infty,\ldots,-2,-1,0,1,2,\ldots,\infty$. Since all but the n=0 transformations are realized non-linearly (the n=0 transformation can be identified as an A_μ^a-dependent local gauge transformation), the $n \neq 0$ transformations will be spontaneously broken, i.e. whereas $\langle 0 \mid \Delta_0^a A_\mu \mid 0 \rangle = 0$; for $n \neq 0$, then $\langle 0 \mid \Delta_n^a A_\mu \mid 0 \rangle \neq 0$. In perturbation theory, the eight fundamental vector potentials $A_\mu^a(x)$ remain exactly massless because the unbroken SU(3) symmetry is a local gauge invariance. In the non-perturbative theory, the (non-degenerate) bound states will form an infinite dimensional representation of the unbroken generators of the affine algebra. The addition of quark matter may also explicity but weakly break the affine symmetry. The higher mass resonances will probably be very broad, as in the hydrogen atom, and thus not conflict with current day cosmology. Since the affine symmetry is the affine algebra of the compact SU(N) symmetry, rather than of a non-compact internal symmetry such as SU(1,1) or $E_{7(7)}$ as in the extended supergravity models, the disadvantages of non-compactness will not be present.

The basic unitary representations of the affine symmetry are given in terms of latticized vertex operators of the dual string model. The rank of the associated finite parameter algebra is equal to number of space time dimensions d of the flat space through which the one-dimensional string moving in time traces out a curved two-dimensional surface. The momentum k_μ appearing in the vertex operator V(k,z) is now defined on a d-dimensional lattice. The standard (generalized Veneziano) dual model for fixed d can be expanded in a power series in z to exhibit an affine symmetry with a fixed constant coefficient in the central extension. This coefficient k is the level of the representation.[12] Because of both the string and gauge theory realizations of the affine symmetry, we may now be able to establish a direct connection between the non-abelian local gauge field theory and the string, or hadronic spectrum.

In order to carry out the construction of the composite states, one must construct a function of the non-linearly transforming fundamental fields which carry a linear representation of the symmetry group. This has been carried out in the CP^N sigma models and in the four-dimensional extended supergravity theories, whose scalars also live in sigma models on coset spaces.[6]

A well known example from the literature is N=4 extended supergravity in four dimensions.[13] The symmetry of this Lagrangian is (SU(4) × SU(1,1)) global, × U(4) local. The particle content is 1 graviton, 4 spin-3/2, 6 spin-1, 4 spin-1/2, and 2 spin-0 particles. The spin-0 particles are valued on the coset

$$\frac{SU(4) \times SU(1,1)}{U(4)} \sim \frac{SU(1,1)}{U(1)}.$$

They can be described by a complex field z(x). In the symmetric gauge, the spin-0 kinetic term in the Lagrangian density is given by

$$\mathcal{L}_0 = \frac{1}{2} \frac{\partial_\mu z \partial^\mu \bar{z}}{(1-z\bar{z})^2}, \tag{36}$$

This can be written in standard sigma-model form:

$$\mathcal{L}_0 = -\frac{1}{16} \mathrm{Tr}\, \partial_\mu q \partial^\mu q^{-1} \tag{37}$$

where[14]

$$q(x) = g(x) g_0 g^{-1*}(x). \tag{38}$$

q(x) is an arbitrary element in the two-dimensional representation of the coset space $\frac{SU(1,1)}{U(1)}$; and

$$g(x) = \begin{pmatrix} \alpha & \beta \\ \bar{\beta} & \bar{\alpha} \end{pmatrix} = \exp[\sigma_3\, i\frac{\omega_0}{2} + \sigma_2(\mathrm{Re}\,\omega) + \sigma_1(\mathrm{Im}\,\omega)]$$

with $\alpha\bar{\alpha} - \beta\bar{\beta} = 1$. g(x) is an arbitrary element of SU(1,1). Infinitesimally, $\alpha \sim 1 + \frac{i\omega_0}{2}$ and $\beta \sim -i\omega$. Define $h(x) = \exp[\sigma_3 i\theta(x)]$ as the arbitrary element of U(1).

The star operation is an involutive automorphism, $(g_1^* g_2^*) = (g_1 g_2)^*$ and $(g^{*})^{*} = g$. In (38), choose $g^* = g^{-1\dagger}$ and the fixed element g_0 of SU(1,1) to be the identity $g_0 = \begin{pmatrix} 1 & 0 \\ 0 & 1 \end{pmatrix}$.

$$g_0 h^*(x) g_0^{-1} = h(x). \tag{39}$$

Therefore,

$$g(x) = gg^{\dagger}$$

and if we identify $z = \frac{\bar{\beta}}{\alpha}$ then

$$g(x) = \frac{1}{1-z\bar{z}} \begin{bmatrix} 1+z\bar{z} & 2\bar{z} \\ 2z & 1+z\bar{z} \end{bmatrix}, \tag{40}$$

which is the subset of elements of SU(1,1) which form the coset $\frac{SU(1,1)}{U(1)}$.

The symmetries of $\mathscr{L}_0$ are

$$g(x) \longrightarrow g_1 g(x) \tag{41}$$

and

$$g(x) \longrightarrow g(x)h(x). \tag{42}$$

Under (41),

$$\begin{aligned} q(x) &\longrightarrow g_1 q(x) g_1^{\dagger} \\ z(x) &\longrightarrow \frac{\alpha z(x)+\beta}{\bar{\beta} z(x)+\bar{\alpha}} \sim z(x) + i\omega_0 z - i\omega - i\omega^{*} z^2. \end{aligned} \tag{43}$$

and under (42),

$$\begin{aligned} q(x) &\longrightarrow q(x) \\ z(x) &\longrightarrow z(x). \end{aligned} \tag{44}$$

On the level of the Lagrangian $\mathscr{L}_0$, global SU(1,1) symmetry acting on the left appears. An additional symmetry motivated by (42) can be displayed in the theory in the following way.[2,6] Add one additional scalar and one gauge particle:

$$\mathcal{L}_0' = -\frac{1}{4}\,\mathrm{Tr}\, D_\mu g\, D^\mu g^{-1} \tag{45}$$

where $D_\mu g = \partial_\mu g - gQ_\mu$, Q_μ is an element of the Lie algebra of U(1), i.e. $Q_\mu = i\sigma^3 A_\mu(x)$. As g transforms according to (42), Q_μ shifts as

$$Q_\mu \longrightarrow h^{-1}Q_\mu h + h^{-1}\partial_\mu h. \tag{46}$$

(Under (41), $Q_\mu \longrightarrow Q_\mu$.) If we vary the auxiliary field A_μ while holding g fixed then

$$\Delta\mathcal{L}_0' = -\frac{1}{2}\,\mathrm{Tr}\,[\Delta Q_\mu(g^{-1}\partial^\mu g - Q^\mu)]. \tag{47}$$

Since ΔQ_μ is also an element of the U(1) algebra, then

$$g^{-1}\partial_\mu g - Q_\mu = \mathrm{Re}(\bar{\alpha}\partial_\mu\beta - \beta\partial_\mu\bar{\alpha})\sigma_1 - \mathrm{Im}(\bar{\alpha}\partial_\mu\beta - \beta\partial_\mu\bar{\alpha})\sigma_2 \tag{48}$$

that is

$$A_\mu = -i(\bar{\alpha}\partial_\mu\alpha - \beta\partial_\mu\bar{\beta}) = \frac{i}{2}\,\frac{\bar{z}\partial_\mu z - z\partial_\mu\bar{z}}{1 - z\bar{z}} - i\,\partial_\mu \ln\frac{\alpha}{\bar{\alpha}}. \tag{49}$$

As z transforms non-linearly under SU(1,1) global (43), the combination of the fundamental z's given by A_μ in (49) transforms as a gauge transformation. (In the non-abelian case this will be a linear transformation up to an inhomogeneous term.) Furthermore in the CP^{N-1} model, where a dynamical $\frac{1}{N}$ calculation is available, a similarly constructed gauge field can be shown to be a real physical particle of the spectrum.[6]

The suggestion for Yang-Mills can now be made explicit. The z(x) plays the role of the gauge potentials. The goal is now to solve for a composite function of gauge fields which transform linearly under the affine symmetry. These bound states will also form multiplets under the unbroken affine transformations. So a direct analysis in terms of unitary infinite-dimensional representations of this algebra will give information about the hadronic spectrum.

REFERENCES

[1] J. Ellis, M.K. Gaillard, M. Günaydin, and B. Zumino, Nucl. Phys. B224, 427(1983).

[2] M.K. Gaillard and B. Zumino, Nucl. Phys. B 193, 221(1981).

[3] L. Dolan and M.J. Duff, Phys. Rev. Lett. 52, 14(1984).

[4] L. Dolan, Rockefeller Report RU84/B/89, May 1984.

[5] A. Salam and J. Strathdee, Annals of Physics, 141, 316(1982).

[6] E. Cremmer and B. Julia, Nucl. Phys. B159, 141(1979). See also A. D'Adda, P. diVecchia and M. Lüscher, Nucl. Phys. B146, 63(1978), B152, 125(1979) and E. Witten, Nucl. Phys. B149, 285(1979); H. Haber, I. Hinchcliffe, E. Rabinovici, Nucl. Phys. B172 458(1980).

[7] L. Dolan, Phys. Lett. B113, 387(1982); see also Rockefeller preprint RU83/B/163, August 1983, to appear in Physics Reports.

[8] I. Singer and B. Zumino, private communication.

[9] B. Wybourne, Classical Groups for Physicists, Wiley, (New York) 1974.

[10] L. Dolan, in preparation.

[11] The author is grateful to S. Mandelstam for conversations on this point.

[12] See papers in these Proceedings.

[13] E. Cremmer and J. Scherk, Nuc. Phys. B127, 259(1977); Phys. Lett. B74, 61(1978); A. Das, Phys. Rev. D15, 2805(1977).

[14] E. Brezin, S. Hikami, and J. Zinn-Justin, Nucl. Phys. B165, 528(1980).

†The Rockefeller University
New York, NY 10021

LOCAL CHARGE ALGEBRAS IN QUANTUM CHIRAL MODELS AND GAUGE THEORIES

Itzhak Bars[†]

ABSTRACT

Affine and loop algebras are analogous to local current algebras that occur in many physical field theories. Sugawara's formulation of the chiral model is a particularly interesting case. If this model is extended by the addition of the Wess-Zumino term, then the quantum algebra of the local charge densities $[J_0^a(x), J_0^b(y)]$ becomes anomalous and has a central extension. This is the first example of a theory with such an anomaly. In 2 dimensions, the central extension is a c-number, just as for affine algebras. Gauge theories can be reformulated in terms of string variables, i.e. ordered line integrals of the form $(\exp i\int_S A\cdot dx)_+$. With a particular choice of rigid strings $\{S\}$, the theory can be rewritten in terms of the "strings" which are group elements rather than the gauge potentials A_μ which are in the algebra. In this formulation the canonical quantization of the theory again is in terms of a local non-Abelian charge algebra analogous to the chiral theory, or affine type algebras in higher dimensions.

1. LOOP AND AFFINE TYPE ALGEBRAS AS LOCAL CHARGE ALGEBRAS.

The infinite dimensional affine and loop algebras have generators which may be labeled as X_n^a, with n=integer $-\infty<n<+\infty$. The label "a" corresponds to the generators of any finite dimensional

---

† Partially supported by the National Science Foundation through the Mathematical Sciences Research Institute.

Vertex Operators in Mathematics and Physics - Proceedings of a Conference November 10-17, 1983. Publications of the Mathematical Sciences Research Institute #3, Springer-Verlag, 1984.

Lie algebra. In the mathematics literature "a" is replaced by a Cartan root, but I will use "a" as a more general label to denote any linear combination of generators. In particular, I have in mind physicists' prefered linear combinations of generators that are hermitian and ortho-normalized. See, for example, the normalization provided by the N×N Gell-Mann matrices $X^a \longrightarrow \lambda^a/2$ for SU(N). In this case, the Killing metric is the unit matrix and the structure constants f^{abc} are completely antisymmetric.

The infinite dimensional affine algebra is

$$[X_n^a, X_m^b] = i\, f^{abc} X_{n+m}^c + \gamma\, n\, \delta_{n,-m}\, \delta^{ab} \tag{1.1}$$

The factor of $i = \sqrt{-1}$ is inserted for consistency with hermiticity $(X_n^a)^\dagger = X_{-n}^a$. The number γ is the "central extension", and takes different values in different representations of the algebra (1.1).

Let us construct the Fourier series

$$X^a(\theta) = \frac{1}{2\pi} \sum_{n=-\infty}^{\infty} e^{-in\theta} X_n^a \tag{1.2}$$

Think of θ = "position" on the loop $0 \leq \theta \leq 2\pi$, and n as its conjugate "momentum". In eq. (1.1) "momentum" is conserved on both sides of the equation. We want to obtain the "position" space version of eq. (1.1). Using eq. (1.1) we find

$$[X^a(\theta), X^b(\theta')] = i\, f^{abc} X^c(\theta)\delta(\theta-\theta') + i\frac{\gamma}{2\pi} \frac{\partial}{\partial\theta}\delta(\theta-\theta')\delta^{ab}, \tag{1.3}$$

where the periodic Dirac delta function is given by

$$\delta(\theta-\theta') = \frac{1}{2\pi} \sum_{n=-\infty}^{\infty} e^{-in(\theta-\theta')} \tag{1.4}$$

and is normalized so that $\int_0^{2\pi} \delta(\theta-\theta')d\theta = 1$. The delta "function" is a *distribution* which vanishes everywhere, except at the origin,

$\theta-\theta'=0$, where it diverges. Therefore, we see that the right hand side of (1.3) vanishes unless $\theta=\theta'$. Thus, the commutation rule (1.3) is "local", i.e. $X^a(\theta)$ commutes with $X^b(\theta')$ if $\theta\neq\theta'$.

We compare (1.3) to current algebras which are very familiar in physics[1]. Consider any field theory in d-dimensions with a symmetry group G. Using Noether's theorem[2] one finds a conserved current J^a_μ corresponding to every generator in the algebra G.

$$(1.5) \qquad \sum_{\mu=1}^{d} \frac{\partial J^a_\mu}{\partial x_\mu}(x) = 0; \quad a = \text{label of generator.}$$

If one uses the standard, equal-time, canonical commutation relations for the canonical conjugates in this theory, it is straightforward to derive the commutation rules for the currents. In particular, for the time component of the current, i.e. the "charge density" $J^a_0(x)$, one finds, in any of the usual theories, the *local charge algebra*

$$(1.6) \qquad [J^a_0(\vec{x},t), J^b_0(\vec{y},t)] = if^{abc}J^c_0(\vec{x},t)\delta^{(d-1)}(\vec{x}-\vec{y})$$

where the Dirac delta function in d-1 space dimensions is given by the integral representation.

$$(1.7) \qquad \delta^{(d-1)}(\vec{x}-\vec{y}) = \frac{1}{(2\pi)^{d-1}}\int d^{d-1}_p e^{-i\vec{p}\cdot(\vec{x}-\vec{y})}$$

where $\vec{p}$ is momentum, and the dot product, $\vec{p}\cdot\vec{x} = p_1x_1 + p_2x_2 +\ldots+ p_{d-1}x_{d-1}$, appears in (1.7).

The charge operator $Q^a = \int dx^{d-1}J^a_0(\vec{x},t)$ is time independent because it is the generator of the symmetry of the field theory we started with. It must satisfy the commutation rule

$$(1.8) \qquad [Q^a,Q^b] = if^{abc}Q^c,$$

as indeed it follows from (1.6).

So, we see that there is a definite analogy between the local "charge" algebras of eqs. (1.6) and (1.3). If the field theory is formulated in 1-space, 1-time dimensions (i.e. d-1=1), and if "space" is a loop rather than the real line, then the delta function in eq. (1.7) is replaced by eq. (1.4), so that the analogy becomes precise.

The only term missing in eq. (1.6) is the central extension γ. The local charge algebra $[J_0^a, J_0^b]$, can be extended to a larger algebra, the current algebra, that involves all the components of the current $J_\mu^a(x)$, not just J_0^a. It is known that the space components of the currents $J_i^a(x)$, i=1,2,...d-1, satisfy commutation relation $[J_0^a, J_i^b]$, $[J_i^a, J_j^b]$ with such "central extensions" known as Schwinger terms[1]. The Schwinger terms depend on the details of the theory. However, the space components do not form a closed algebra by themselves, like the J_0^a's do.

The central extension in the local charge algebra $[J_0^a, J_0^b]$ is indeed very unusual. The absence of the central extension in (1.6) had always prevented us from understanding fully the affine algebra in the context of current algebras. However, recent work involving the chiral model with Wess-Zumino term[3,4] has finally led us to the first example of a field theory whose charge algebra $[J_0^a, J_0^b]$ has an "anomaly" or a central extension. Rajeev[4] and Witten[3] have also observed related anomalies. It is interesting that this central extension is related to global topological properties of the theory. This example will be explained in section 3.

Some two dimensional field theories have an infinite number of conserved currents[5]. Their charges Q_n^a have been shown[6] to form affine type algebras. This is a *different* connection between current algebras and affine algebras than the one described above. In the case above we were dealing with the *time dependent local* zeroth component of *one* current $J_0^a(x)$, in the case of refs. 5,6 we are dealing with the *time independent* integrals $Q_n^a = \int dx J_{on}^a(x)$ of the zeroth components of an *infinite number* of currents. The two dimensional chiral model including the Wess-Zumino term exhibits *both* types of relations to affine algebras, as outlined in Section 3.

From the local charge algebra (1.6) in many dimensions we suggest generalizations of affine type algebras which the mathematicians have not studied yet.* To make it nice mathematically, let us replace the (d-1) dimensional Euclidean space R^{d-1} of eq. (1.6) by a compact space, e.g. the (d-1) dimensional torus, or the (d-1) dimensional sphere, etc.

As a definite example, take d-1=2, and a torus described by two angles θ and ϕ. Then, it is natural to define

(1.9)
$$X^a(\theta,\phi) = \left[\frac{1}{2\pi}\right]^2 \sum_{m,n=-\infty}^{\infty} X^a_{mn} e^{-i(m\theta+n\phi)}$$
$$\delta(\theta-\theta')\delta(\phi-\phi') = \frac{1}{(2\pi)^2} \sum_{m,n=-\infty}^{\infty} e^{-i[m(\theta-\theta')+n(\phi-\phi')]}$$

Substituting in eq. (1.6) we obtain the algebra

(1.10)
$$[X^a(\theta,\phi),X^b(\theta',\phi')] = if^{abc}X^c(\theta,\phi)\delta(\theta-\theta')\delta(\phi-\phi') + \text{"anomaly"}$$
$$[X^a_{mn},X^b_{k\ell}] = if^{abc}X^c_{m+k,n+\ell} + \text{"anomaly"}$$

which generalizes eq. (1.1) to 2-dimensions. An appropriate central extension or anomaly can be cooked up by adding in (1.10) a linear combination of single derivatives $\left[\alpha\frac{\partial}{\partial\theta} + \beta\frac{\partial}{\partial\phi}\right]$ of the delta functions in eq. (1.9).

(1.11)
$$\text{"anomaly"} = \left[\alpha\frac{\partial}{\partial\theta} + \beta\frac{\partial}{\partial\phi}\right]\delta(\theta-\theta')\delta(\phi-\phi')$$

If the two dimensional space in eqs. (1.6) and (1.7) is a sphere instead of a torus, the appropriate expansions of $X^a(\theta,\phi)$ and the spherical delta functions, are in terms of spherical harmonics $Y_{\ell m}(\theta,\phi)$, with ℓ=0,1,2,3,... and $-\ell\leq m\leq\ell$. Then we arrive at

---

* Remarks made by I. Singer during this lecture indicate that he has started work along a direction similar to the one described here.

$$[X^a(\theta,\phi),X^b(\theta,\phi)] = if^{abc}X^c(\theta,\phi)\delta(\cos\theta-\cos\theta')\delta(\phi-\phi') + \text{"anomaly"}$$

(1.12)

$$[X^a_{\ell_1 m_1},X^b_{\ell_2 m_2}] = if^{abc}\sum_{\ell,m} F(\ell_1 m_1,\ell_2 m_2;\ell m)X^c_{\ell m} + \text{"anomaly"}$$

where the sum extends over

(1.13) $$|\ell_1-\ell_2|\leq\ell\leq\ell_1+\ell_2 \qquad -\ell\leq m\leq\ell$$

and, the structure constant $F(\ell_1 m_1;\ell_2 m_2;\ell m)$ is given by

(1.14)

$$F(\ell_1 m_1;\ell_2 m_2;\ell m) = \left[\frac{(2\ell_1+1)(2\ell_2+1)(2\ell+1)}{4\pi}\right]^{1/2}\begin{bmatrix}\ell_1 & \ell_2 & \ell\\ 0 & 0 & 0\end{bmatrix}\begin{bmatrix}\ell_1 & \ell_2 & \ell\\ m_1 & m_2 & -m\end{bmatrix}.$$

Here $\begin{bmatrix}\ell_1 & \ell_2 & \ell\\ m_1 & m_2 & -m\end{bmatrix}$ is Wigner's 3-j symbol related to Clebsh-Gordan coefficients that appear in the sum of angular momentum

(1.15) $$\begin{bmatrix}\ell_1 & \ell_2 & \ell\\ m_1 & m_2 & -m\end{bmatrix} = \frac{(-1)^{\ell_1-\ell_2+m}}{\left[2\ell+1\right]^{1/2}} < \ell_1 m_1 \ell_2 m_2 | \ell m > .$$

The last two equations in (1.12) are related to ach other via the identity

(1.16)

$$\sum_{\substack{\ell_1 m_1\\ \ell_2 m_2}} F(\ell_1 m_1;\ell_2 m_2;\ell m)Y_{\ell_1 m_1}(\theta,\phi)Y_{\ell_2 m_2}(\theta',\phi') = Y_{\ell m}(\theta,\phi)\delta(\cos\theta-\cos\theta')\delta(\phi-\phi').$$

Again, a central extension would follow from additional terms in (1.2) with a linear combination of first derivatives $\left[\alpha\frac{\partial}{\partial\phi} + \beta\frac{\partial}{\partial\cos\theta}\right]$ applied on the spherical delta function $\delta(\cos\theta-\cos\theta')\delta(\phi-\phi')$. First derivatives provide an antisymmetric expression and will always satisfy the Jacobi identities. I have not investigated yet higher odd powers of derivatives.

Generalizations such as these are clearly possible for higher dimensions, and many compact manifolds that generalize the torus, sphere, etc.

2. CHIRAL MODEL AND THE SUGAWARA CURRENT ALGEBRA

Consider an N×N matrix U(x) which represents a map from Minkowski (or other) space in d-dimensions, x^{μ}, $\mu=1,2,\ldots d$, into a group G. For definiteness I will take G=SU(N), $U^{-1}=U^{+}$, detU=1, but other groups may be considered just as well. Form the Cartan connection

(2.1) $$A_{\mu} = -iU^{+}(x)\frac{\partial U(x)}{\partial x_{\mu}},$$

and define the action for the chiral model

(2.2) $$S = \frac{1}{g^{2}}\int dx^{d}\mathrm{Tr}\left[A_{\mu}(x)A^{\mu}(x).\right].$$

The Euler-Lagrange equations reduce to

(2.3) $$\partial_{\mu}A^{\mu} = 0$$

Note that from (2.1) we can write

(2.4) $$\partial_{\mu}A_{\nu} - \partial_{\nu}A_{\mu} + i[A_{\mu}A_{\nu}] = 0$$

This action is invariant under left and right multiplication of U(x) by x-independent unitary N×N matrices g_L and g_R^{-1}, with $\det g_L=\det g_R=1$:

$$U(x) \longrightarrow g_L U(x) g_R^{-1}$$

(2.3)

$$S \longrightarrow S$$

Since the left and right transformations commute, the invariance is described by the group

$$G = SU(N)_L \otimes SU(N)_R \tag{2.4}$$

Thus, by Noether's theorem we find two conserved currents $J_{\mu L} = \frac{i}{g^2}(\partial_\mu U)U^\dagger$ and $J_{\mu R} = \frac{i}{g^2}(\partial_\mu U^\dagger)U = \frac{1}{g^2}A_\mu$. Their conservation can be verified using (2.3). The algebras of these two currents commute because they correspond to the generators of $SU(N)_L \otimes SU(N)_R$. It is sufficient to concentrate only on $J_{\mu R} = A_\mu$.

The hamiltonian that follows from (2.2) is

$$H = \frac{1}{g^2}\int d_x^{d-1}\, \mathrm{Tr}(A_0^2 + A_i^2) \tag{2.5}$$

The canonical variables are computed in the standard way

$$\Lambda_j^i(x) = \frac{\delta S}{\delta \partial_0 U_i{}^j(x)} = -\frac{2}{g^2}(U^\dagger \partial_0 U\ U^\dagger)_j^i \tag{2.6}$$

However, the naive Poisson brackets are not consistent with the constraints $U^\dagger U=1$, $\det U=1$. The Dirac brackets must be computed with the methods of ref (7). These are not simple, however it turns out that the combinatiion $\Lambda U = -\frac{2}{g^2}U^\dagger \partial_0 U$, which can be recognized as being proportional to A_0, has simple commutation rules if we write

$$A_0 = \sum_{a=1}^{N^2-1} A_0^a \lambda^a/2 = \Lambda U \tag{2.7}$$

where $\lambda^a/2$ is a complete set of N×N *traceless* hermitian matrices. Then

$$\left[U_\alpha{}^\beta(x)\ ,\ U_\gamma{}^\delta(y)\right] = 0$$

(2.8a)

$$\left[A_0^a(x)\ ,\ U_\alpha{}^\beta(y)\right] = g^2(U(x)\lambda^a/2)_\alpha{}^\beta \delta^{d-1}(x-y)$$

(2.8b) $$\left[A_0^a(x), A_0^b(y)\right] = if^{abc}g^2A_0^c(x)\delta^{d-1}(x-y).$$

We recognize in this last equation the non-Abelian local charge algebra. It can be verified that the equation of motion (2.3) follows from the quantum rules (2.8) and the Hamiltonian (2.5). It turns out that it is convenient to compute the commutators of A_i with A_0 and with itself. This can be done using (2.8)

$$\left[A_0^a(x), A_i^b(y)\right] = if^{abc}g^2A_i^c(x)\delta^{d-1}_{(x-y)} + ig^2\frac{\partial}{\partial x^i}\delta^{d-1}_{(x-y)}$$

(2.9)

$$\left[A_i^a(x), A_j^b(y)\right] = 0$$

Thus, the algebra of the conserved currents A^a_μ, together with the Hamiltonian completely define the quantum theory. The physics problem is to find the eigenstates and other properties of this Hamiltonian. In 2-dimensions it is possible to rewrite A_0 and the Hamiltonian in terms of canonical variables reminiscent of a gauge field and its canonical conjugate electric field, as in the Appendix of ref. (8).

The current algebra of the A_0, and A_i is due to Sugawara[(9)]. Here note the central extension $\partial_i\delta(x-y)$ known as a Schwinger term. This is an example of a quantum theory which can be defined purely in terms of its current algebra. As in all known theories, except the next section, the local charge algebra $[A_0, A_0]$ has no anomalies.

3. CHIRAL MODEL WITH WESS-ZUMINO TERM AND THE ANOMALY IN THE ALGEBRA OF LOCAL CHARGE DENSITIES

The Wess-Zumino term (WZ) in the chiral model has recently been discussed by Witten[3]. It is related to anomalies caused by fermions in quantum field theory and has a topological significance. To write an action that includes the WZ term it is convenient to increase the dimension of space-time by 1. Then it can be expressed as a volume integral in d+1 dimensions. Let us write the action in d=2 and d=4 dimensions

$$S_2 = \int d^2x \frac{\mathrm{Tr}(A_\mu A^\mu)}{g^2} - \frac{i}{3\lambda^2}\int d^3x\ \mathrm{Tr}(A_\mu A_\nu A_\lambda)\epsilon^{\mu\nu\lambda}$$

(3.1)

$$S_4 = \int d^4x\ \frac{\mathrm{Tr}(A_\mu A^\mu)}{g^2} - \frac{1}{5\lambda^2}\int d^5x\ \mathrm{Tr}(A_\mu A_\nu A_\lambda A_\sigma A_\rho)\epsilon^{\mu\nu\lambda\sigma\rho}.$$

A_μ was defined in eq. (2.1) as the Cartan connection; $\epsilon^{\mu\nu\lambda}$ and $\epsilon^{\mu\nu\lambda\sigma\rho}$ are the Levi-Civita completely antisymmetric tensors in 3 and 5 dimensions respectively; and $1/\lambda^2$ is a dimensionless constant in any dimension, while $1/g^2$ is dimensionless in d=2 but has dimensions of $(\mathrm{mass})^2$ in 4 dimensions.

It can be shown that the new term is a total divergence so that it can be rewritten as a surface term back in d-dimensions which remains ambiguous up to a constant. This can be understood via the homology groups $H_{d+1}(G)=Z$ for d=2,4 etc.[3]. For the quantum mechanics to be unambiguous Witten argues that $1/\lambda^2$ must be quantized in units of a pure number $1/\lambda_0^2$

$$1/\lambda^2 = n/\lambda_0^2,\ n=\text{any integer} \tag{3.2}$$

so that the ambiguity in the action is $2\pi n$ and does not matter in the path integral $\int e^{iS}$.

The action (3.1) is invariant under the same transformation laws of (2.3). Therefore, we expect conserved currents. We shall concentrate again on the right-hand side current. The WZ term in the action makes a contribution, so that the equation of motion for the matrix U(x) is now different than (2.3). It takes the form

$$\partial_\mu J^\mu = 0$$

(3.3)
$$J\mu = \frac{1}{g^2}A\mu - \frac{1}{\lambda^2}\epsilon_{\mu\nu}A^\nu \quad \text{for } d=2$$

$$J\mu = \frac{1}{g^2}A_\mu - \frac{i}{\lambda^2}\epsilon_{\mu\nu\lambda\sigma}A^\nu A^\lambda A^\sigma \quad \text{for } d=4$$

From now on we concentrate on d=2. In this case, using eq. (2.4), which is still valid, we derive via (3.3) that

(3.4)
$$(d=2) \quad \partial_\mu J_\nu - \partial_\nu J_\mu + ig^2[J_\mu J_\nu] = 0$$

So that we may write $J\mu$ as a conserved Cartan connection

(3.5)
$$(d=2) \quad J\mu = -\frac{i}{g^2}V^{-1}\partial_\mu V$$

$$\partial\mu J^\mu = 0$$

where V(x) is a N×N matrix in SU(N). Thus, the original matrix U(x) is related now to V(x).

From eqs. (3.4) and (3.5) we can easily derive an infinite number of conserved currents, with the iterative methods of ref. 5. Therefore, the new action, including the WZ term, corresponds to a completely integrable system.

We may proceed with the canonical quantization as in section 2. However, now the canonical conjugate $\Lambda^i_j = (\delta S/\delta\partial_0 U^j_i)$ is a more complicated expression in terms of $A\mu$ or U. The simple quantities are the conserved currents which we may write explicitly as

(3.6)
$$J^a_0 = \frac{1}{g^2}A^a_0 - \frac{1}{\lambda^2}A^a_1; \quad J^a_1 = \frac{1}{g^2}A^a_0 - \frac{1}{\lambda^2}A^a_0$$

The Hamiltonian takes the form

$$H = \frac{1}{2g^2}\int dx \left[(A_0^a)^2 + (A_1^a)^2\right]$$

(3.7)

$$= \int dx \left\{ \frac{\lambda^4 g^2 (\lambda^4 + g^4)}{2(\lambda^4 - g^4)^2} \left[J_0^a J_0^a + J_1^a J_1^a\right] + \frac{\lambda^6 g^4}{(\lambda^4 - g^4)^2} \left[J_1^a J_0^a + J_0^a J_1^a\right] \right\}$$

which is written entirely in terms of the *conserved currents.*

The canonical commutation rules are now similar to eq. (2.8) with J_0^a replacing A_0^a, but also with a *surprise*:

(3.8a) $$[U_\alpha^\beta(x), U_\gamma^\delta(y)] = 0$$

(3.8b) $$[J_0^a(x), U_\alpha^\beta(y)] = \left(U(x)\frac{\lambda^a}{2}\right)_\alpha^\beta \delta(x-y)$$

(3.8c) $$\left[J_0^a(x), J_0^b(y)\right] = if^{abc} J_0^c(x)\delta(x-y) - \frac{2i}{\lambda^2}\delta^{ab}\partial_x \delta(x-y)$$

Note now the *central extension* in the commutator of the zeroth component of the current. This is the first example of a theory with this kind of anomaly. This is exactly as in affine algebras, eq. (1.3). If we take Witten's[3] quantization condition $1/\lambda^2 = n/\lambda_0^2$ we see that the central extension is quantized.

The necessity of the anomaly can be verified by requiring that the quantization rules (3.8) together with the Hamiltonian (3.7) reproduce the equations of motions 3.4 and 3.5. To do so we must derive from (3.8) the commutation rules for $J_1^a(x)$. This can be done by eliminating the time derivatives $U^{-1}\partial_0 U$ from the expression of J_1 in favor J_0, via eq. (3.6). Then we find the remaining commutators of the current algebra

$$\left[J_0^a(x), J_1^b(y)\right] = if^{abc} J_1^c(x)\delta(x-y) + i\left(\frac{1}{g^2} + \frac{g^2}{\lambda^4}\right)\partial_x \delta(x-y)\delta^{ab}$$

(3.9)

$$\left[J_1^a(x), J_1^b(y)\right] = if^{abc}\left(\frac{g^4}{\lambda^4}J_0^c + \frac{2g^2}{\lambda^2}J_1^c\right)\delta(x-y) - \frac{2i}{\lambda^2}\partial_x (x-y)\delta^{ab}$$

The system of equations (3.7), (3.8c), (3.9) provide a new theory of currents in the spirit of Sugawara, but with anomalous commutators. It is interesting that the equations of motion of $J\mu$, (3.4) and (3.5), are classically the same as the old theory of currents. However, quantum mechanically, the new matrix V of eq. (3.5) has non-commuting operator matrix elements as can be seen from the second equation in (3.9).

In addition to the charge algebra of eq. 3.8c, the infinite number of conserved charges of this theory (alluded to following eq. 3.5) form a further affine algebra in the sense of refs. 5,6. It is possible that it differs from the old theory by a central extension which was missing in previous analyses[6].

Rajeev[4] has also pointed out an unusual quantization of the chiral model with Wess-Zumino term in 4-dimensions. However, he concentrated on the commutation rules of $[A_0^a, A_0^b]$ rather than the conserved currents $[J_0^a, J_0^b]$. The algebra of the currents $J\mu$ is more complicated in 4-dimensions. The $[J_0 J_0]$ commutator has an operator anomaly rather than a c-number anomaly.

Witten[3] has also discovered a central extension in the *lightcone* (rather than equal time) quantization of a special case of this theory, namely the special value of the coupling $\lambda^2 = \pm g^2$. Our quantum current algebra with arbitrary $\lambda^2 \neq g^2$ is more general and reveals the anomaly in $[J_0, J_0]$ which is most unusual. An anomaly $\partial_- \delta(x-y)$ in lightcone commutators of currents $[J_\pm, J_\pm]$ in any theory is not unusual since the well known Schwinger term of $[J_0, J_1]$ must make a contribution.

To see how $g^2 \neq \lambda^2$ describes a more general algebra we take the special value $g^2 = \pm\lambda^2$ in eq. (3.6). We find that $J_0 = \mp J_1 = -iU^\dagger \partial_\mp U$ and we recover the lightcone combinations. The eq. of motion simplifies $\partial_0 J_0 - \partial_1 J_1 = 0 = \partial_\pm (U^\dagger \partial_\mp U)$ so that the currents are functions either of x_- or x_+ respectively. We can substitute x_1 by x_+ or x_-, as needed, in our equal time commutation rules since the current is a function of a single variable x_+ or x_-. Then, the 3 equations 3.8c and 3.9 degenerate to a single one and reproduce the special commutation rules of Witten.

In this theory there is, of course, a conserved left-handed

current $\tilde{J}_\mu$ in addition to the conserved right-handed current J_μ discussed above. It can be rewritten in terms of J_μ as follows:

$$\tilde{J}_\mu = \frac{i}{g^2}\partial_\mu U\,U^\dagger + \frac{i}{\lambda^2}\epsilon_{\mu\nu}\partial^\nu U\,U^\dagger$$

(3.10)

$$= -U\left\{\frac{\lambda^4+g^4}{\lambda^4-g^4}J_\mu + \frac{2\lambda^2 g^2}{\lambda^4-g^4}\epsilon_{\mu\nu}J^\nu\right\}U^\dagger$$

Its consistent commutation rules are obtained from eqs. (3.10), (3.8) and (3.9). It can be verified that $[\tilde{J}^a_0, \tilde{J}^b_0]$ has an anomaly with opposite sign to $[J^a_0, J^b_0]$ and that $\tilde{J}^a_0$ commutes with J^b_0, etc. Thus the *zeroth* components of the left and right-hand currents form two commuting affine algebras. The algebra including the space components J_1 and $\tilde{J}_1$ does not split into commuting left versus right sets because of the Schwinger terms.

Can these observations now help us solve the quantum chiral model? Here is my suggestion: A vertex representation of the affine algebra, which is an exponential of harmonic oscillators, can be used to construct Virasoro-like operators L_n. In this conference I learned that they are called Segal operators. These operators turn out to be quadratic in the harmonic oscillators. In particular L_0 is the Hamiltonian because it acts precisely like the time translation operator. So, in order to solve our chiral model we must find the "Vertex representation" of the current algebra in eqs. (3.8-3.9). This will give a Hamiltonian which is quadratic in oscillators and hence solves the problem. Progress in this direction will be reported elsewhere.

4. GAUGE THEORY

Gauge theories can be reformulated, in any dimension, in terms of string, i.e., ordered line integrals of the form

$$\left[\exp\, ig \int A\cdot dx\right]_+ \qquad (4.1)$$

where the line is a "string" S embedded in d-dimensions. If we take a set of rigid strings of a definite shape it is possible to write the action completely in terms of such strings. This was done in refs. 7, 10, 11, where it was also shown how to canonically quantize the theory in terms of such string variables. The path integral quantization is briefly given in ref. 12. The canonical quantization is described by equations analogous to the ones above that apply to the chiral model (without the WZ term), and thus involve a local non-Abelian charge algebra in higher dimensions. Hence, the relevance of affine type algebras to gauge theories. The unitary matrix U is replaced by the string, while the zeroth component of the current $J_0(x)$ is replaced by a local "glue" density $G(\vec{x})$. The gauge theory requires several variable like U and J_0 for a complete description in d-dimensions, as outlined below.

Following ref. 7, we briefly describe the reformulation. Let $A_\mu(x)$, $\mu=1,2,\ldots,d$ be the SU(N) gauge potentials (connections) written in the form of N×N hermitian, traceless matrices in the algebra of SU(N). For <u>every</u> μ, introduce a unitary matrix $B_\mu(x)$, such that

$$A_\mu = \frac{i}{g} B_\mu^\dagger \partial_\mu B_\mu$$

(4.2) $$B_\mu^\dagger B_\mu = 1$$

$$\det B_\mu = 1$$

If we imagine writing $B_\mu = \exp i\,\theta_\mu$, with θ_μ=traceless and hermitian, we see that the N×N matrices A_μ and B_μ are described by the same number of independent variables. Indeed, given $A_\mu(x)$, $B_\mu(x)$ can be determined from the d differential equations (for each μ)

(4.3) $$\partial_\mu B_\mu = -igB_\mu A_\mu ,$$

as

(4.4) $$B_\mu^{(x)} = \left[e^{-ig\int_{-\infty}^{x_\mu} dx'_\mu A_\mu(x')}\right]_+$$

where the path ordered integral is along a straight line in the μ direction from $-\infty$ up to the point X_μ.

(4.5) 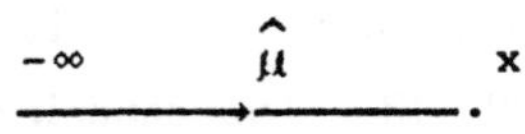

$B_\mu(x)$

Using (4.2) we compute the covariant field strength (curvature) $F_{\mu\nu}$ and find

$$
\begin{aligned}
F_{\mu\nu} &= \partial_\mu A_\nu - \partial_\nu A_\mu - ig[A_\mu, A_\nu] \\
&= \frac{i}{g} B_\mu^\dagger [\partial_\mu (B_{\mu\nu} \partial_\nu B_{\nu\mu})] B_\mu \qquad (4.6) \\
\mathrm{Tr}(F_{\mu\nu} F_{\mu\nu}) &= -\frac{1}{g^2} \mathrm{Tr}[\partial_\mu (B_{\mu\nu} \partial_\nu B_{\nu\mu})]^2
\end{aligned}
$$

where $B_{\mu\nu} = B_\mu B_\nu^\dagger$ is the string that starts at $-\infty$, comes along the μ direction to the point x where it turns sharply into the ν direction and goes off to $-\infty$ again, as in the figure.

(4.7) 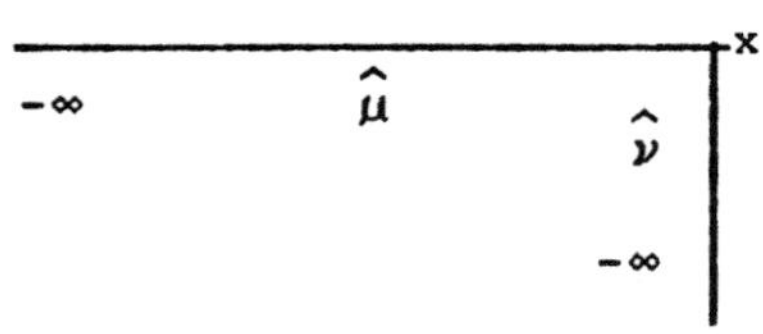

$B_{\mu\nu}(x)$

Thus, the action is completely rewritten in terms of such rigid strings.

We no longer need to consider the nonlocal dependence of $B_\mu(x)$ on A_μ since its appearance in the action is naturally local. If we apply the canonical formalism, as in ref. 7, we find the Hamiltonian density

$$
H(x) = g^2 \sum_{I=1}^{d-1} \mathrm{Tr}\left[\Pi_I(x)\right]^2 - \frac{1}{g^2} \sum_{I>J}^{d-1} \mathrm{Tr}\left[\partial_I (B_{IJ} \partial_J B_{JI})\right]^2 \qquad (4.8)
$$

where the first term is the usual "electric" term $\mathrm{Tr}(F_{0I})^2$ and the second the "magnetic" term $\mathrm{Tr}(F_{IJ})^2$. The "glue" density G(x) we mentioned above is

$$G_I(x) = -\partial_I \Pi_I(x), \quad \text{no sum on } I, \tag{4.9}$$

or

$$\Pi_I(x) = -\frac{1}{\partial_I} G_I(x) = \int_{x_I}^{\infty} dx_I' \, G_I(x') \tag{4.10}$$

This set of canonical variables $(G_1, G_2, G_3 \ldots, B_{12}, B_{13}, B_{23} \ldots)$ may be described via a set of axial gauges, as they were first discovered in ref. 11. Thus, in the gauge $A_1=0$ the electric field F_{01} is a complicated function of the canonical sets of variables (A_2, F_{02}), (A_3, F_{03}), etc. Its commutation relations are determined by this fact. Π_1 is identified as F_{01} in the $A_1=0$ gauge. Note that $Tr(\Pi_1^2)$ is gauge independent. Similarly in the gauge $A_2=0$, Π_2 is the dependent variable $F_{02}=\Pi_2$, etc. The local gauge transformations from the gauge $A_1=0$ to the gauge $A_2=0$ is identified as $B_{12}(x)$, and from $A_2=0$ to $A_3=0$ as $B_{23}(x)$, and so on. This is illustrated in the diagram (for d-1=3).

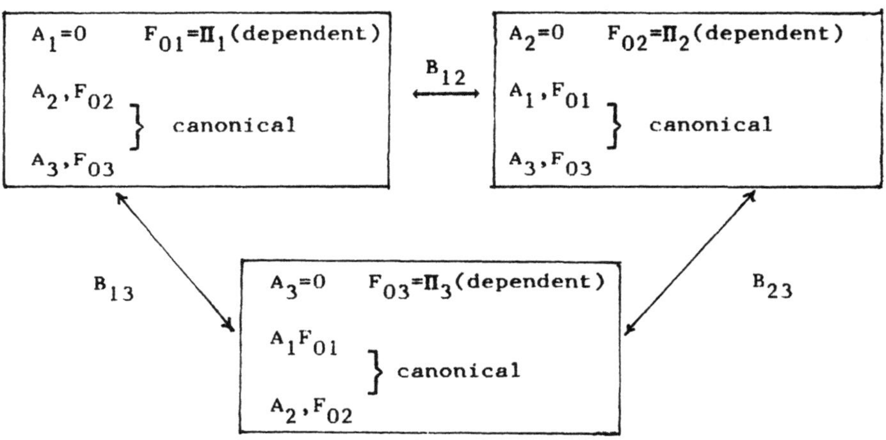

The quantum algebra of these variables are reminiscent of the SU(N)×SU(N) chiral model except that now instead of just the left and right sides of a single matrix U, we will consider the 1-side, the 2-side, the 3-side etc. of the unitary matrices matrices B_{12}, B_{13},

B_{23}, etc. The role of the left and right handed local current charge densities $\tilde{J}_0^a$ and J_0^a will now be played by the local "glue" densities G_1^a, G_2^a, G_3^a, etc. that act on the 1-side, 2-side, 3-side etc. Thus, we will have a local charge algebra corresponding to

$$SU(N)_1 \otimes SU(N)_2 \otimes SU(N)_3 \otimes \ldots$$

The quantum algebra can be written compactly

$$[\left(B_{IJ}(x)\right)_\alpha^\beta, \left(B_{KL}(y)\right)_\gamma^\delta] = 0$$

$$(4.11) \quad [G_I^a(x), B_{KL}(y)] = \left(-\frac{\lambda^a}{2}B_{KL}\delta_{IK}+B_{KL}\frac{\lambda^a}{2}\delta_{IL}\right)\delta^{d-1}(x-y)$$

$$[G_I^a(x), G_J^b(y)] = if^{abc}G_I^c(x)\delta^{d-1}(x-y)\delta_{IJ}$$

We see that a generalization of the affine or loop algebras to higher dimensions is relevant for studying quantum gauge theories. The physics problem is to find the eigenstates (and other properties) of the Hamiltonian $H=\int d^{d-1}xH(x)$, given the above commutation rules.

A simplified Hamiltonian, namely the first term of (4.8) corresponding to the strong coupling limit ($g\to\infty$), can immediately be solved using the algebra in (4.11). The solution is given in refs. (10,11) in the presence of a short distance cut-off. At fixed time, the eigenstates correspond to closed static strings of any shape embedded in d-1 dimensions, representing closed electric flux tubes, with quantized flux. The more difficult problem including the magnetic term can be dealt with in the strong coupling expansion. More sophisticated methods, perhaps involving generalizations of techniques presented in this conference, are needed for a more satisfactory non-perturbative treatment of this very challenging problem.

REFERENCES

[1] See e.g. the books on Current Algebras by S. Adler and R. Dashen, and by D. Gross, R. Jackiw and S. Treiman.

[2] See e.g. J.D. Bjorken and S. Drell, Relativistic Quantum Field Theory, Addison Wesley.

[3] E. Witten, Princeton preprints (1983), "Global Aspects of Current Algebra", "Current Algebra, Baryons, and Quark Confinement", "Non Abelian Bosonization in Two Dimensions".

[4] S.G. Rajeev, Syracuse preprint SV-4222-266 (1983), "Fermions from Bosons in 3+1 Dimensions Through Anomalous Commutators".

[5] H. Lusher and K. Pohlmeyer, Nucl. Phys. B137, 46(1978), E. Brezin, et. al. Phys. Lett. 82B, 442(1979).

[6] L. Dolan, Phys. Rev. Lett. 47, 1371(1981). B.Y. Hou, M.L. Ge, Y.S. Wu, Phys. Rev. D24, 2238(1981). For a review, see L. Dolan, Rockefeller preprint RU83/B/63.

[7] I. Bars and F. Green, Nucl. Phys. B148, 445(1979). See also Appendix of ref. 8.

[8] I. Bars, M. Günaydin, S. Yankielowicz, Nucl. Phys. B219, 81(1983).

[9] H. Sugawara, Phys, Rev. 170, 1659(1968). See also K. Bardakci and M. Halpern, Phys, Rev. 172, 1542(1968).

[10] I. Bars, Nucl. Phys. B149, 39(1979).

[11] I. Bars, Phys. Rev. Lett. 40, 688(1978), and in New Frontiers in High Energy Physics, P. 611, Eds. B. Kursunoglu, A. Perlmutter, L.F. Scott (Plenum).

[12] I. Bars, Phys. Lett. 116B, 57(1982).

Department of Physics
University of Southern California
Los Angeles, CA 90089-0484

SUPERGEOMETRY AND KAC-MOODY ALGEBRAS

B. Julia†

ABSTRACT

This exposition is meant to be a sequel to a previous review (to appear in the proceedings of the AMS-SIAM summer seminar on "Applications of Group Theory in Physics and Mathematical Physics" Chicago - July 1982). The main observation and conjectures date back to 1980-81 but further progress will require a concrete realization of affine and hyperbolic Kac-Moody groups and homogeneous spaces beyond the Lie algebraic construction and other formal or algebraic results. The first section summarizes the main features of supersymmetry needed for the construction of a gauge theory thereof. In the next the dimensional reduction of the bosonic part of eleven-dimensional supergravity is brought under the mathematical microscope. We review the impressive computations of General Relativists in section III and the emergence of the loop algebras of so(2,1) and su(2,1). Finally we discuss some links with the main approaches to completely integrable systems. This modest attempt to bring into the same framework such widely different subjects must be superficial, let us hope that the subsequent frustration will generate fruitful work.

1. SUPERSYMMETRY

In ref. [1] the connections between dual models of spinning strings and supersymmetric theories in 10 dimensions were reviewed. The precise correspondence is still under investigation [2]. It

---

† Partially supported by the National Science Foundation through the Mathematical Sciences Research Institute.

Vertex Operators in Mathematics and Physics - Proceedings of a conference November 10-17, 1983. Publications of the Mathematical Sciences Research Institute #3, Springer-Verlag, 1984.

involves an intermediate step namely the construction of the Scattering matrix and the problem is to construct a local field theory from it; the S-matrix describes only solutions of the quantum equations of motion whereas a field theory is usually constructed from a variational principle for unconstrained wave functions (one speaks of off-shell formulation before the equations of motion are implemented). The only information needed to construct these 10 dimensional field theories is actually their linearized form and a guess for their "global" symmetries (in this case invariance under a finite dimensional superalgebra.) Let us explain this directly in one time plus ten space dimensions because as it turns out the field theory describing the zero-mass states of the string model is the dimensional reduction of an 11 dimensional theory to 10 dimensions.

The first step is to obtain an irreducible linear representation of the Poincaré superalgebra. This can be done by a Wigner-Mackey induction from the so called "little superalgebra". The even part of the Poincaré superalgebra is the semi-direct product $P = \mathbb{R}^{11} \times so(1,10)$; it is a standard result that if one takes the stabilizer s_p of a character of $\mathbb{R}^{11}$, $p = (p_0, p_1 \dots p_{10})$ (where the Casimir $p^2 = -p_0^2 + p_1^2 + \dots + p_{10}^2 = 0$ for massless states) the induced representation consists of "functions" on the light cone ($p \neq 0$, $p^2 = 0$) with values in a representation σ of s_p. The odd part of the Poincaré superalgebra is the representation of P corresponding to $p = 0$ with σ the 32-component representation of so(1,10). The anticommutation relations are:

$$\{Q_{\underline{A}}, (Q^+\Gamma_0)^{\underline{B}}\} = (\Gamma^M)_{\underline{A}}^{\ \underline{B}} P_M \tag{1}$$

($\underline{\Gamma}^M$ are Clifford-Dirac matrices of so(1,10).)

So we must enlarge s_p to s'_p which includes the 32 odd-generators and find an irreducible representation σ' of it. For a fixed null vector (1) reduces to a Clifford algebra which together with $\mathbb{R}^9 \times so(9)$ form s'_p. In physics one considers usually trivial actions of $\mathbb{R}^9$, if we take the short simple root last the representations of B_4 that arise have highest weight (2,0,0,0) for the graviton, (1,0,0,1) for the "gravitino" and (0,0,1,0) for a bosonic field that accounts for the equal number of boson (even) states and of fermion (odd) states 84 + 44 = 128.

The second step is to find reducible "field" representations namely representations induced by the Lorentz subgroup SO(1,10) that contain the "state" representations. More precisely one requires the irreducible representation to be a subquotient of the reducible one. One obtains it by imposing equations of motion (variational Euler-Lagrange equations) and eventually (i.e. for gauge theories) by quotienting out by a residual gauge subrepresentation which cannot be eliminated by the usual gauge conditions; this last phenomenon is called "on-shell" gauge invariance, to preserve it the non-linear coupling (which one adds most conveniently at this stage by rendering the equations of motion non-linear) must involve conserved currents. This second step is implemented for the double-cover of the Poincaré group but has not always been implemented for the "Super-Poincaré group." Namely supersymmetry is only realized on the solutions of the field equations and it is an urgent challenge to find a "superfield" formulation for the most interesting supersymmetric theories.

The third step is to introduce interactions i.e. non linear terms in the equations or higher order terms in the Lagrangian. An algorithm for deforming linearized gravitation into its non-linear (General Relativity) generalization was first proposed in ref. [3] see also [4], let us explain it for the simpler case of Yang-Mills theories. A careful discussion of the gravitational interactions is deferred to [5]. We start with a local field theory of g vector gauge fields: the gauge theory of $\mathbb{R}^g$. The Lagrangian density is g times that of electromagnetism:

$$L^0 = \sum_{i=1}^{g} - \tfrac{1}{4}(\partial_M A^i_N - \partial_N A^i_M)^2 \tag{2}$$

(implicit sum over M and N = 0,1,2,...,d-1)

It possesses an invariance under the "local" gauge group:

(3) $$\delta^0(\epsilon)A^i_M = -\partial_M\epsilon^i(x)$$

and under a global (or rigid) compact group G having a g dimensional representation T with invariant quadratic form δ_{ij} (the Kronecker symbol). The transformation law reads:

(4) $$\delta^1_{(\alpha)}A^j_M = \sum_{j,k}(\alpha^i T_i)^j{}_k A^k_M.$$

It is associated to a conserved current by Noether's theorem:

(5) $$J^{M1}_i = \frac{\delta L^0}{\delta\partial_M\alpha^i} = -\sum_{j,k}\left[\partial_M A^j_N - \partial_N A^j_M\right]T_{ijk}A^k_N$$

(6) $\partial_M J^{M1}_i = 0$ if the equations of motion are satisfied.

Let us ignore the subtleties of the Legendre transformation in gauge theories. The conservation equation (6) which implies the conservation of the total charges:

(7) $$\frac{dQ_i}{d\Gamma} = 0 \quad \text{with} \quad Q_i = \int d^{d-1}x J^0_i$$

translates into an action on phase space of the group $\mathbb{R}^g \times \mathbb{R}_t$ where the direct factor $\mathbb{R}_t$ represents the time evolution. The Q_i's are the Hamiltonian functions for the $\mathbb{R}$ factors, they are obtained by the so-called moment map [6].

If the representation T is the adjoint representation of G one may try to deform the gauge group $\mathbb{R}^g$ into G; in terms of trivial bundles one replaces by a principal G-bundle the vector bundle of its tangent spaces at a global section. (This procedure can be generalized to include other representations of G: "matter fields"). The idea is to consider the transformation $\delta^0(\epsilon) + \delta^1(e\epsilon) = \delta(\epsilon(x))$ where e is the deformation parameter usually called coupling constant. Obviously

(8) $$\delta L^0 = \delta^1 L^0 = e\sum_i J^M{}_i{}^1\partial_M\epsilon^i \neq 0.$$

So one tries to cancel δL order by order in e by adding higher order terms. In our example we have to solve $\delta^1 L^0 + \delta^0 L^1 = 0$ or:

$$(9) \qquad \frac{\partial L^1}{\partial A^i_M}\left(A^i_M, \partial_M A^i_N - \partial_N A^i_M\right) = eJ^M{}_i{}^1.$$

The derivatives are partial Fréchet derivatives, in the general case one only requires invariance of the action (the space-time integral of the Lagrangian density.) Let us consider $\partial_M A^i_N - \partial_N A^i_M = F_{MN}^{\ i}$ and A^i_M as independent variables, (9) implies the Helmholtz [7] consistency conditions:

$$(10) \qquad \frac{\partial J^M{}_i(x)}{\partial A_N{}^i(y)} = \frac{\partial J^N{}_j(x)}{\partial A_M{}^i(y)} .$$

We must remark that (9) does not determine L^1 uniquely; non minimal couplings can be added which involve only $F_{MN}^{\ i}$ furthermore the current can be modified by the addition of the divergence of any antisymmetric tensor if we require only invariance of the action:

$$(11) \qquad J^M_{mod.i} = J^M{}_i + \partial_N K^{NM}{}_i \quad (\partial_M J^M_{mod.i} = 0 \text{ too.})$$

(One requests also global invariance of L^1 under G to have $\delta^1 L^1 \sim \partial\epsilon$). Because (10) holds one can find a simplest $L^1 = -\frac{e}{2} F_{MN}^{\ j} T_{ijk} A^i_M A^k_N$. The next order is similar and the procedure stops because the Yang-Mills theory has the required invariance.

The algorithm we just described is sometimes called the Noether method because the Noether current is coupled to the gauge field. It has been the most powerful method for the construction [8] of supergravity theories. The formulation of consistency conditions (10) is new however [5], one expects also nontriviality conditions of the deformation... But what is supergravity?

Let us go back to step 2 and try to find a field representation for the state representation with 44 components; the latter is nothing else but a symmetric traceless second rank tensor of so(9). It is natural to look for a symmetric tensor field on the eleven dimensional manifold P/So(1,10), but a metric has 66 components or 22 too many. The equations of motion should impose the masslessness but also reduce the number of fields. This requires some gauge invariance and a ("harmonic") gauge condition:

$$\partial_M\left[(|g_{PQ}|)^{1/2}g^{MN}\right] = 0 \tag{12}$$

then the linearized Hilbert-Einstein Lagrangian density leads to the appropriate equations of motion, it is a function of $h^{MN} = \frac{1}{\kappa}(g^{MN} - g^{MN}_{Flat})$:

$$L^0 \sim -\frac{1}{4}\Big[(\partial_M h_{NP})^2 - 2(\partial_M h^M{}_N)^2 + 2(\partial_M h^{MN})(\partial_N h^P{}_P) - (\partial_M h^N{}_N)^2\Big] \tag{13}$$

where the contractions are done with g_{Flat} of signature $(-1,+1,+1,\ldots,+1)$.

Using the linearized form of (12) h_{MN} must be harmonic i.e. $p^2 = 0$ and 11 of its components are killed. Eleven more components are arbitrary, in the nonlinear theory they decouple from the conserved current; the latter is a symmetric 2-tensor which is equal to a modified form (11) of the Noether current of translations for solutions of the equations of motion. The decoupling of these last 11 components is the result of the on-shell gauge invariance: $h^{(p)}_{MN} \longrightarrow h^{(p)}_{MN} + p_M X_N + p_N X_M$ $(p^2=0)$.

Similar arguments lead to a three form A(x) with "gauge" invariance: $A \longrightarrow A+dB$ with B an arbitrary 2-form and to a spinor valued 1 form $\psi^{\underline{A}}$ with linearized gauge invariance: $\psi^{\underline{A}} \longrightarrow \psi^{\underline{A}} + d\epsilon^{\underline{A}}$ where $\epsilon^{\underline{A}}(x)$ are the 32 parameters of local supersymmetry transformation. This local supersymmetry is abelian and plays the role of the $\mathbb{R}^g$ gauge invariance above. It is to be deformed into a local "analog" of the Poincaré superalgebra [8][5]. We know a one parameter family of deformations of this eleven dimensional supergravity, let us write the action:

$$\kappa^2 S_{11} = \int_{\mathcal{M}_{11}} d^{11}x\left[|g_{PQ}|\right]^{1/2}[\tfrac{1}{4}R(g)] - \int_{\mathcal{M}_{11}} [\frac{1}{2\cdot 7!}dA'\wedge {}^*(dA') - \frac{2}{(12)^4}\, dA'\wedge dA'\wedge A']$$

(14)

$$+ \tfrac{1}{2}\int_{\mathcal{M}_{11}} d^{11}x\left[|g_{PQ}|\right]^{1/2}\ \psi'^{+\underline{A}}_{M}\cdot S^{MN}_{\underline{AB}}\cdot\psi'^{\underline{B}}_{N}$$

where $A' = \kappa A$, R is the scalar curvature, $\psi' = \kappa\psi$ and the gravitino wave operator is defined by

$$(\Gamma^0\cdot S^{MN}) = i\Gamma^{[M}\Gamma^{P}\Gamma^{N]}(\partial_p + \tfrac{1}{4}\omega''_{PRS}\Gamma^{[R}\Gamma^{S]})$$

(15)

$$+ \frac{1}{12}\partial_{[P}A''_{QRS]}\left\{\Gamma^{[M}\Gamma^{P}.\Gamma^{Q}.\Gamma^{R}.\Gamma^{S}.\Gamma^{N]} + 12g^{MP}\Gamma^{[Q}\Gamma^{R]}g^{SN}\right\}$$

Γ^0, Γ^1, $\Gamma^2 ... \Gamma^{10}$ are Clifford matrices, the bracket indicates antisymmetrization and A" and ω" are equal to A' and to the "Lorentz connection" on the spin bundle up to appropriate quadratic terms in ψ [9]. The coupling constant κ is related to Newton's constant of gravitation. Supergeometry has not been properly understood yet in 11(Bose) + 32(Fermi) dimensions. So we shall retreat to the purely bosonic sector by setting ψ equal to zero.

II THE E SERIES.

1. Reconstruction of symmetries

In reference [1] we indicated the general ideas of dimensional reduction on a small torus. To summarize one supposes an abelian algebra of (D-d) independent Killing vector fields which we take to be $\partial/\partial x^i$ i = d, d+1..., D-1. Tensor fields in the original D dimensional space-time split into several tensor fields differing in d-dimensional character, for example one vector field gives one vector field plus (D-d) scalar functions all depending on d residual coordinates. We reduce the diffeomorphism group in D dimensions to that in d dimensions; but d dimensional quantities of the same tensor character can arise from different sources and internal symmetry (i.e. symmetries commuting with the Poincaré group) grows faster than expected as will be illustrated below.

What is expected on moral grounds [1] is an internal symmetry $SL(D-d,\mathbb{R})$ and in most cases $GL(D-d)$. For example the 10 dimensional reduction of (14) involves one scalar "field" (= function) namely $\Delta = g_{11,11}$ but its kinetic term depends only on $\partial\Delta/\Delta$ and one can rescale the other fields in such a way that $GL(1)$ or $\mathbb{R}^+$ be a symmetry of the full Lagrangian and of the equations of motion. In [1] we also discussed the emergence of $SL(2,\mathbb{R})$ in the 3 dimensional reduction of the equations of motion of pure gravitation in 4 dimensions. We should insist that to this day there is no conceptual understanding of this group of invariance, it acts on a space including the original tensor fields but also some "dual" field.

Let us proceed to the exciting discovery of E_7 from the equality 70 + 63 = 133. We may distinguish 2 steps: a) the enlargement of $SL(7)$ to $SL(8)$ and b) showing that $SL(8)/SO(8)$ is actually a truncation of $E_7(+7)/SU(8)$. It goes as follows: after dimensional reduction from 11 to 4 dimensions let us consider those terms in the action that involve g_{ij} (28 of them) and $A_{\mu\nu i}$ (7 of them). By some (discrete) "duality" one can define 7 dual potentials φ^i to $A_{\mu\nu i}$ (i = 4,5,...10) and a dual Lagrangian that leads to equivalent equations of motion as (14) after reduction. This Lagrangian can be written as:

$$(16) \qquad L_+ = \frac{e}{16}\partial_\mu g^{ij}\partial^\mu g_{ij} + \frac{e}{8\Delta}\partial_\mu\varphi^i\partial^\mu\varphi^j g_{ij} - \frac{e}{32}\frac{\partial_\mu\Delta}{\Delta}\frac{\partial^\mu\Delta}{\Delta}$$

where $\mu = 0,1,2,3$, $g^{ij} = (g_{ij})^{-1}$ and $\Delta = \det g_{ij}$. An important feature is the polynomial dependence of L_+ in φ. On the other hand 28 + 7 = 35 is the dimension of the symmetric space $SL(8)/SO(8)$ which can be canonically parametrized by a symmetric matrix of determinant one with positive eigenvalues, it is therefore tantalizing to restore manifest $SL(8)$ invariance in the $GL(7)$ invariant equation (16); we guessed a symmetric form of L_+:

$$(17) \qquad L_+ = \frac{e}{16} \partial_\mu S^{i'j'} \partial^\mu S_{i'j'} \qquad a', i', j' = 4,5...11$$

and the corresponding parametrization:

$$(18) \qquad S^{i'j'} = \Delta^{-3/4} \begin{bmatrix} \Delta g^{ij} + \varphi^i\varphi^j, & \varphi^j \\ \varphi^i & , 1 \end{bmatrix} = (V_+^t)^{i'}{}_{a'} V_+{}^{a'j'}$$

$$S_{i'j'} = (S^{i'j'})^{-1}$$

V_+ is partially triangular *hence* both V_+ and V_+^{-1} are polynomial in φ^i.

Let us now collect also those terms involving the 35 "pseudo-scalar" fields A_{ijk} and try to restore SL(8) invariance there too. Actually motivated by lowest order terms in κ [10] one would like to restore an invariance under a global SU(8) as well. But the couplings of the 28 vector fields and of their dual (magnetic) vector potentials suggest that one should consider the 28 + 28 dimensional representation of SO(8):

$$8\wedge 8 \oplus \text{contragredient } (8\wedge 8).$$

After some gymnastics one finds the universal occurrence of a 56 × 56 matrix $V = V_- V_+$ where V_+ is just the appropriate representative of V_+ and V_- has the following interesting form after some reordering of the basis:

$$(19) \qquad V_- = \exp T_- , \; T_- = \begin{bmatrix} 0 & A_1 & 0 & 0 \\ 0 & 0 & {}^*A & 0 \\ 0 & 0 & 0 & A_2 \\ 0 & 0 & 0 & 0 \end{bmatrix}$$

where $A_1 = A^F_{[i][jk]}$, ${}^*A = \frac{1}{6} \epsilon^{[ij][k\ell]mno} A^F_{mno}$ and $A_2 = A^F_{[ij][k]}$; a bracket counts as one index and A^F_{ijk} are the components of A_{ijk} in an orthonormal basis for g_{ij}. Once again we were led to (19) by the polynomiality of the Lagrangian in A_{ijk}.

Actually one checks easily that $\mathcal{V} = \mathcal{V}_+\mathcal{V}_-^c$ where $\mathcal{V}_-^c$ has the same form as $\mathcal{V}_-$ (19) but expressed in terms of the components $A^c_{i\,jk} = A_{ijk11}$ in the "curved" basis for $S_{i'j'}$ (see [11]).

2. A generalized Iwasawa decomposition.

Let us now recall the Iwasawa decomposition; given a connected real Lie group G there is an analytic diffeomorphism of the product K.A.N. of its subgroups K, A, N, respectively compact, Cartan and nilpotent (generated by the positive root vectors) onto G. In particular K\G is uniquely parametrized by pairs (a,n) in A×N. We can apply the Iwasawa decomposition to SL(8$\mathbb{R}$) and $\mathcal{V}_+$, and by using the gauge invariance under SO(8): $\mathcal{V}_+ \cong \sigma\mathcal{V}_+$ (σ may depend on x^μ and on the spinless fields) we find $\mathcal{V} \cong \alpha\mathfrak{n}_+\mathcal{V}_-^c$. This parametrization of $\mathcal{V}$ was found in [11] where is was remarked that near the identity this is the Iwasawa decomposition of $(Su(8)/\mathbb{Z}_2)\backslash E_7(+7)$. In fact the proof of such a generalized Iwasawa decomposition is very easy even globally. We only have to find the product decomposition of an element $\mathfrak{n}$ of N (in E_7) into a product $\mathfrak{n}_+ \exp T_-$ and to use the fact that the exponential map is one to one in connected nilpotent groups. (The Lie algebra of N is 63 dimensional and decomposes into one dimensional subspaces by a $\mathbb{Z}$-grading.) In the linear 56 representation the element $\mathfrak{n}$ can be obtained uniquely in the form of a product $\mathfrak{n} = \prod_i \exp T_i$ where T_i belongs to the ith subspace of any vector space decomposition of the Lie algebra of N. The exponents can be adjusted one by one and uniquely by using the matrix algebra grading of N (the distance to the diagonal) in the basis [11] where $\mathfrak{n}$ is triangular.

This establishes that the 70 spinless fields parametrize (take values in) the symmetric space $A_7\backslash E_7$. The equations simplify considerably if instead of using a global section of the principal bundle $H \longrightarrow G \longrightarrow H\backslash G$ (here $A_7 \longrightarrow E_7 \longrightarrow A_7\backslash E_7$) as in (18) one restores a (local) gauge invariance under A_7 (A_7 contains SO(8)). The generalization of (18) is:

(20) $$L = -\frac{e}{48\kappa^2}\,\mathrm{Tr}\Big(\big[(\partial_\mu V)V^{-1}\big]^{\perp}\Big)^2$$

with V an arbitrary element of E_7 in its fundamental representation and $\perp$ meaning the Killing-orthogonal projection along a_7 onto its orthogonal complement in e_7. Clearly (20) and its Euler-Lagrange equations are invariant under A_7^{Left} (local) $\times$ E_7^{Right} (global); it turns out that the complete set of *equations* of motion is *invariant* [11,12], this E_7 defies our understanding as the (Ehlers) $SL(2,\mathbb{R})$ of stationary pure gravity mentioned above does.

3. A spinorial symmetric space and E_8

Elementary counting arguments [11] suggested E_6 (resp. E_8) as symmetry groups for the reduction of (14) to 5 (resp. 3) dimensions. Assuming E_6 invariance the authors of [13] constructed a consistent five dimensional supergravity theory. The emergence of E_8 (+8) was established in [12] up to a Clebsch-Gordan computation left as an exercise (we considered only the Bose fields). To convince the frightened reader that the group E_8 is not such a monster, we present a sketch of a simplified proof that the maximal supergravity (14) reduced to 3 dimensions is classically equivalent to a symmetric space model $D_8\backslash E_8$ on a curved background 3-space-time.

We start from 4 dimensions with E_7 symmetry 70 scalar fields in V with 28 vector fields and their 28 dual potentials and with a pseudo-Riemannian 4-metric $g_{\mu\nu}$. The latter decomposes into a 3-metric a vector field and a scalar field Δ, the vector is dual to a (classically equivalent) scalar field $\underline{P}$ (called B in [1]); both Δ and $\underline{P}$ are invariant under E_7. The 28 vectors give 28 3-vector gauge fields and 28 scalars, by duality one again converts the vector into scalars and the 56 scalars S(x) transform irreducibly under the E_7 internal symmetry. The equations of motion are the variational equations for a Lagrangian which is E_7 invariant:

(21) $$\kappa^2 L_3 =$$

$$\frac{e}{4}R - \frac{e}{8\Delta^2}(\partial\Delta)^2 - \frac{e}{48}\,\mathrm{Tr}(\partial V.V^{-1})^{\perp 2} - \frac{e}{4\Delta}(V.\partial S)^2$$

$$- \frac{e}{8\Delta^2}(\partial P - S^t\Omega\partial S)^2$$

Ω is the symplectic form invariant under E_7 and ∂ stands for $\partial/\partial x^\mu$. If we set $S(x) = 0$ we obtain a Lagrangian of the form (20) for the subspace $(A_7\backslash E_7)\times(U_1\backslash A_1$ of $(D_8\backslash E_8)$; this implies the unique relative normalization of (21) and has been checked by computing the quadratic Casimir operators of A_1 and E_7 in their fundamental and adjoint representations [12]. Once again $D_8\backslash E_8$ appears in a partially triangular form and we must check that

(22) $$\kappa^2 L_3 = -\frac{e}{240}\,\mathrm{Tr}\left[(\partial g\cdot g^{-1})^{\perp}\right]^2.$$

The splittings are: 248 gives the (133×1) + (56×2) + (1×3) representations of $E_7\times A_1$ and $128 = 1_2 + 1_{-2} + 28_1 + \overline{28}_{-1} + 70_0$ where n_λ is the n-representation of A_7 and has (helicity) character $e^{i\lambda\theta}$ for U_1.

Let us now use the gradation corresponding to the highest root of E_8. P corresponds to the highest root (grade 2) and the S's correspond to the 56 roots of grade 1 (the latter form pairs of sum equal to the highest root) and the last term of (21) fixes their relative normalization. We recall that (20) should hold for $D_8\backslash E_8$. But $(\partial g\cdot g^{-1})^{\perp}$ contains 2 terms of grade 2: necessarily

$$\partial P\cdot T_2 + \tfrac{1}{2}[S_i T_1^i,\ \partial S_j T_1^j]_- = (\partial P - S^t\Omega\partial S)T_2$$

and

(23) $$[T_1^i,\ T_1^j]_- = -2\Omega^{ij}T_2$$

It remains only to check the normalization of the penultimate term of

(21). Given $g = V_{248}(V,\Delta)\cdot\exp(PT_2 + S_i T_1^i)$ the normalization is just the value of the Killing form on T_1^i. So we can avoid all Clebsch-Gordan computations and use instead the standard formula:

$$[T_\alpha, T_\beta] = \pm\left[\frac{q(1-p)}{2}\langle H_\alpha | H_\alpha\rangle\right]^{1/2} T_{\alpha+\beta} \tag{24}$$

if $(\beta+n\alpha)$ are roots for $p \leq n \leq q$. Here we must also use a formula of [14]: $\langle H_\alpha | H_\alpha\rangle = h^{-1}$ (h = 30 = Coxeter number). But (24) holds for normalized T_α's whereas $tr(T_2 T_2^t) = 60$ by (21). We deduce from all this that $Tr(T_1^i T_1^{i\,t}) = 120$ for each i and that (22) holds.

III LOOP ALGEBRAS

If one extrapolates the above phenomenon to dimension 2 one is led to conjecture an internal symmetry E_9 there (see [1] and references [10] and [11] therein). Of course E_9 is $E_8^{(1)}$ and similar extrapolations suggested already in July 1980 that the affine Kac-Moody algebras (of type I) associated to ordinary finite dimensional Lie groups by the usual extension of the Dynkin diagram should also correspond to them by dimensional reduction; precisely a symmetry group G of a 3-dimensional supergravity should grow into the group $G^{(1)}$ of Lie-Bäcklund transformations leaving invariant the equations of motion (and set of solutions) of this supergravity theory reduced to 2 dimensions. This is to be contrasted with the "K.P. hierarchy" [15] for which GL(∞) acts on solutions of an infinite number of equations in infinitely many variables, there the dimensional reduction reduces the symmetry group to an affine Kac-Moody group ($A_1^{(1)}$ in the case of the "K.d.V. hierarchy").

The author succeeded actually during the winter 1981 in recognizing this affinization of the group G in the cases of General Relativity and of General Relativity coupled to electromagnetism. This statement requires explanations and qualifications; first of all most of the hard work had been done by general relativists who found an action of the loop algebras (not groups) of $\mathfrak{sl}(2,\mathbb{R})$ and $\mathfrak{su}(2,1)$ on a space lying over the space of solutions [16]; of course at that time Kac-Moody theory was just at its beginning and these loop algebras were not recognized as such. Let us note that "non-compact" real forms appear here. Secondly motivated by the Serre presentation and the extrapolation of the E series the author argued that in fact the full affine Kac-Moody algebra was at work on the fields already introduced and he showed that the central charge had been overlooked because it would act only on some components of the 4 by 4 metric.

Loosely speaking R. Geroch discovered in 1972 [17] the Serre presentation of $A_1^{(1)}$. One $s\ell(2,\mathbb{R})$ (actually $g\ell(2,\mathbb{R})$) is the expected consequence of reparametrization invariance in four dimensions after reduction to two dimensions. The other $s\ell(2,\mathbb{R})$ (Ehlers) is acting also naturally on the space of stationary solutions (3-dimensions subsist instead of 2) we mentioned it above and described it in detail in [1]. Our observation was that the two Cartan generators are not equal, they are only almost equal, namely their difference δ acts only on the least interesting components of the metric which have been traditionally neglected. This is consistent with a central charge action on a homogeneous space $G^{(1)}/H$! Let us explain this.

In the theories of chiral fields (harmonic maps from $\mathbb{R}^2$ into a coset space) the group appears in two ways: as a set of transformations and also in the homogeneous space. For infinite dimensional groups the second point of view had not been considered; in fact there is a third object to study; a (sub)set of solutions might be also a coset space of this group (the Geroch conjecture is such a statement for stationary axially symmetric solutions of Einstein's equations). In [18] two candidates for H were considered in order to reformulate the Geroch equations as a chiral model for $A_1^{(1)}/H$. The first is the "maximal compact" subgroup K of $A_1^{(1)}$ and we identified the basic fields of the theory with the 4 coordinates on $\frac{G\ell(2,\mathbb{R})}{So(2)} \times \mathbb{R}^+$; the first factor describes the internal part of the metric and the second term is the conformal factor of the 2 by 2 metric along the remaining (non cyclic) directions. Clearly $G\ell(2,\mathbb{R}) \times \mathbb{R}^+$ acts on the left and $\mathbb{R}^+$ is central but it acts on itself by left multiplication, this is precisely the action of δ (the central charge in the group of transformations) on the conformal factor (i.e. the coordinate along the central generator in G/K). In [18] we also identified the next coordinates on G/K corresponding to level 1 and an Iwasawa decomposition of $A_1^{(1)}$. The second candidate for the subgroup H was found against most expectations to be of finite codimension.

The proof of the equivalence between such a chiral model and the theory of 2-"dimensional" gravitational solutions will involve a lot of analysis. We need to define properly non compact Kac-Moody groups and coset spaes, this is the price to pay for solving partial differential equations derived from Einstein's equations. We refer to our forthcoming review for more information [19]. We should say however that since our remarks a lot of progress has been made in understanding the hidden symmetries of non-linear systems. In particular in theories where one does not expect a Kac-Moody algebra of symmetries but a loop algebra only (like the chiral model SL(2)/SO(2), there is no conformal factor there) the full Kac-Moody algebra acts on the τ-functions [20] so there could be a new differential geometric interpretation of these objects in the case of the Einstein equations.

In [1] we also conjectured the emergence of hyperbolic Kac-Moody algebras acting on spaces lying over the finite dimensional space of homogeneous ($\vec{x}$ independent) solutions to Einstein's equations. This would be of course a nice model for these very abstract constructs but the proof demands that one simplifies the 2 dimensional case first. Let us just make 2 remarks. For N = 8 supergravirty after E_9 comes E_{10} as a natural candidate, for pure gravity after $A_1^{(1)}$ comes a group that contains it as well as the obvious group $S\ell(3)$ such that an $S\ell(2)$ subgroup of the latter commutes with the Ehlers $S\ell(2)$. Secondly we must realize that hyperbolic Kac-Moody algebras have no subalgebras of finite codimension (V. Kac and D. Peterson unpublished) so our second candidate for H becomes less likely.

Some work has also been devoted to the study of symplectic symmetries but the situations is still unclear so we shall not report on it.

In conclusion let us pray that more analysts will join us in the study of these amazing models.

REFERENCES

[1] B. Julia, Kac-Moody Symmetry of Gravitation and Supergravity Theories, in Chicago 1982-AMS-SIAM Proceedings (to appear) (1984).

[2] See for example J.H. Schwarz, Phys. Reports 89 no. 3 (1982).

[3] S.N. Gupta, Phys. Rev. 96 (1954) 1693 and Proc. Phys. Soc. (London) A65 (1952) 608.

[4] S. Deser, Gen. Rel. and Grav. 1 (1970) 9.

[5] B. Julia, to appear in the Proceedings of the Conference on Group Theoretical Methods in Physics, College Park (1984) (and work in progress).

[6] J.M. Souriau, Structure des systèmes dynamiques (1970) Dunod-Paris.

[7] Helmholtz J. reine und ang. Mathematik 100 (1887) 137.

[8] P. van Nieuwenhiuzen, Phys. Reports 68 (1981) 189.

[9] E. Cremmer, B. Julia and J. Scherk, Phys. Lett. 76 B (1978) 409.

[10] B. de Wit and D. Freedman, Nucl. Phys. B130 (1977) 105.

[11] E. Cremmer and B. Julia, Phys. Lett. 80B (1978) 48 and Nucl. Phys. B159 (1979) 141.

[12] B. Julia, Application of supergravity to gravitation theory, in "Unified Field Theories of more than 4 dimensions" ed. V. de Sabbata World Scientific 1983.

[13] E. Cremmer, J. Scherk and J.H. Schwarz, Phys. Lett. 84B (1979) 83.

[14] B. Kostant, Adv. in Math. 20 (1976) 179.

[15] See for example M. Jimbo and T. Miwa: Solitons and infinite dimensional Lie algebras (March 1983 R.I.M.S. preprint).

[16] W. Kinnersley and D.M. Chitre, J. Math. Phys. 18 (1977) 1538.

[17] R. Geroch, J. Math. Phys. 13 (1972) 394.

[18] B. Julia, Proceedings of the Johns Hopkins Workshop on Particle Theory (May 81).

[19] B. Julia, Physics Reports, in preparation.

[20] G. Segal and G. Wilson, to be published by I.H.E.S.: Loop groups and equations of KdV type.

Mathematical Sciences Research Institute
Berkeley, CA 94720
Laboratoire de Physique Théorique de
l'Ecole Normale Superieure
24 Rue Lhomond
75231 Paris Cedex 05
France

A PROOF OF THE NO-GHOST THEOREM USING THE KAC DETERMINANT

Charles B. Thorn[†]

ABSTRACT

A new proof of the no-ghost theorem for dual resonance models using results of Kac on infinite dimensional Lie algebras is presented.

* * *

The physical states of the Dual Resonance Model (DRM)[1] satisfy

$$L_n \mid \text{phys} > = 0 \qquad n > 0$$

(1)

$$(L_0 - 1) \mid \text{phys} > = 0$$

where L_n are the generators of the conformal algebra

(2) $$[L_n, L_m] = (n - m)L_{n+m} + \frac{D}{12}(n^3 - n)\,\delta_{n,-m}$$

and where D is the dimension of space-time. The representation of the L_n's occurring in the DRM is given in terms of operators $a_n^\mu = a_{-n}^{\dagger\mu}$, $n = 0,\pm1,\pm2,\ldots$; $\mu = 0,1,2,\ldots, D-1$ satisfying the commutation relations

(3) $$[a_n{}^\mu, a_m{}^\nu] = n\delta_{n,-m}\, g^{\mu\nu},$$

with $g_{\mu\nu} = 0$ for $\mu \neq \nu$ and $-g_{00} = g_{11} = \cdots = g_{D-1,D-1} = 1$:

---

† Partially supported by the U.S. Department of Energy under contract No. DE-AS-05-81-ER40008.

Vertex Operators in Mathematics and Physics - Proceedings of a Conference November 10-17, 1983. Publications of the Mathematical Sciences Research Institute #3, Springer-Verlag, 1984.

$$L_n = \frac{1}{2} \sum_m a_m^{\mu} a_{n-m}^{\nu} g_{\mu\nu} \qquad \text{for } n \neq 0,$$

(4)

$$L_0 = \frac{a_0^{\mu} a_0^{\nu} g_{\mu\nu}}{2} + \sum_{m=1}^{\infty} a_{-m}^{\mu} a_m^{\nu} g_{\mu\nu}.$$

The L_n's act on a vector space generated by the states

(5) $$a_{-n_1}^{\mu_1} \; a_{-n_2}^{\mu_2} \cdots a_{-n_k}^{\mu_k} \; | \; 0, a_0'^{\mu} >$$

where the state $| \; 0, a_0'^{\mu} >$ satisfies

$$a_n^{\nu} \; | \; 0, a_0'^{\mu} > \; = \delta_{n,0} \, a_0'^{\mu} \; | \; 0, a_0'^{\mu} > \qquad \text{for } n \geqslant 0.$$

(6)

$$< 0, a_0'^{\mu} \; | \; 0, a_0'^{\mu} > \; = 1.$$

In the DRM $a_0'^{\mu}$ is $\sqrt{2}$ x (energy momentum vector of a particle), and in the particle's rest frame it is just (M,0,0,..., 0) with M the mass of the particle. Since $g^{00} = -1$, the vector space generated by the states (5) contains negative norm states, called ghosts, e.g.

$$< 0, a_0' \; | \; a_n^0 \, a_{-n}^0 \; | \; 0, a_0' > \; = n \, g^{00} = -n.$$

The no-ghost theorem, proved over a decade ago by Brower[2] and by Goddard and Thorn[3], asserts that all the solutions to Eqs. (1) have non-negative norm for $D \leqslant 26$.

Before those proofs appeared, a direct algebraic analysis of Eqs. (1) was attempted by Brower and Thorn[4]. This analysis established a sufficient condition for the absence of ghosts. To state this condition Brower and Thorn defined the $p(m) \times p(m)$ matrix ($p(m)$ = the number of partitions of the integer m)

(7) $$\mathcal{M}^m_{\{\lambda'\}, \{\lambda\}} = < h \; | \; L_m^{\lambda_m'} \cdots L_1^{\lambda_1'} \; L_{-1}^{\lambda_1} \cdots L_{-m}^{\lambda_m} \; | \; h >$$

where $\{\lambda'\}$, $\{\lambda\}$ are partitions of m: $\sum_k k\lambda' = \sum_k k\lambda_k = m$, and the state $| \; h >$ satisfies the conditions

$$\langle h \mid h \rangle = 1$$

$$\text{(8)} \qquad L_0 \mid h \rangle = h \mid h \rangle$$

$$L_n \mid h \rangle = 0 \qquad \text{for } n > 0.$$

Note that $\mathcal{M}^m(h)$ may be evaluated using only the algebra (2) and properties (8). A sufficient condition for the absence of ghosts in the solution space of Eqs. (1) is that the signature of the matrix $\mathcal{M}^m(h)$ is the same as that of the matrix

$$\text{(9)} \qquad T^m_{\{\lambda'\},\,\{\lambda\}} = \delta_{\lambda_1'\lambda_1} \cdots \delta_{\lambda_m'\lambda_m} (-1)^{\sum_k \lambda_k}$$

for $m \geqslant 1$ and for $h = 1-m-k$, $k = 0,1,2,\ldots$.

Ref. 4 actually states a slightly weaker sufficient condition, which allows any of the negative eigenvalues of $\mathcal{M}^m(1-m)$ to be replaced by zero. In Ref. 4 it was further shown that the signature of $\mathcal{M}^m(h)$ as $h \longrightarrow -\infty$ for fixed m was precisely that of T^m. Thus, the above conditions will certainly be satisfied if

$$\text{(10)} \qquad \det \mathcal{M}^m(h) \neq 0 \qquad \text{for all } h < 1 - m,$$

although this is by no means necessary.

A calculation of det $\mathcal{M}^m(h)$ was not available at the time of our work. But since that time Victor Kac has obtained a closed form expression for det $\mathcal{M}^m(h)$:[5]

$$\det \mathcal{M}^m(h) = \text{const} \prod_{rs \leqslant m} (h - h_{r,s})^{p(m-rs)}$$

where

$$h_{r,s} =$$

$$\text{(11)} \qquad = \frac{1}{48}[(13-D)(r^2+s^2) - 24rs + 2D - 2 + (D^2-26D+25)^{1/2}(r^2-s^2)]$$

and $p(k)$ is the number of partitions of the integer k ($p(0) \equiv 1$).

Now for $1 < D < 25$, the only real zeroes of det $\mathcal{M}^m$ are at

$$(12) \qquad h^*_{r,r} = \frac{D-1}{24}(1 - r^2).$$

Since $r^2 \leqslant m$ for det $\mathcal{M}^m$, the real zeroes h^{*m} of det $\mathcal{M}^m$ satisfy

$$(13) \qquad h^{*m} \geqslant \frac{D-1}{24}(1 - m) \geqslant 1 - m$$

for $m \geqslant 1$ and $1 < D \leqslant 25$. Thus, for all $D \leqslant 25$, Kac's calculation completes the direct proof of the no-ghost theorem initiated by Brower and Thorn.

We conclude with a comparison of these results with those of Brower, Goddard and Thorn. The latter authors established the absence of ghosts in the dual resonance model for all dimensions $D \leqslant 26$. Since the proof relied on the explicit representations of the L_n's in that model, D was necessarily an integer. The direct proof presented here would establish the absence of ghosts in a more general class of models (as yet to be discovered) for which the D of Eq. 2 could be any real number $1 \leqslant D \leqslant 25$. On the other hand, our new proof cannot be easily extended to $D = 26$, because, for $D > 25$, $h_{r,s}$ is real for all r, s, and for some $r \neq s$, $h_{r,s} < 1-m$. Thus the proofs are complementary in a certain sense.

In a more general class of models, the second of Eqs. (1) might be generalized to

$$(L_0 - \alpha_0) \mid \text{phys} > = 0 \qquad \text{for } \alpha_0 < 1.$$

(For $\alpha_0 > 1$ there are always ghosts.) It is clear that the proof presented in this article immediately extends to this more general case for $D \leqslant 25$. The older proofs can also be extended to this case, for integers $D \leqslant 25$, through the expedient of dimensional reduction: apply the 26 dimensional no-ghost theorem on the subspace of states generated by only the first D components of a^μ_n, but with the 26th component of $a_0'^\mu$ nonvanishing: $1 - \alpha_0 = (a_0'^{26})^2/2$.

Let us review our results. By combining the present proof with older proofs, we can say that in the framework of models labeled by continuous α_0, D, the no-ghost theorem is established for

$$1 \leqslant D \leqslant 25 \qquad \text{and} \qquad \alpha_0 \leqslant 1$$

and for

$$D = 26 \qquad \text{and} \qquad \alpha_0 = 1.$$

By simple inspection there are ghosts present in models with $\alpha_0 > 1$ and any D. The situation for general $\alpha_0 \leqslant 1$ and $D > 25$ is complicated except for the special case $\alpha_0 = 1$, $D = 26$. Of course it is a straightforward matter to exhibit ghosts for certain ranges of these parameters by examining the signature of the matrices $\mathcal{M}^m(h)$ for low values of m. For example, inspection of $\mathcal{M}^1$ establishes ghosts for all $\alpha_0 > 1$; inspection of $\mathcal{M}^2$ establishes ghosts for all

$$D > \frac{26(1 + \frac{8}{13}(1-\alpha_0))(2-\alpha_0)}{3 - 2\alpha_0} ,$$

etc. It would be interesting to establish the regions of the (D, α_0) plane for which ghosts are present.

REFERENCES

[1] For a review of these models see, e.g. S. Mandelstam, Physics Reports 13C 261 (1974).

[2] R.C. Brower, Phys. Rev. D6, 1655 (1972).

[3] P. Goddard and C.B. Thorn, Phys. Letters 40B, 235 (1972).

[4] R.C. Brower and C.B. Thorn, Nuclear Physics B31, 163 (1971).

[5] V.G. Kac, in *Group Theoretical Methods in Physics*, edited by W. Beiglböck, A. Böhm, and E. Takasugi (Springer-Verlag, New York, 1979), p. 441. For a proof of the Kac formula, see B.L. Feigin and D.B. Fuchs, Functs. Anal. Prilozhen. 16 (1982), 47-63.

APPENDIX

Here we outline the Goddard-Thorn proof of the no-ghost theorem[2].

Let $V^D(a_0')$ be the vector space generated by the states in Eq. 5, and let k^μ be a light-like vector $(k\cdot k \equiv g^{\mu\nu}k_\mu k_\nu = 0)$ satisfying $k\cdot a_0' \equiv g^{\mu\nu}k_\mu a_{0\nu}' \neq 0$. We then define the subspaces:

$$V_\ell^D(a_0') = \{|\psi\rangle \in V^D(a_0'): \ (L_0-\ell)|\psi\rangle = 0\}$$

$$S_\ell^D(a_0') = \{|s\rangle \in V_\ell^D(a_0'): \ |s\rangle = L_{-1}|\psi_1\rangle + L_{-2}|\psi_2\rangle$$

$$\forall |\psi_1\rangle \in V_{\ell-1}(a_0'), \ |\psi_2) \in V_{\ell-2}(a_0')\}$$

$$K_\ell^D(a_0') = \{|\phi\rangle \in V_\ell^D(a_0'): \ k\cdot a_n|\phi\rangle = 0 \ \forall n > 0\}$$

It is not hard to establish the following facts for $D \geqslant 2$ (see ref. 2):

(a) Any $|\psi\rangle \in V_\ell^D(a_0')$ has a decomposition

$$|\psi\rangle = |s\rangle + |\phi\rangle \text{ with } |s\rangle \in S_\ell^D(a_0'), |\phi\rangle \in K_\ell^D(a_0')$$

and states in S_ℓ are linearly independent of those in K_ℓ i.e.

$$V_\ell^D(a_0') = S_\ell^D(a_0') \oplus K_\ell^D(a_0').$$

(b) If $|\phi\rangle \in K_\ell^D$, $L_n|\phi\rangle \in K_{\ell-n}^d$ for $n > 0$.

(c) If $|\phi\rangle \in K_\ell^D$, $\langle\phi|\phi\rangle \geqslant 0$.

__Lemma__: If $D = 26$, L_1 map S_1^{26} into S_0^{26} and $\tilde{L}_2 \equiv L_2 + \frac{3}{2}L_1^2$ maps S_1^{26} into S_{-1}^{26}.

__Proof__. If $|s\rangle \in S_1^{26}$, then by definition

$$|s\rangle = L_{-1}|\psi_1\rangle + L_{-2}|\psi_2\rangle \quad \text{with} \quad |\psi_1\rangle \in V_0^{26}, |\psi_2\rangle \in V_{-1}^{26}.$$

By acting on $|s\rangle$ in turn with L_1 and $\tilde{L}_2$, and making use of the conformal algebra of Eq. (2) one finds

$$L_1|s\rangle = |s'\rangle + 2L_0|\psi_1\rangle$$

and

$$\tilde{L}_2|s\rangle = |s''\rangle + 3(L_0 + 1)|\psi_2\rangle + 3(L_0 + 1)L_1|\psi_1\rangle$$

with $|s'\rangle \epsilon S_0^{26}$, $|s''\rangle \epsilon S_{-1}^{26}$. The Lemma follows by noting that $L_0|\psi_1\rangle = (L_0 + 1)|\psi_2\rangle = (L_0 + 1)L_1|\psi_1\rangle = 0$.

<u>No-ghost</u> <u>Theorem</u>. If $D = 26$, $L_n|\psi\rangle = (L_0 - 1)|\psi\rangle = 0 \;\forall\; n>0$ implies $\langle\psi|\psi\rangle \geqslant 0$.

<u>Proof</u>: By (a), write $|\psi\rangle = |s\rangle + |\phi\rangle$ with $|s\rangle \epsilon S_1^{26}$, $|\phi\rangle \epsilon K_1^{26}$. Then $L_n|\psi\rangle = 0 \Longrightarrow L_1|s\rangle = L_1|\phi\rangle = 0$ and $\tilde{L}_2|s\rangle = \tilde{L}_2|\phi\rangle = 0$ where we have used (a), (b) and the Lemma. Since L_n for $n > 0$ is an element of the Lie Algebra generated by L_1 and L_2 it follows that

$$L_n|s\rangle = L_n|\phi\rangle = 0 \qquad \forall\; n > 0.$$

Thus $\langle s|s\rangle = \langle s|\phi\rangle = 0$ and $\langle\psi|\psi\rangle = \langle\phi|\phi\rangle \geqslant 0$ by (c).

<u>Corollary</u>: The no-ghost theorem applies for all integers D satisfying $2 \leqslant D \leqslant 26$.

<u>Proof</u>: $v_\ell^D(a_0')$ for dimension D is a subspace of $V_\ell^{26}(a_0')$. Furthermore L_n for $D < 26$, $n \geqslant 0$ differs from L_n for $D = 26$, $n \geqslant 0$ by an operator which is zero on $V_\ell^D(a_0')$, provided the 26th component of a_0' is zero. Thus a physical state in $D < 26$ dimensions is also a physical state in 26 dimensions.

Department of Physics
University of Florida
Gainseville, Florida 32611

CONFORMAL INVARIANCE, UNITARITY AND TWO DIMENSIONAL CRITICAL EXPONENTS†

Daniel Friedan, Zongan Qiu, and Stephen Shenker

ABSTRACT

We show that conformal invariance and unitarity severely limit the possible values of critical exponents in two dimensional systems by finding the discrete series of unitarisable representations of the Virasoro algebra. The realization of conformal symmetry in a given system is parametrized by a real number c, the coefficient of the trace anomaly. For $c<1$ the only values allowed by unitarity are $c=1-6/m(m+1)$, $m=2,3,4\cdots$. For each of these values of c unitarity determines a finite set of rational numbers that must contain all possible critical exponents. These finite sets account for the known critical exponents of the following two dimensional models: Ising(m=3), tricritical Ising(m=4), 3-state Potts(m=5), and tricritical 3-state Potts(m=6).

1. INTRODUCTION

One of the most intriguing features of statistical mechanical systems and of their analogs, euclidean field theories, is the existence of special critical points where the systems are scale invariant. Correlation functions of the fluctuating fields mirror this lack of scale by transforming very simply under dilations of space. The two point function $\langle\phi(\vec{r})\phi(0)\rangle$, for example, is proportional to r^{-2x}. The number x is called the scaling dimension of ϕ (the anomalous dimension in field theory). The scaling dimensions of the fields determine the critical exponents of the system.

† This work was supported in part by the U.S. Department of Energy contract DE-Ac02-81ER-10957, National Science Foundation grant NSF-DMR-82-16892 and the Alfred P. Sloan Foundation.

Vertex Operators in Mathematics and Physics - Proceedings of a Conference November 10-17, 1983. Publications of the Mathematical Sciences Research Institute #3, Springer-Verlag, 1984.

The renormalization group has provided a satisfying conceptual framework for understanding the occurrence of scale invariant points. They are fixed points of the system under renormalization group transformations. The scaling dimensions describe the linearized behavior of the renormalization group near the fixed point. Although this viewpoint is elegant and compelling, it provides no general procedure for classifying or constraining possible fixed points. Calculations are done within the context of specific families of models.

Many known examples of fixed points display a richer symmetry than simple scale invariance. They are invariant under local rescalings: transformations of space that preserve angles but change lengths differently at different points. These are the conformal transformations. The question arises whether conformal invariance can be used to constrain or construct possible fixed point theories. This possibility was suggested by Polyakov[1].

Two dimensions is an especially promising place to apply notions of conformal invariance, because there the group of conformal transformations is infinite dimensional. Any analytic function mapping the complex plane to itself is conformal. Belavin, Polyakov and Zamolodchikov (BPZ)[2] have shown how the rich structure of the conformal group in two dimensions can be used to analyze conformally invariant field theories.

Many two dimensional statistical mechanical systems can also be interpreted as 1+1 dimensional quantum field theories. The distinguishing feature of the quantum theories is unitarity, equivalent to the property of reflection positivity in the statistical systems. We shall see that the quantum mechanical condition of unitarity, in the presence of the large conformal transformation group, puts a powerful constraint on the allowed physical systems.

2. CONFORMAL INVARIANCE IN TWO DIMENSIONS

Conformal invariance is simplest to describe in complex coordinates $z=x+iy$, $\bar{z}=x-iy$. The infinitesimal conformal transformations are $z \longrightarrow z+v(z)$, $\bar{z} \longrightarrow \bar{z}+\bar{v}(\bar{z})$. A simple basis is $v(z)=\epsilon z^{n+1}$, $\bar{v}(\bar{z})=\bar{\epsilon}\bar{z}^{n+1}$. When $n=-1,0,1$ these generate the group of fractional linear transformations $z \longrightarrow w(z)=(az+b)/(cz+d)$, which are the only globally defined conformal transformations on the Riemann sphere. This group is abbreviated SL_2.

Conformally invariant systems are described by correlation functions of a collection of conformal fields. Conformal fields are tensors of the form $\phi(z,\bar{z})dz^h d\bar{z}^{\bar{h}}$, which transform infinitesimally as

$$(1) \qquad \phi \longrightarrow \phi + \left[v(z)\frac{\partial}{\partial z} + hv'(z) + \bar{v}(\bar{z})\frac{\partial}{\partial \bar{z}} + \bar{h}\bar{v}'(\bar{z})\right]\phi.$$

The correlation functions are invariant under the SL_2 transformations. The two point function, for example, obeys

$$(2) \qquad <\phi(z_1)\phi(z_2)> = w'(z_1)^h \bar{w}'(\bar{z}_1)^{\bar{h}} w'(z_2)^h \bar{w}'(\bar{z}_2)^{\bar{h}} <\phi(w(z_1))\phi(w(z_2))>.$$

From the invariance under Euclidean and scale transformations we get $<\phi(re^{i\theta})\phi(0)>=r^{-2(h+\bar{h})}e^{-2i\theta(h-\bar{h})}$, so $h+\bar{h}$ is x, the scaling dimension of ϕ, and $h-\bar{h}$ is its spin.

The consequences of conformal variations with n not equal -1,0,1 are summarized by saying that there is a field $T_{\mu\nu}(x,y)$, called the stress energy tensor, which generates local scale transformations. Translation invariance and global scale invariance require $T_{\mu\nu}$ to be conserved and traceless; in complex notation we can write $T_{\mu\nu}dx^\mu dx^\nu = T(z)dz^2 + \bar{T}(\bar{z})d\bar{z}^2$. The requirement that T generate local scale transformations is expressed in the operator product of T with conformal fields:

$$(3) \qquad T(w)\phi(z,\bar{z}) \sim h(w-z)^{-2}\phi(z,\bar{z}) + (w-z)^{-1}\partial_z\phi(z,\bar{z})$$

and similarly for $\bar{T}$. The action of conformal transformations on correlation functions can then be expressed as

$$(4) \qquad \frac{1}{2\pi i}\oint_C dw\ v(w)T(w)\ \phi(z,\bar{z}) = \left[v(z)\frac{\partial}{\partial z} + hv'(z)\right]\phi(z,\bar{z})$$

where the contour surrounds z. A similar equation holds for $\bar{T}$. The splitting of the z and $\bar{z}$ dependence is a noteworthy simplification.

The action of conformal transformations on T itself is given by the operator product:

$$(5) \qquad T(w)T(z) \sim \tfrac{1}{2}c(w-z)^{-4} + 2(w-z)^{-2}T(w) + (w-z)^{-1}T'(w)$$

The form of (5) is forced by equation (3) and SL_2 invariance. $\bar{T}$ satisfies the same relation, assuming the theory is $z \leftrightarrow \bar{z}$ invariant.

The number c measures the deviation of T from transforming as a true conformal field for $n \neq -1,0,1$. It also describes the lack of invariance of the ground state to curving the underlying space - an effect referred to as the trace anomaly[3].

3. HILBERT SPACE INTERPRETATION

There exist general conditions on a system that allow it to be reconstructed as a theory of operator fields acting on a space of states with a hermitian inner product. The correlation functions are interpreted as vacuum expectation values of time-ordered products of operator fields. Some of the most important models have the property that the inner product on the space of states is positive definite, i.e. the state space is a Hilbert space. This unitarity property allows such models to be interpreted as quantum field theories. Models where a hermitian transfer matrix construction can be made provide concrete examples.

The positivity of the metric in the Hilbert space of states is equivalent to the property of reflection positivity in the correlation functions. Reflection positivity is imposed by singling out a hypersurface and requiring that correlations of products of fields on one side of the hypersurface with their reflections on the other side be nonnegative. The hypersurface is typically interpreted as constant in time. In two dimensions the hypersurface would be a line and the reflection would be $z \leftrightarrow \bar{z}$.

The infinitesimal conformal transformations $\delta z = \epsilon z^{n+1}$ are singular for $n \neq -1,0,1$ either at the origin or at infinity, so it is useful to define operators on an alternate hypersurface, the unit circle, where z^{n+1} is always well-behaved. The reflection operation is

$z\leftrightarrow 1/\bar{z}$. Because correlation functions are SL_2 invariant, and because the unit circle can be mapped by an SL_2 transformation to a line, reflection positivity through a line is equivalent to reflection positivity through the unit circle.

We can express correlation functions for quantum field theories as operator expectation values in this nonstandard Hilbert space. Dilation takes the place of time translation and radial ordering takes the place of time ordering. Alternatively, using coordinates $z=e^{\tau+i\theta}$, this is quantum field theory in Euclidean "time" τ and a periodic one dimensional space θ.

We define the operators that implement conformal transformations by

(6) $$T(z)=\sum_{n=-\infty}^{\infty} z^{-n-2}L_n \quad \bar{T}(\bar{z})=\sum_{n=-\infty}^{\infty} \bar{z}^{-n-2}\bar{L}_n.$$

It follows from (3), (5) and hermiticity of the inner product that the stress energy tensor is self-adjoint, i.e. $[T(1/\bar{z})d(1/\bar{z})^2]^{+}=T(z)dz^2$, and the equivalent for $\bar{T}$. This translates to

(7) $$L_n^{+}=L_{-n} \quad \bar{L}_n^{+}=\bar{L}_{-n}.$$

Formula (3) rewritten in operator language reads

(8) $$[L_n,\phi] = z^{n+1}\frac{\partial}{\partial z}\phi + h(n+1)z^n\Phi.$$

Dilations are generated by $L_0+\bar{L}_0$, rotations by $L_0-\bar{L}_0$, and translations by L_{-1} and $\bar{L}_{-1}$. The SL_2 invariance of the correlation functions implies SL_2 invariance of the vacuum: $L_n|0\rangle=0=\bar{L}_n|0\rangle$, for n=-1,0,1.

Equation (5) is equivalent to the commutation relations

(9) $$[L_m,L_n] = (m-n)L_{m+n} + \frac{c}{12}m(m^2-1)\delta_{m,-n}.$$

The operators $\bar{L}_n$ satisfy the same algebra and commute with all the L's. The algebra (9) is called the Virasoro algebra[4]. It was first encountered in string theory where conformal invariance is a residue of full reparametrization invariance in a certain (conformal) gauge[5].

The central term with coefficient c in (9) describes the particular realization of conformal symmetry in the theory. We see that the product of two Virasoro algebras each with central charge c acts on the space of states of the conformal field theory. Reflection positivity tells us that this is a unitary representation.

If a field ϕ satisfies the transformation law (8), at least for n=0, then $L_0\phi(0)|0\rangle=h\phi(0)|0\rangle$. Thus the problem of finding critical exponents is reduced to understanding the allowed eigenvalues of L_0. To constrain these values we must remember that the states of the Hilbert space are not only eigenstates of L_0; they also form a unitary representation of the Virasoro algebra.

4. REPRESENTATIONS OF THE VIRASORO ALGEBRA

We focus on representations of one Virasoro algebra since representations of the product will just be tensor products of representations of the factors. First, the L_n for n>0 are lowering operators for L_0, i.e. $L_0L_n=L_n(L_0-n)$. The vacuum must have the lowest eigenvalue of L_0, so it is annihilated by all the L_n for n>0, in addition to the SL_2 generators n=-1,0,1. Each conformal operator ϕ can be associated with a state $|h\rangle=\phi(0)|0\rangle$. By (8), this state satisfies $L_0|h\rangle=h|h\rangle$ and $L_n|h\rangle=0$ for n>0. A state such as $|h\rangle$ which is annihilated by all the lowering operators is called a highest weight vector. There is a one to one correspondence between the highest weight vectors and the conformal fields of the theory.

Once we have a highest weight vector we can build a representation of the Virasoro algebra by applying the L_{-n}, $n\geq 1$. These states can be classified by L_0 eigenvalue. A state is in the nth level if its L_0 value is h+n. A basis of states at the nth level is given by

$$(10) \qquad \{L_{-k_1}L_{-k_2}\cdots L_{-k_m}|h\rangle: \Sigma k_i=n,\ k_1\geq k_2\geq\cdots\geq k_m>0\}.$$

There are P(n) such states, where the classical partition function P(n) is the number of ways of writing n as a sum of positive integers. The tower of such levels is called a highest weight representation of the algebra.

In terms of field theory, these higher level states correspond to operators of higher scaling dimension, obtained by applying products of stress-energy tensors to some conformal field. We should think of each conformal field as carrying such a conformal family along with it. The organization of all the fields of the theory into conformal families, each associated with a conformal field, can be accomplished by making repeated operator products of T(z) with an arbitrary field. The operator coefficient of the most singular term will eventually obey the defining relation (3) of a conformal field.

The inner product of any two states in the span of basis (10) can be computed from the hermiticity condition (7) and the commutation relations (9). The unitarity constraint is that the matrix of inner products should have no negative eigenvalues. We can impose the positivity constraint level by level because different levels have different L_0 eigenvalues and hence are orthogonal. A state $|\psi\rangle$ in the span of basis (10) with $\langle\psi|\psi\rangle$ negative is called a "ghost." If a ghost is found on any level the representation cannot occur in any unitary theory.

At level 1 there is a single state, $|1\rangle=L_{-1}|h\rangle$, and $\langle 1|1\rangle=2h$. Therefore positivity at level 1 rules out all $h<0$. At level n the state $|n\rangle=L_{-n}|h\rangle$ has $\langle n|n\rangle=2nh+cn(n^2-1)/12$. If $c<0$ this is a ghost state for large n. So we can limit our attention to the region $c\geq 0$, $h\geq 0$.

5. THE KAC DETERMINANT AND UNITARITY

To proceed we employ a version of Kac's formula for the determinant of the nth level matrix of inner products[6]:

$$\det M_{(n)}(c,h) = \prod_{k=1}^{n} \psi_k(c,h)^{P(n-k)} \tag{11}$$

where

$$\psi_k(c,h) = \prod_{pq=k} (h-h_{p,q}(c)) \tag{12}$$

where p,q range over the positive integers and

$$h_{p,q}(c) = \frac{[(m+1)p-mq]^2-1}{4m(m+1)} \quad (13)$$

The parameter m is given by

$$c = 1 - \frac{6}{m(m+1)}. \quad (14)$$

An overall positive constant has been suppressed in formula (11). The determinant has no zeroes in the region $1<c$, $0<h$, and the matrix of inner products for a given level is positive definite for h large enough, so there are no ghosts in the region $1\leq c$, $0\leq h$. Moreover a point is ghost free up to level n if it can be continuously connnected to $1<c$, $0<h$, without crossing a vanishing curve $h=h_{p,q}(c)$, $pq\leq n$. The region $-\infty<c<1$ corresponds to $0<m<\infty$.

The determinant is negative in some regions and positive in others. We can immediately eliminate all negative regions because they necessarily contain an odd number of ghosts. A straightforward examination of the $h_{p,q}$ curves (see the figures) shows that any point in the c,h plane for $c<1$ will have a negative determinant at some level unless that point is actually on one or more of the curves $h_{p,q}$. Then its determinant will be zero on all levels $n\geq pq$.

We already see that unitarity has limited the possible scaling dimensions for a given c (or m) to the discretely infinite set of numbers $h_{p,q}(c)$.

We mention that if the determinant vanishes at a point c,h at level n then there is a null state at that level, i.e. a state in the span of (10) orthogonal to every state including itself. Such states give rise to linear differential equations on the correlation functions[2]. The above argument tells us that correlation functions of all unitary models for $c<1$ satisfy such differential equations.

In fact the unitary theories are far more limited. We find that there are ghosts everywhere on the vanishing curves except, possibly, at a certain set of isolated points. We give a sketch of the argument here; details can be found in reference 7.

First let us examine what happens on the lowest levels. The vanishing curves for level 2 are shown in figure 2. We have already explained that region I and its border are ghost free through the second level because they are connected to $c>1$, $h>0$. The level 2 determinant is negative in region II, so there is a ghost there.

Now look at level 3 (figure 3). Region I and its border are ghost free. The level 3 determinant is negative in regions III and II_b. Region II_b was already eliminated on level 2 so the new information is that region III has a ghost. The question is whether there is a ghost on the vanishing curve $h_{2,1}$ on the border between regions II_a and III.

To settle the question we examine the pattern of subrepresentations at the intersection point A ($m=3, h=1/2$). There is a null state $|2,1\rangle$ at level 2 because $h=h_{2,1}$. There are no null states at lower levels, i.e. levels 0 and 1, so $|2,1\rangle$ is a highest weight vector, i.e. is annihilated by all the lowering operators L_n for $n>0$. It weight is given by $L_0|2,1\rangle=(h+2)|2,1\rangle$. The raising operators acting on $|2,1\rangle$ generate a subrepresentation consisting entirely of null states. In particular, $L_{-1}|2,1\rangle$ is a null state on level 3. There is also a null state $|3,1\rangle$ on level 3 because $h=h_{3,1}$ at A. The question is whether $|3,1\rangle$ is in the subrepresentation generated from $|2,1\rangle$. If it were, then Kac's determinant formula for the subrepresentation $\det M_{(3-2)}(c,h+2)=2(h+2)$ would have to vanish. Since it does not vanish, the ghost in region III is distinct from the states in the subrepresentation and eliminates all of the vanishing curve $h_{2,1}$ to the left of A. The curve $h_{3,1}$ is an example of what we will call a "cutting" curve.

Now we sketch the general argument. We consider one curve $h_{p,q}$ at a time, in order of increasing $n=pq$. We will describe how successive segments of $h_{p,q}$ are eliminated by cutting curves. The curve $h_{p,q}$ first appears as a vanishing curve at level n and remains as a vanishing curve on all levels greater than n. We define the first intersection on $h_{p,q}$ at level $n'\geqslant n$ to be the intersection closest to $c=1$ of $h_{p,q}$ with another vanishing curve. We will show that there are ghosts at all points of $h_{p,q}$ except, possibly, the first intersections. When $h_{p,q}$ appears at level n it has a first intersection (except for the case p=n, q=1). Along $h_{p,q}$ as c decreases from this first intersection the curve enters the region already known by the determinant argument to contain ghosts on previous levels, except for points where it intersects other vanishing curves from previous levels. By assumption we have already considered the points on those curves. So we need consider only the interval on $h_{p,q}$ from c=1 to the first intersection (or its whole length if q=1).

We will show that only the first intersections can be ghost free by arguing that, when a new vanishing curve $h_{p',q'}$ appears at level n'=p'q' and makes a new first intersection by crossing $h_{p,q}$ between c=1 and what was the first intersection on level n'-1, the interior of the interval between the two first intersections acquires a ghost. The curve $h_{p',q'}$ thus acts as a cutting curve. Figures 4-12 show the vanishing curves on levels 4-12. Open dots mark the intervals eliminated by cutting curves.

The cutting pattern is clearest if we organize the $h_{p,q}$ curves into clusters indexed by k=p-q. An elementary argument shows that the first intersections are exactly the intersections between curves in cluster k and curves in cluster -1-k (taking the intersection closest to c=1 if there are two). These are the points

$$c = 1 - \frac{6}{m(m+1)} \qquad m=2,3,4,\cdots$$

(15)

$$h = h_{p,q}(c) = \frac{[(m+1)p-mq]^2 - 1}{4m(m+1)}$$

for p=1,2,$\cdots$,m-1 and q=1,2,$\cdots$,p.

We now generalize the argument we used for level 3. Consider the new first intersection $h=h_{p,q}=h_{p',q'}$. At level n, since $h=h_{p,q}$ and there are no vanishings at lower levels, there is a null state $|p,q\rangle$ which is a highest weight vector. It generates a highest weight representation consisting of null states. At level n' there is a null state $|p',q'\rangle$ because $h=h_{p',q'}$. It is simple to check that $\det M_{(n'-n)}(c,h+n)\neq 0$ at all the first intersections (15). It follows that

$|p',q'\rangle$ lies outside the subrepresentation generated from $|p,q\rangle$. Therefore there is a ghost on $h_{p,q}$ on one side or the other of $h_{p',q'}$. It remains only to show that the ghost is on the side away from c=1. We do this by following the successive first intersections the cutting curve $h_{p',q'}$ makes with curves in cluster $k=-1+q'-p'$ as it leaves the c=1 axis. At the first one the ghost is on the side away from c=1 because the other side can be connected to c>1. The ghost continues to exist along the cutting curve on the side farther away from c=1 at each successive first intersection the cutting curve makes in cluster k because there are no intervening intersections with curves not in cluster k. This serves to eliminate all the intervals on curves $h_{p,q}$ between first intersections, and leaves only the first intersections as possibly ghost free.

We have proved that all unitary representations are contained in the list (15) of first intersections, but not that all representations on the list are in fact unitary. We have verified numerically that all first intersections are ghost free through level 12. We have a heuristic argument that they remain ghost-free to all levels. Assume that the matrix of inner products can be diagonalized analytically throughout the whole region of interest (h>0,c>0 say), so that the norm squared of each state is a product of factors of the form $(h-h_{p,q})(h-h_{q,p})$, $p\neq q$ or $(h-h_{p,p})$. The pattern of subrepresentations[6,8] at c=1 implies that states whose norms vanish on the curve $h_{p,q}$ will also have zero norm on all the curves $h_{p-k,q-k}$, $k\geq 1$. At the first intersections m is an integer. Then the symmetry $h_{p,q}=h_{p+m,q+m+1}$ ensures that whenever a factor $(h-h_{p,q})(h-h_{q,p})$ or $(h-h_{p,p})$ is negative, another factor $(h-h_{p-k,q-k})(h-h_{q-k,p-k})$ or $(h-h_{p-k,p-k})$ is zero. Thus the absence of ghosts at the first intersections would be proved if we could verify the diagonalization assumption.

We have carried out the analytic diagonalization explicitly for levels 1-5. It might be possible to study the barrier to such a diagonalization from inconsistencies in subrepresentation patterns around polygons of vanishing curves in the c,h plane. In principle we can check for these. There are no such barriers in the region of interest for levels 1-12. Alternatively we may be able to use techniques developed by Feigin and Fuks[9], Zamolodchikov[10] and

Kadanoff and Nienhuis[11] to exhibit analytic deformations of correlation functions away from c=1.

6. CONCLUDING REMARKS

Unitarity has restricted the possible values of scaling dimensions to the simple list of rational numbers (15) when $c<1$. We display the allowed h values for m=3,4,5,6 in the right half of table I. Remember that the scaling dimension of a field is $x=h+\bar{h}$ and its spin is $h-\bar{h}$. We find[12] the Ising model is described by m=3 representation, tricritical Ising by m=4, 3-state Potts by m=5, and tricritical 3-state Potts by m=6. The known scaling indices for these models and the $h,\bar{h}$ candidates are displayed in the left half of table I. All of the known exponents are accounted for. Note that systems with continuously variable critical exponents like the Baxter, Ashkin-Teller and gaussian models live at c=1 where unitarity allows all $h \geqslant 0$.

We have used unitarity to provide strong constraints on possible reperesentations of conformal invariance, but there are additional requirements on a sensible theory: closure of the operator product expansion and crossing symmetry of correlation functions[1,2]. BPZ[2] have developed differential equation techniques which allow implementation of these conditions when the space of states is made up of representations with null states, i.e. $h=h_{p,q}(c)$. For the special values of c corresponding to m rational, $m=r/(s-r)$, $r<s$, $3r \geqslant 2s$, they have found[13] finite sets of conformal fields that must close under the operator product expansion. The scaling dimensions are given by $h=h_{p,q}(c)$, $1 \leqslant p<r$, $1 \leqslant q<s$. Note that the unitary representations (15) correspond to $s=r+1$. Dotsenko[14] has used the differential equation technique to find a closed operator algebra for the 3-state Potts model and to construct some of its correlation functions. Notice that in table I the 3-state Potts and tricritical 3-state Potts models do not exhaust all the possibly unitary representations. For 3-state Potts, the representations that are used make up the closed subalgebra found by Dotsenko. For the tricritical model the representations used form a closed subalgebra of the same type. It is an interesting question whether there exist larger models for m=5 and 6 using all possible

representations and containing the Potts models as sub-models.

Unitarity tells us that the only allowed representations for $c<1$ have null states, so the differential equation techniques apply. It should now be straightforward to sort out all possible unitary conformally invariant models with $c<1$. The problem is a finite one because only a finite number of representations are allowed at a given m. Such a systematic construction of scale invariant models would partially realize the bootstrap program initiated by Kadanoff[15] and Polyakov[16].

The new models with $m\geq 7$ are particularly tantalizing. Do they exist? What are they? What are their symmetries? If any of these models had a continuous internal symmetry generated by the line integral of a conserved local current then Euclidean invariance would require the two components of the current to have $h=1$, $\bar{h}=0$ and $h=0$, $\bar{h}=1$. But $h=1$ is never unitary, so there is no possibility of continuous internal symmetry.

There exists a supersymmetric extension of the Virasoro algebra called the Ramond-Neveu-Schwarz algebra[17]. Kac has written a determinant formula for this algebra as well[6]. The methods of section 5 apply; only first intersections can be ghost-free. The allowed representations, for $\frac{2}{3}c<1$, are

$$\frac{2}{3}c = 1 - \frac{8}{m(m+2)} \qquad m = 2,3,4,\cdots$$

(16)

$$h = h_{p,q}(c) = \frac{\left[\frac{1}{2}(p-q)m+p\right]^2-1}{2m(m+2)}$$

were p,q are integers, both even or both odd, $0<p<m$, $0<q\leq p$. Note that $h=1/2$ does not occur, so there are no internal supercurrents.

The representations $c=7/10$, $h=0$, $1/10$ of the super conformal algebra are composed of the representations $c=7/10$, $h=0$, $3/2$, $1/10$, $3/5$ of the Virasoro algebra. These are the representations which occur in the Z_2 invariant subalgebra of the tricritical Ising model, so it seems likely that a $c=7/10$ supersymmetric model forms a subsector of the tricritical Ising model[7]. Presumably the fermionic partners are combinations of order and disorder variables. If this identification of models is justified then physical systems described by the tricritical Ising model, such as helium adsorbed on Krypton-plated graphite[18], are supersymmetric.

Models made from representations with $m \geq 4$, i.e. $1 \leq c < 3/2$, would all be new, with the possible exception $m=4$, $c=1$. For $c=1$ the allowed representations are $h=0$, $1/16$, $1/6$ and 1. Models using $h=0$, $1/6$ and 1 or $h=0$, $1/16$ and 1 can tentatively be identified with special points in the gaussian model[7]. The $h=1$ representation would correspond to a marginal operator breaking the supersymmetry. The representations of the super algebra might also be of interest in superstring theory[19].

From a general point of view the striking feature of our result is the extreme rigidity of the conformal bootstrap. For $c<1$ it seems that unitarity is the crucial constraint. For $c \geq 1$ unitarity is only a mild requirement, but the possibility exists that crossing symmetry, the other element of the bootstrap, will play a crucial role in limiting the possible realizations of conformal invariance.

ACKNOWLEDGEMENTS

We thank L.P. Kadanoff and M. dens Nijs for a number of helpful conversations. We would like to express our appreciation to D. Arnett for providing access to the Chicago Astronomy and Astrophysics computer, and to M. Crawford, P. Schinder and G. Toomey for their generous help in using it.

D.F. is grateful to the U.S. National Academy of Science and to the Academy of Sciences of the U.S.S.R for their support of the exchange which led to this work, to the L.D. Landau Institute of Theoretical Physics for its hospitality, to A.A. Migdal for interesting conversations and to A.B. Zamolodchikov, A.M. Polyakov and Vl.S. Dotsenko for many patient explanations of their ideas and much discussion.

REFERENCES

[1] A.M. Polyakov, Pisma ZhETP 12, 538 (1970) [JETP Letters 12, 381 (1970)] and ZhETP 66, 23(1974) [JETP 39, 10(1974)].

[2] A.B. Zamolodchikov, A.M. Polyakov and Vl.S. Dotsenko, private communication. After our work was completed we received references 13 and 14.

[3] D. Friedan in Common Trends in Field Theory and Statistical Mechanics (Les Houches, Summer 1982), J.B. Zuber ed., to be published.

[4] M.A. Virasoro, Phys. Rev. D1, 2933 (170).

[5] See review articles in Dual Theory, M. Jacob ed. (North Holland, Amsterdam 1974).

[6] V.G. Kac, Proceedings of the International Congress of Mathematicians, Helsinki, 1978 and Lecture Notes in Physics 94, 441 (1979); B.L. Feigin and D.B. Fuchs, Functs. Anal. Prilozhen. 16, 47(1982) [Funct. Anal. and Appl. 16, 114(1982)].

[7] D. Friedan, Z. Qiu, S.H. Shenker, in preparation.

[8] G. Segal, Comm. Math. Phys. 80, 301(1981); I.B. Frenkel and V.G. Kac, Invent. Math. 62, 23(1980).

[9] B.L. Feigin and D.B. Fuchs, unpublished.

[10] A.B. Zamolodchikov, private communication.

[11] L.P. Kadanoff and B. Nienhuis, unpublished.

[12] L.P. Kadanoff has confirmed our assignments of c using four point funtions of energy operators in the q-state Potts model calculated by himself and B. Nienhuis. He finds $c=(3t-1)(3-4t)/t$ where $-\sqrt{q}=2\cos(\pi/2t)$.

[13] A.A. Belavin, A.M. Polyakov, and A.B. Zamolodchikov, STATPHYS 15, Edinburgh 1983 (J. Stat. Phys., to be published).

[14] Vl.S. Dotsenko, L.D. Landau Institute preprint, September 1983.

[15 L.P. Kadanoff, Phys. Rev. Letters 23, 1430 (1969).

[16] A.M. Polyakov, ZhETP 57, 271 (1969) [JETP 30, 151 (1970)].

[17] P. Ramond, Phys. Rev. D3, 2415 (1971); A. Neveu and J.H. Schwarz, Nucl. Phys. B31, 86(1971) and Phys. Rev. D4, 1109 (1971).

[18] M.J. Tejwani, O. Ferreira, and O.E. Vilches, Phys. Rev. Letters 44, 152(1980); W. Kinzel, M. Shick, and A.N. Berker in Ordering in Two Dimensions, S.K. Sinha ed. (Elsevier North Holland, Amsterdam 1980).

[19] M.B. Green and J.H. Schwarz, Phys. Lett. 109B, 444(1982).

[20] M. dens Nijs, University of Washington preprint, September 1983.

[21] D.A. Huse, J. Phys. A, to be published. We thank Daniel S. Fisher for bringing this reference to our attention.

Enrico Fermi and James Franck Institutes
and Department of Physics
University of Chicago, Chicago, IL 60637

TABLE I COMPARISON WITH KNOWN CRITICAL EXPONENTS[a]

model	field	x	h	$\bar{h}$	unitary highest weights $h_{p,q}(c)$[b]			
	σ	$\frac{1}{8}$	$\frac{1}{16}$	$\frac{1}{16}$	<u>m=3</u>			
Ising	ϵ	1	$\frac{1}{2}$	$\frac{1}{2}$				$\frac{1}{16}$
	ψ	$\frac{1}{2}$	$\frac{1}{2}$	0	0			$\frac{1}{2}$
	σ	$\frac{3}{40}$	$\frac{3}{80}$	$\frac{3}{80}$				
tricritical	ϵ	$\frac{1}{5}$	$\frac{1}{10}$	$\frac{1}{10}$	<u>m=4</u>			$\frac{1}{10}$
Ising		$\frac{7}{8}$	$\frac{7}{16}$	$\frac{7}{16}$		$\frac{3}{80}$		$\frac{3}{5}$
	t	$\frac{6}{5}$	$\frac{3}{5}$	$\frac{3}{5}$	0	$\frac{7}{16}$		$\frac{3}{2}$
		3	$\frac{3}{2}$	$\frac{3}{2}$				
	σ	$\frac{2}{15}$	$\frac{1}{15}$	$\frac{1}{15}$				
	ϵ	$\frac{4}{5}$	$\frac{2}{5}$	$\frac{2}{5}$	<u>m=5</u>			$\frac{1}{8}$
3-state		$\frac{4}{3}$	$\frac{2}{3}$	$\frac{2}{3}$			$\frac{1}{15}$	$\frac{2}{3}$
Potts	CH	$\frac{9}{5}$	$\frac{7}{5}$	$\frac{2}{5}$		$\frac{1}{40}$	$\frac{21}{40}$	$\frac{13}{8}$
	ϵ'	$\frac{14}{5}$	$\frac{7}{5}$	$\frac{7}{5}$	0	$\frac{2}{5}$	$\frac{7}{5}$	3
		3	3	0				

(continued)

TABLE I (continued)

model	field	x	h	$\bar{h}$	unitary highest weights $h_{p,q}(c)$				
	σ	$\frac{2}{21}$	$\frac{1}{21}$	$\frac{1}{21}$					
	ϵ	$\frac{2}{7}$	$\frac{1}{7}$	$\frac{1}{7}$	<u>m=6</u>				$\frac{1}{7}$
tricritical		$\frac{20}{21}$	$\frac{10}{21}$	$\frac{10}{21}$				$\frac{5}{56}$	$\frac{5}{7}$
3-state	t	$\frac{10}{7}$	$\frac{5}{7}$	$\frac{5}{7}$			$\frac{1}{21}$	$\frac{33}{56}$	$\frac{12}{7}$
Potts	CH	$\frac{17}{7}$	$\frac{12}{7}$	$\frac{5}{7}$		$\frac{1}{56}$	$\frac{10}{21}$	$\frac{85}{56}$	$\frac{22}{7}$
		$\frac{8}{3}$	$\frac{4}{3}$	$\frac{4}{3}$	0	$\frac{3}{8}$	$\frac{4}{3}$	$\frac{23}{8}$	5
		$\frac{23}{7}$	$\frac{22}{7}$	$\frac{1}{7}$					
		5	5	0					

[a] See reference 20 and references therein. For the operator of dimension 3 in the tricritical Ising model see reference 21. Note that reference 20 gives infinite sequences of possible irrelevant operators, not all of which are known to occur. We list only those which our results allow as conformal operators. The spin assignments, $h-\bar{h}$, are consistent with what is known. The CH operators can also be interpreted as derivatives of other operators.

[b] The weights $h_{p,q}$ are listed with p running horizontally from 1 to m−1 and q running vertically from 1 to p.

KEY TO THE FIGURES

The vanishing curves for level n are shown in figure n. These are the curves $h=h_{p,q}(c)$, $pq \leq n$. All of the curves in figure n are also present in figure n+1. Note that the vertical axis is $\sqrt{h}$ and that the scale changes from graph to graph.

The individual curves can be identified by their behavior for c near 1:

$$h_{p,q}(1-6\epsilon) \sim \frac{1}{4}(p-q)^2 + \frac{1}{4}(p-q)(p+q)\sqrt{\epsilon} \quad p \neq q$$

$$h_{p,p}(1-6\epsilon) \sim \frac{1}{4}(p^2-1)\epsilon.$$

The region connected to $c>1$, $h>0$ is ghost free through level n, as are the solid lines and the points marked with a circle and cross. The marked points, and also the ends of solid lines, are first intersections.

A region bounded by two solid lines and a line of open dots was ghost free on the previous level but has negative determinant and therefore a ghost on the present level.

A line of open dots represents a segment of a vanishing curve which was ghost free through the previous level but which acquires a ghost on the current level, by the cutting mechanism described in section 5.

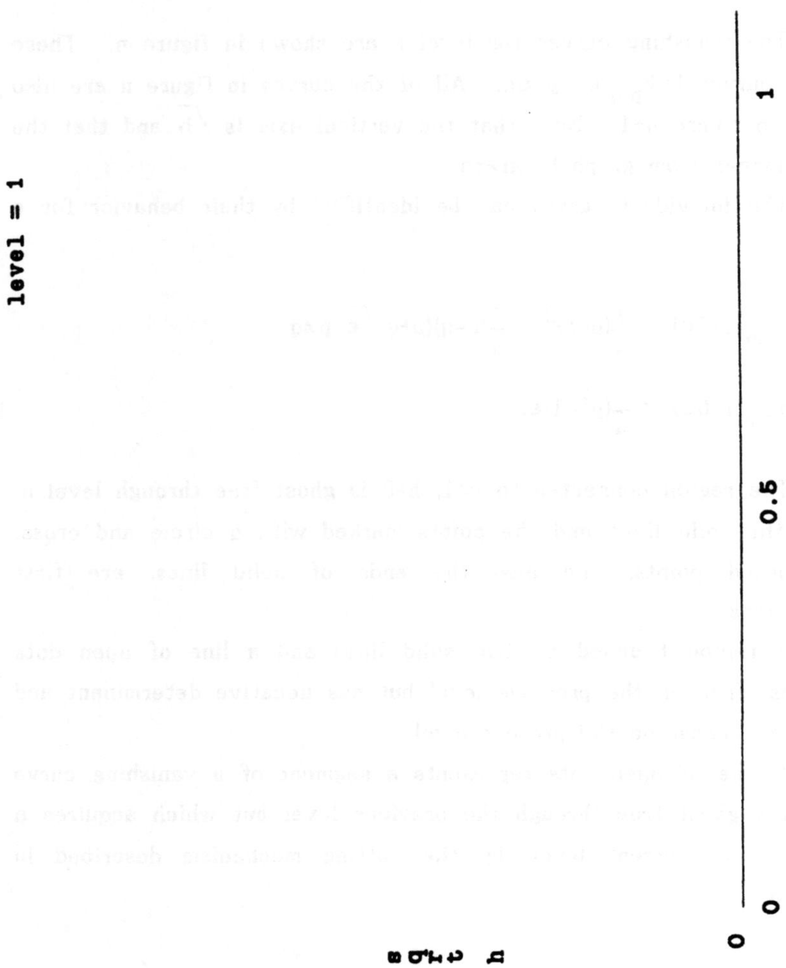

Figure 1

level = 2

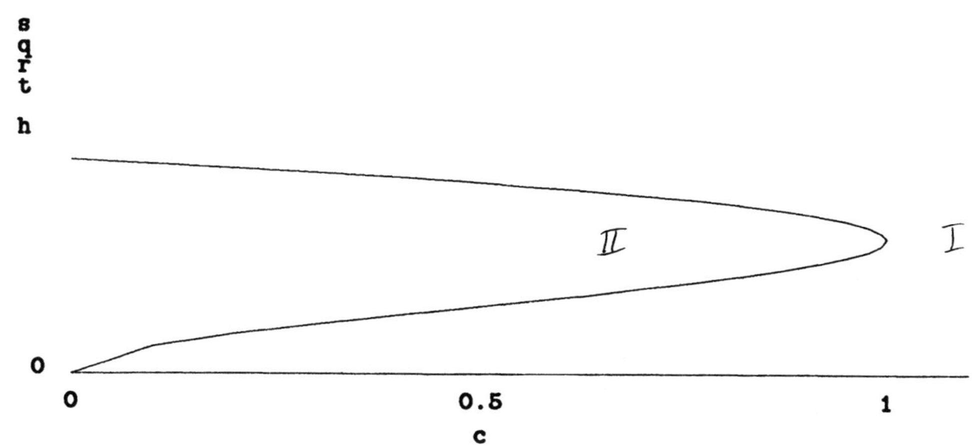

Figure 2

level = 3

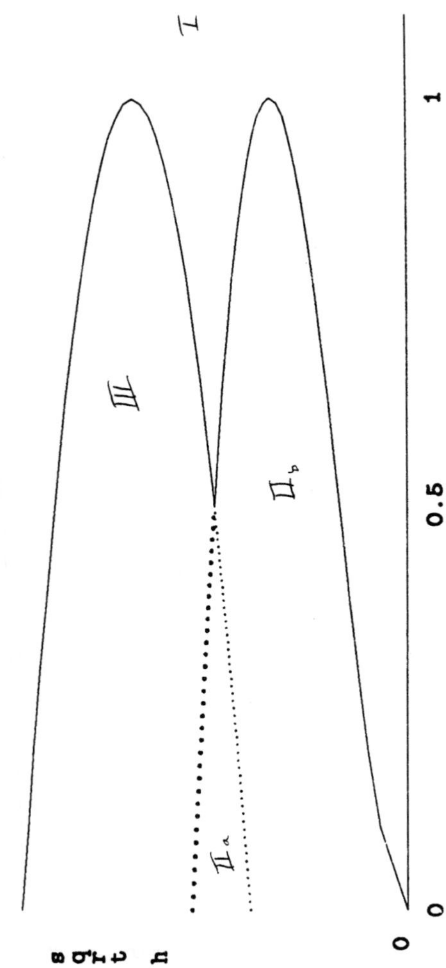

Figure 3

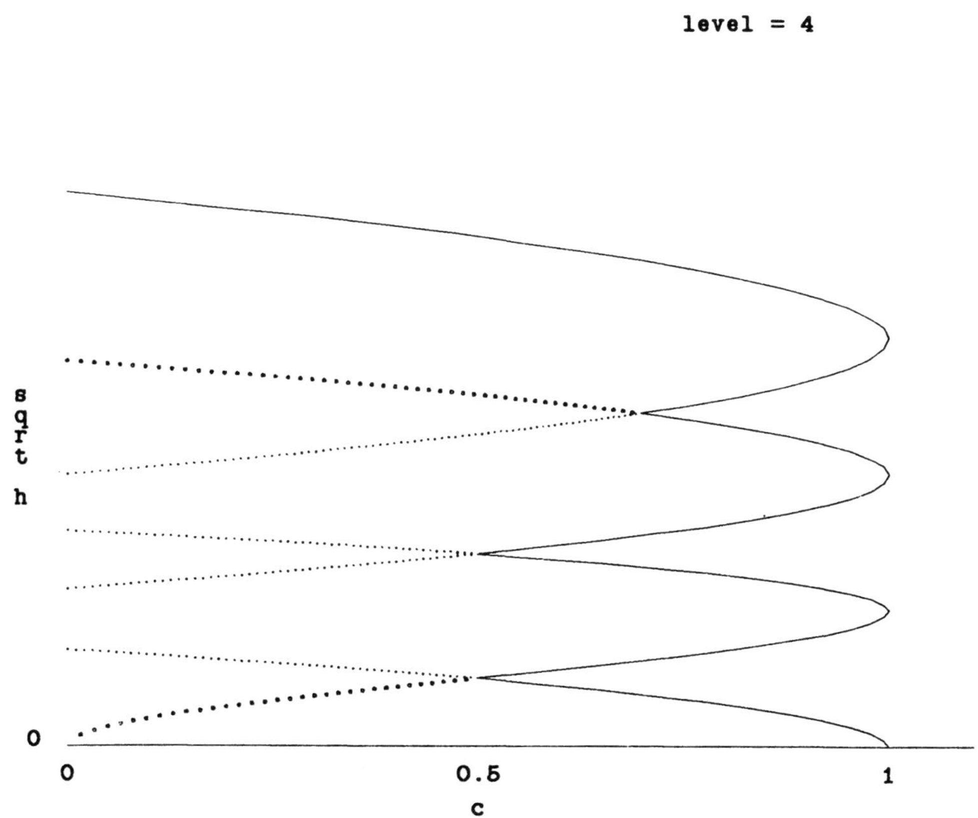

Figure 4

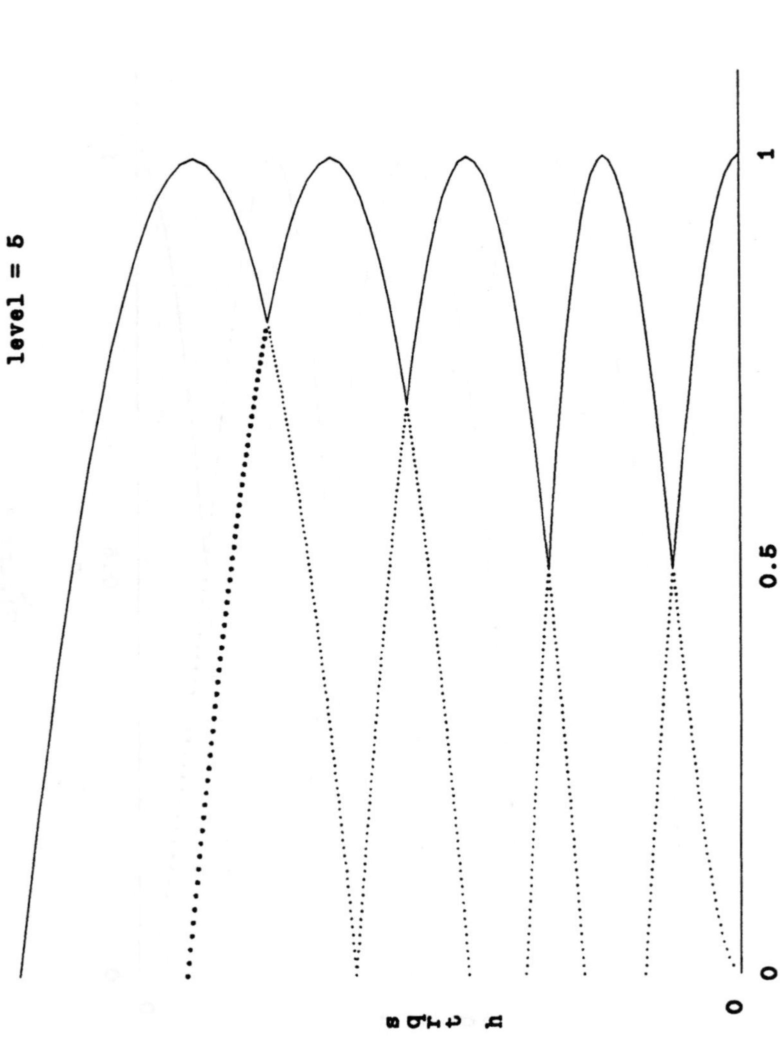

Figure 5

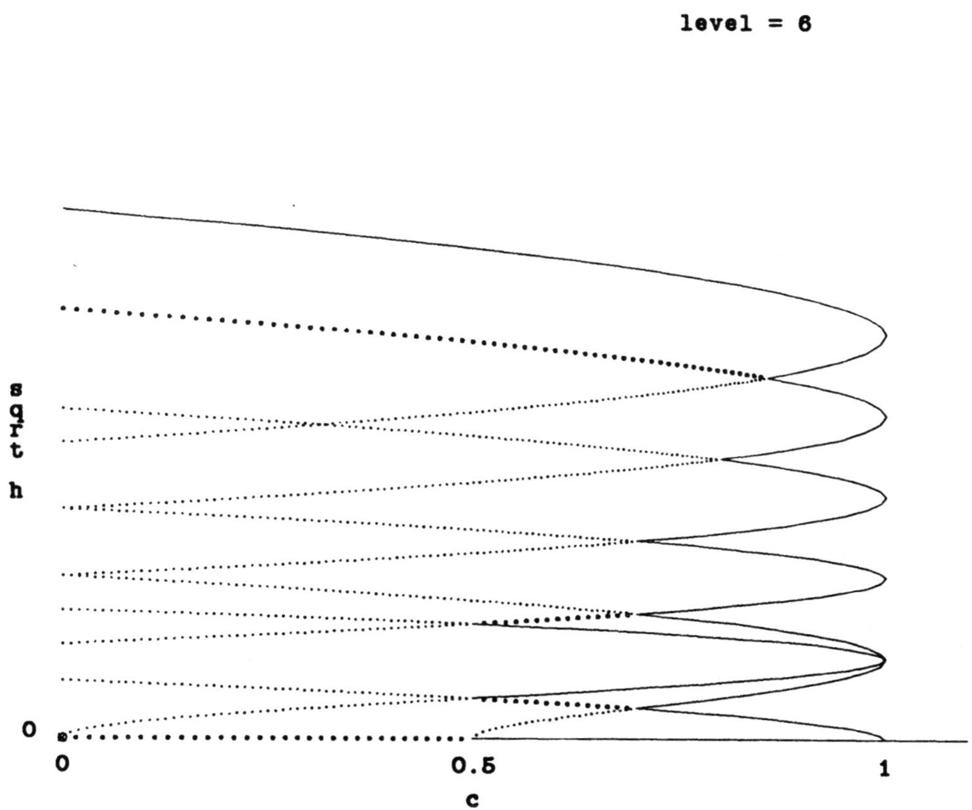

Figure 6

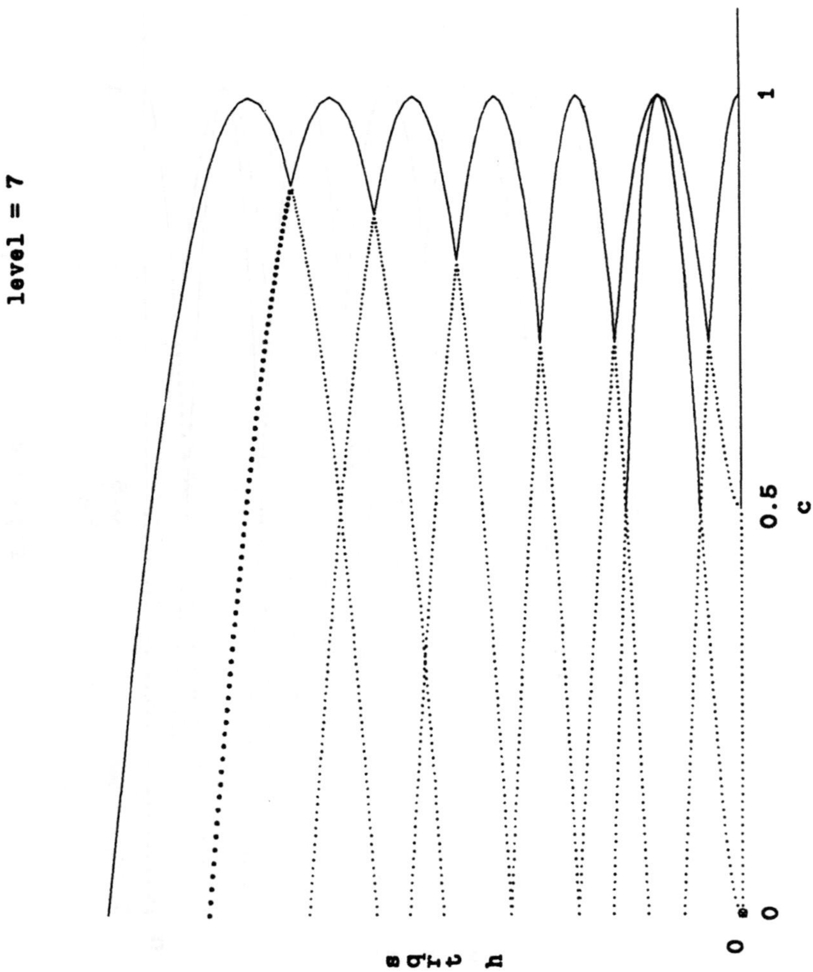

Figure 7

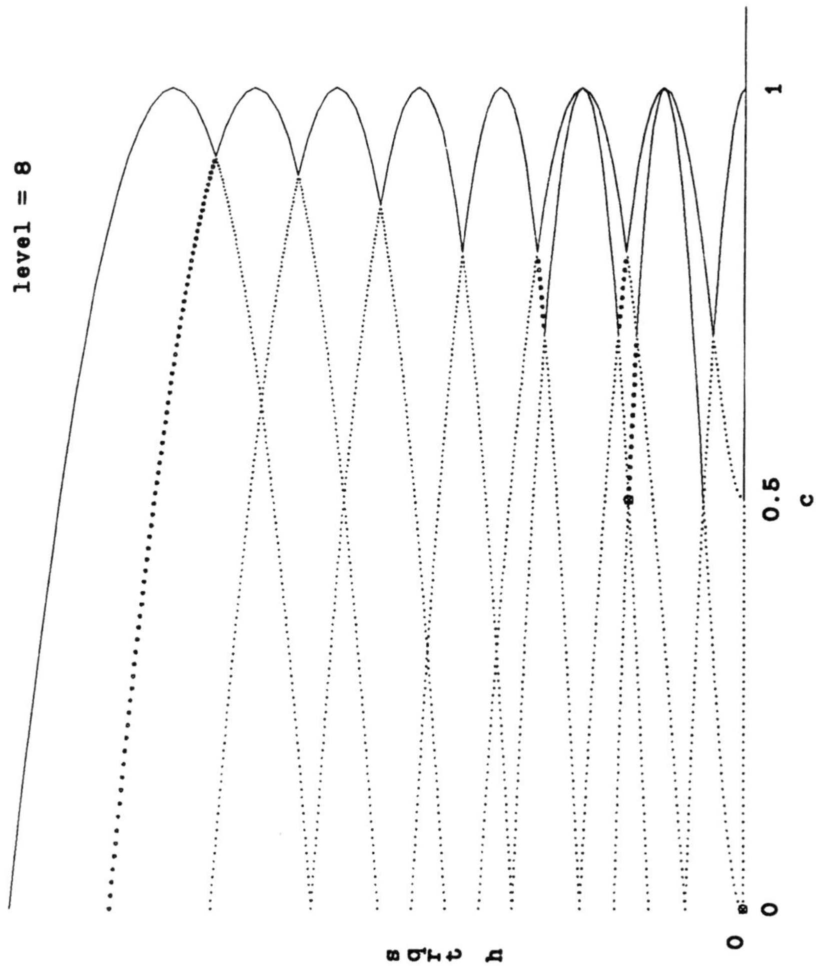

Figure 8

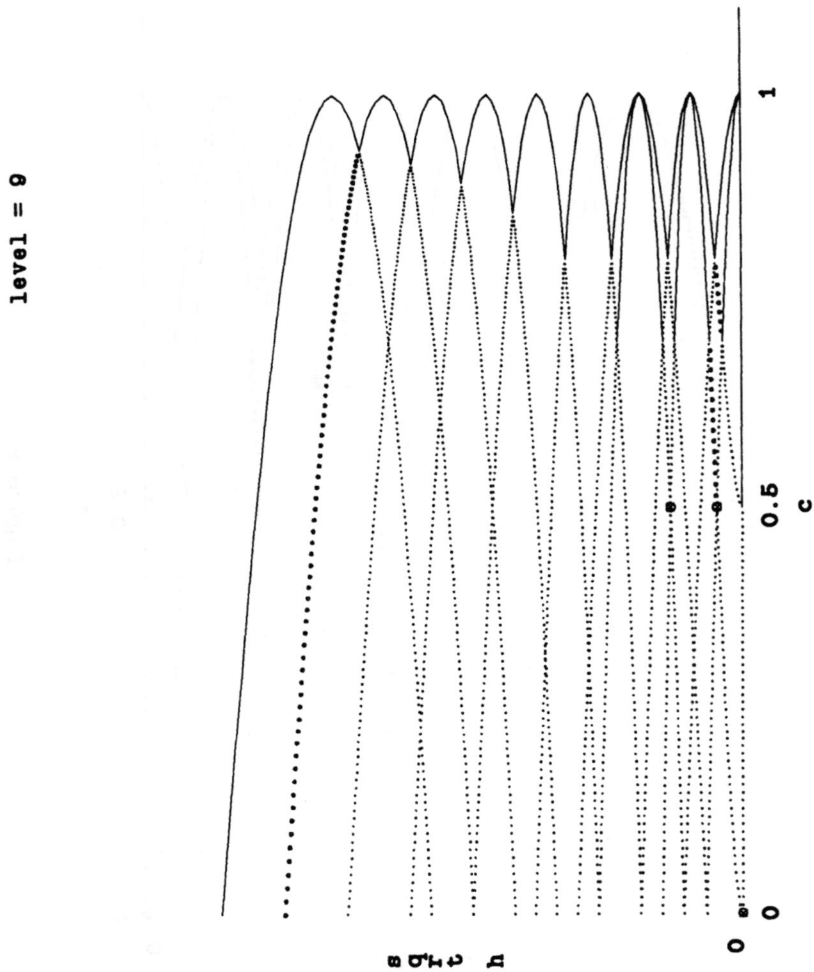

Figure 9

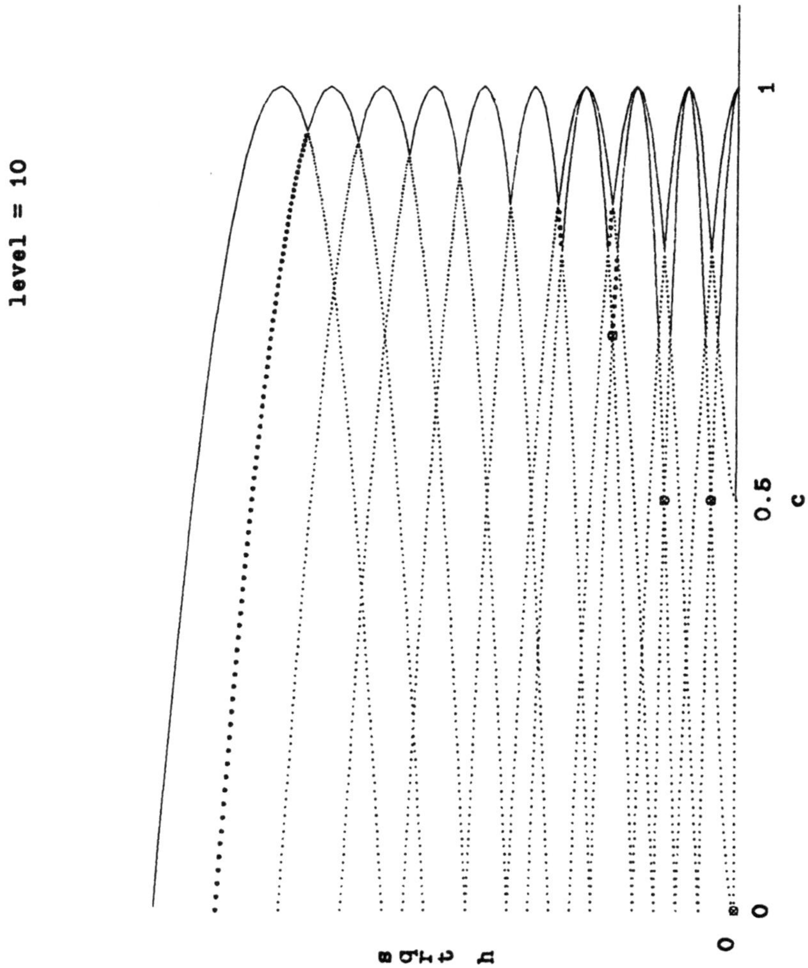

Figure 10

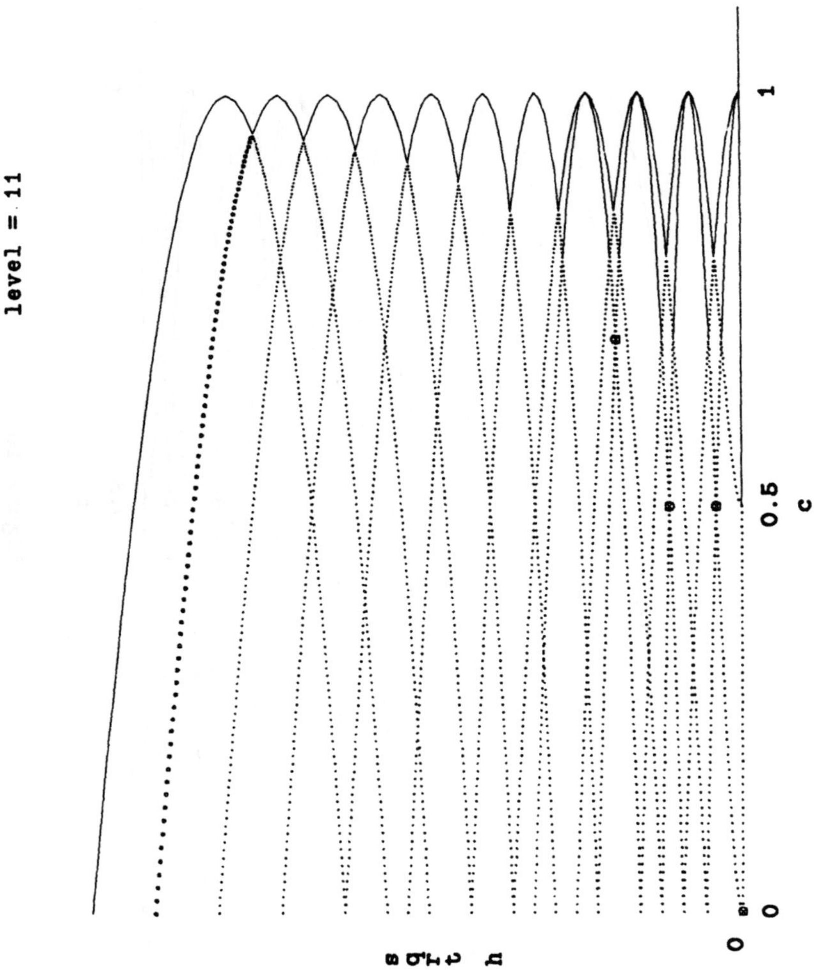

Figure 11

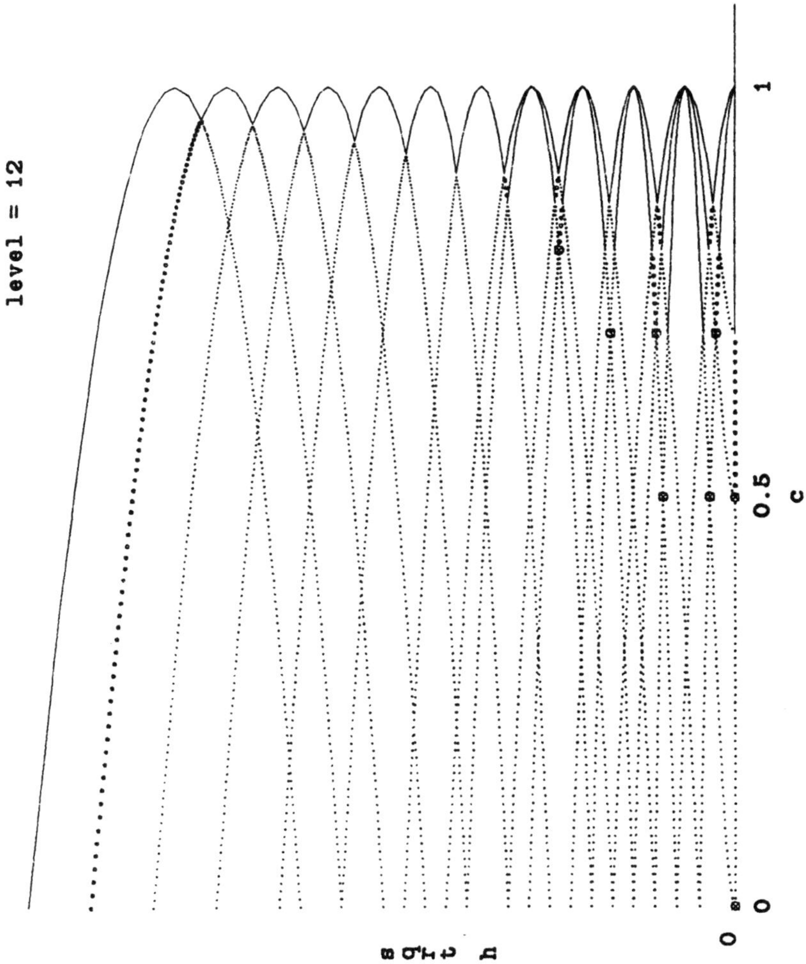

Figure 12

VACUUM VECTOR REPRESENTATIONS OF THE VIRASORO ALGEBRA

A. Rocha-Caridi†

1. INTRODUCTION

A lot of attention has been focused lately on certain infinite dimensional Lie algebras for their importance in some physical theories as well as the richness of their mathematical theories. One of these algebras is the Virasoro algebra. The Virasoro algebra is known to physicists in the theory of dual string models (cf. [25]). The first mathematical reference on the Virasoro algebra that is known to us is by Gelfand and Fuchs [9]. They proved that the second cohomology of the Lie algebra v of polynomial vector fields on the circle is one-dimensional. Using this one can show that the Virasoro algebra is the universal central extension $\hat{v}$ of v (see §4 below). The Virasoro algebra was later realized as an algebra of operators on the representation space of a Kac-Moody algebra (cf. [5, 3, 11, 17]), in a way reminiscent of its earlier introduction in dual models.

Among the first significant results on the representation theory of the Virasoro algebra was Kac's formula for the determinant of the contravariant form [14,15] (see also [1]). One of the main problems in representation theory is to determine the characters of the irreducible highest weight representations. Explicit character formulas for some of these representations have been determined according to the action of the center. In the case where the center acts by the scalar $c = 1$, the formulas were obtained by Kac [14,15], using the above mentioned realization of the Virasoro algebra as an algebra of operators. Wallach and the author [22,23] developed very general methods for the determination of irreducible characters, which were

† Partially supported by National Science Foundation Grant MCS-8201260.

Vertex Operators in Mathematics and Physics - Proceedings of a Conference November 10-17, 1983. Publications of the Mathematical Sciences Research Institute #3, Springer-Verlag, 1984.

applied to the cases c = 0, 25 and 26. In particular, the irreducible characters of the Lie algebra of vector fields on the circle were computed in [22,23].

Recently, Feigin and Fuchs [2][1] announced the results that completely describe the submodules of a Verma module. From their findings one can infer what the irreducible characters look like. There has also been an important development in Physics, where the representation theory of the Virasoro algebra plays a crucial role. In [6,7], Friedan, Qiu and Shenker introduced an infinite family of representations $L((h_i,c_i))$ with (h_i,c_i) in the domain $0 \leqslant c < 1$. The importance of these representations lies in the fact that the complementary set of the weights (h_i,c_i) in the domain $0 \leqslant c < 1$ corresponds to nonunitary representations [6,7]. The representations $L((h_i,c_i))$ that are unitary are intimately related to the unitary, conformally invariant, models in statistical mechanics [6,7]. The characters of these representations seem to be of interest in Physics. In this paper we calculate them explicitly.

In §2 we review the basics on the representation theory of the Virasoro algebra, including Kac's formula. In §3 we present the results of Goodman and Wallach [12] on the Segal operators and use them to indicate how to derive the irreducible characters in the case c = 1. In §4 we discuss the case of the Lie algebra of polynomial vector fields on the circle (c = 0). Although we only concern ourselves with the characters for certain special values of c, we feel that the methods presented in the exposition outlined above are of independent interest. This part of the paper is based on the lecture presented by the author at the Workshop. In §5 we state and discuss the implications of the results announced in [2]. In particular, we show how to derive the irreducible characters. In §6 we derive, using the results of [2], explicit formulas for the characters of the representations introduced in [6,7].

---

[1]We thank I. Frenkel for kindly translating the work [2] into English for us.

ACKNOWLEDGMENT

It is a pleasure to thank D. Friedan for many stimulating discussions.

2. REPRESENTATION THEORY

The *Virasoro algebra* is, by definition, the complex Lie algebra g with basis $\{d_k, d_0'\}_{k\in\mathbb{Z}}$, and bracket relations:

(1) $$[d_k, d_\ell] = (\ell-k)d_{k+\ell} + \delta_{k,-\ell}\frac{k(k^2-1)}{12}d_0'.$$

(2) $$[d_k, d_0'] = 0, \quad \text{for all } k \in \mathbb{Z}.$$

Let $h = \mathbb{C}d_0 \oplus \mathbb{C}d_0'$ and $\lambda \in h^*$ such that $\lambda(d_0) = h$, $\lambda(d_0') = c$. Given any h-module M we define its λ-weightspace:

$$M_\lambda = \{v \in M \mid d_0 v = hv,\ d_0'v = cv\}.$$

If $M = \bigoplus_{\lambda\in h^*} M_\lambda$ and $\dim M_\lambda < \infty$ for all $\lambda \in h^*$, we define the *formal character* of M, ch M by

$$\text{ch } M = \sum_{\lambda\in h^*} (\dim M_\lambda) z^\lambda$$

A *highest weight* (or vacuum vector) *representation* M of g, with highest weight λ, is one generated by a vector v_λ such that:

(3) $$\begin{cases} d_m v_\lambda = 0, & m \geq 1 \\ d_0 v_\lambda = hv_\lambda \text{ and } d_0' v_\lambda = cv_\lambda \end{cases}$$

v_λ is called a highest weight vector of M. Next we recall the definition of the universal highest weight representation with highest weight λ. Let $n = \bigoplus_{i\in\mathbb{N}} \mathbb{C}d_i$, $n^- = \bigoplus_{i\in\mathbb{N}} \mathbb{C}d_{-i}$, $b = h\oplus n$, and let $\mathbb{C}(\lambda)$ denote the one-dimensional b-module with trivial n-action and with h-action given by λ. Let $M(\lambda) = U(g)\otimes_{U(b)}\mathbb{C}(\lambda)$. (Here, we denote by $U(\mathfrak{a})$ the universal enveloping algebra of $\mathfrak{a}$, where $\mathfrak{a}$ is any subalgebra of $\mathfrak{g}$). $M(\lambda)$ is the *Verma module* associated with $\mathfrak{g}$, $\mathfrak{h}$, $\mathfrak{n}$ and λ. We denote by $L(\lambda)$ its unique irreducible quotient. It is clear that $M(\lambda) = \bigoplus_{m\in\mathbb{Z}_+} M(\lambda)_{\lambda-m}$. Hence

$$\text{(4)} \qquad \operatorname{ch}(M(\lambda) = z^{\lambda} \sum_{i\in\mathbb{Z}_+} p(i)z^{-i} = z^{\lambda} \prod_{i=1}^{\infty} (1-z^{-i})^{-1}$$

where $p(i)$ is the number of partitions of i.

Let σ denote the linear antiautomorphism of $U(\mathfrak{g})$ such that $\sigma(d_m) = d_{-m}$, $m \in \mathbb{Z}$, and $\sigma(d_0') = d_0'$. Let β be the projection of $U(\mathfrak{g}) = U(\mathfrak{h})\oplus(\mathfrak{n}^-U(\mathfrak{g}) + U(\mathfrak{g})\mathfrak{n})$ onto $U(\mathfrak{h})$. We write v_λ for $1\otimes1$ in $M(\lambda)$ and set $(Xv_\lambda,Yv_\lambda)_\lambda = (\lambda\circ\beta)(\sigma(X)Y)$, for all $X, Y \in U(\mathfrak{g})$. This defines a symmetric bilinear form on $M(\lambda)$ which is *contravariant* in the following sense:

$$\text{(5)} \qquad (Xv,w)_\lambda = (v,\sigma(X)w)_\lambda$$

for all $v,w \in M(\lambda)$, $X \in U(\mathfrak{g})$. If $v \in M(\lambda)_\mu$, $w \in M(\lambda)_\nu$, with $\mu \neq \nu$, then $(v,w)_\lambda = 0$, by (5). It is easy to see that the radical of $(\ ,\ )_\lambda$, Rad $(\ ,\ )_\lambda$, is the unique maximal submodule of $M(\lambda)$, i.e., $L(\lambda) = M(\lambda)/\text{Rad}\,(\ ,\ )_\lambda$.

We identify $\lambda \in \mathfrak{h}^*$ with $(h,c) \in \mathbb{C}^2$, where $\lambda(d_0) = h, \lambda(d_0') = c$ and let $(\ ,\ )_{h,c,m}$ denote the restriction of $(\ ,\ )_\lambda$ to $M(\lambda)_{\lambda-m}$. The following formula is due to Kac [14,15]:

$$\text{(6)} \qquad \det(\ ,\ )_{h,c,m} = \prod_{i=1}^{m} \left[\prod_{\substack{rs=i\\ r\leqslant s}} \psi_{r,s}(h,c)\right]^{p(m-i)}$$

where $\psi_{r,s}(h,c) = (h-\alpha^+_{r,s}(c))(h-\alpha^-_{r,s}(c))$, $r,s \in \mathbb{N}$, $r \neq s$, $\psi_{r,r}(h,c) = h + \frac{1}{24}(r^2-1)(1-c)$, and

$$\alpha^{\pm}_{r,s}(c) = -\frac{1}{48}[(13-c)(r^2+s^2)\pm\left[c^2-26c+25\right]^{1/2}(r^2-s^2) - 24rs - 2 + 2c].$$

Remarks: 1) Formula (6) is valid up to nonzero constant.

2) A proof of (6) is provided by Feigin and Fuchs [1].

Following [22], we let $T_\lambda: U(n^-) \longrightarrow M(\lambda)$ be the linear isomorphism such that $T_\lambda(X) = Xv_\lambda$. Set $V = U(n^-)$. If we let $\tau_\lambda(X)v = T_\lambda^{-1}(XT_\lambda(v))$ for $X \in g$, $v \in V$, then (τ_λ, V) is a representation of g. For each $t \in \mathbb{C}$, we let (π_t, V) be the representation $(\tau_{\lambda+t}, V)$ for some fixed $\lambda \in \mathbb{C}^2$. Let $B_t(v,w) = (T_{\lambda+t}(v), T_{\lambda+t}(w))_{\lambda+t}$ for all $v,w \in V$. Then $B_t(\pi_t(X)v,w) = B_t(v,\pi_t(\sigma(X))w)$, for all $X \in g$, $v,w \in V$. Let $V_\eta = \{X \in V \mid [H,X] = \eta X$, for all $H \in h\}$. Clearly $V = \bigoplus_{-\eta \in \mathbb{N}} V_\eta$. Let $\mathcal{O}(V)$ (resp. $\mathcal{O}(\mathbb{C})$) denote the space of germs $\mathbf{f}$ of holomorphic functions f at 0 with values in a finite sum of weightspaces V_η of V (resp. with values in $\mathbb{C}$). For each $k \in \mathbb{Z}_+$, set

$$\mathcal{O}(V)_{(k)} = \{\mathbf{f} \in \mathcal{O}(V) \mid B_t(f(t),w) \in t^k\mathcal{O}(\mathbb{C}), \text{ for all } w \in V\}.$$

Now, put $V_{(k)} = \{f(0) \mid \mathbf{f} \in \mathcal{O}(V)_{(k)}\}$ and $M(\lambda)_{(k)} = T_\lambda(V_{(k)})$.

Proposition 1 [22]. $M(\lambda) = M(\lambda)_{(0)} \supset M(\lambda)_{(1)} \supset \ldots$ is a g-module filtration satisfying:

(i) $M(\lambda)_{(1)} = \text{Rad}\ (\ ,\ )_\lambda$.

(ii) $M(\lambda_{(k)}/M(\lambda)_{(k+1)}$ carries a nondegenerate (symmetric) contravariant form

(iii) $\sum_{k>0} \text{ch}\ M(\lambda)_{(k)} = \sum_{m \in \mathbb{Z}_+} z^{\lambda-m}\ \text{ord}_0(\det\ (\ ,\ )_{\lambda+t,m})$

Let $\varphi_i(h,c) = \prod_{\substack{rs=i \\ r \leq s}} \psi_{r,s}(h,c)$ and set $\mathbb{N}_+(\lambda) = \{i \in \mathbb{N} \mid \varphi_i(h,c) = 0\}$.

Corollary 2: In the notation of Proposition 1,

$$\sum_{k>0} \operatorname{ch} M(\lambda)_{(k)} = \sum_{m \in \mathbb{N}_+(\lambda)} \operatorname{ch} M(\lambda - m).$$

We write $\mu \uparrow \lambda$ if there are $m_1, \ldots, m_r \in \mathbb{N}$ such that $m_i \in \mathbb{N}_+(\lambda - m_1 - \ldots - m_{i-1})$, $2 \leq i \leq r, m_1 \in \mathbb{N}_+(\lambda)$, and $\mu = \lambda - m_1 - \ldots - m_r$.

Repeated applications of Corollary 2 give

Theorem 3 [14,15]. Let $\lambda, \mu \in \mathfrak{h}^*$. If $L(\mu)$ is a subquotient of $M(\lambda)$, then $\mu \uparrow \lambda$.

3. THE SEGAL OPERATORS[2] AND THE CASE c = 1

In this section we recall, following Goodman and Wallach [12], a realization of the Virasoro algebra. The difference in the statements below and the ones of [12] is explained by the use of a different invariant bilinear form.

Let $\mathfrak{a}$ be a simple Lie algebra over $\mathbb{C}$, and let $(\, , \,)$ be a symmetric, invariant form on $\mathfrak{a}$, normalized so that the square length of a long root is equal to 2. Let $\mathfrak{t} \subset \mathfrak{a}$ be a Cartan subalgebra and let Δ (resp. $\overset{\sim}{\Delta}$) denote the root system (resp. dual root system) of $(\mathfrak{a}, \mathfrak{t})$. Let $\tilde{\alpha}$ be the highest root of Δ. We write $\tilde{\alpha} = \sum_{i=1}^{\ell} g_i \alpha_i$ where $\alpha_1, \ldots, \alpha_\ell$ are the simple roots of Δ. Set $g = 1 + \sum_{i=1}^{\ell} g_i$. We denote by $B(\, , \,)$ the Killing form of $\mathfrak{a}$. Clearly, $B(\, , \,)$ is a nonzero scalar multiple of $(\, , \,)$. The following result determines this scalar.[3]

---

[2] These operators first appeared in an unpublished manuscript by G. Segal (cf. [3]).

[3] We thank I. Frenkel for bringing this result to our attention.

Lemma 1 [18, Lemma 1.2]. $(X,Y) = \frac{1}{2g}B(X,Y)$, for every $X,Y \in \mathfrak{a}$.

Given $\gamma \in \mathfrak{t}^*$ let h_γ (resp. t_γ) be the unique element of $\mathfrak{t}$ such that $\gamma(H) = B(h_\gamma,H)$ (resp. $\gamma(H) = (t_\gamma,H)$) for all $H \in \mathfrak{t}$. If $\lambda, \mu \in \mathfrak{t}^*$, set $B(\lambda,\mu) = B(h_\lambda,h_\mu)$ and $(\lambda,\mu) = (t_\lambda,t_\mu)$.

Next is another version of Lemma 1:

Lemma **1'**: $(\lambda,\mu) = 2gB(\lambda,\mu)$, for all $\lambda,\mu \in \mathfrak{t}^*$.

Proof: We have $(t_\gamma,H) = \frac{1}{2g}B(t_\gamma,H)$ for all $H \in \mathfrak{t}$, $\gamma \in \mathfrak{t}^*$, by Lemma 1. Hence

$\frac{1}{2g}B(t_\gamma,H) = B(h_\gamma,H)$ for all $H \in \mathfrak{t}$, $\gamma \in \mathfrak{t}^*$. This implies that $t_\gamma = 2gh_\gamma$ for all $\gamma \in \mathfrak{t}^*$. Therefore, $(\lambda,\mu) = 4g^2(h_\lambda,h_\mu) = 2gB(h_\lambda,h_\mu) = 2gB(\lambda,\mu)$, for all $\lambda, \mu \in \mathfrak{t}^*$. q.e.d.

We let $\hat{\mathfrak{a}} = \mathfrak{a}\otimes\mathbb{C}[t,t^{-1}]\oplus\mathbb{C}Z$ and set $X(p) = X\otimes t^p$, for $X \in \mathfrak{a}$, $p \in \mathbb{Z}$. We define

$$[X(p),Y(q)] = [X,Y](p+q) + p\delta_{p,-q}(X,Y)Z$$

for all $X,Y \in \mathfrak{a}$, $p,q \in \mathbb{Z}$, and $[\hat{\mathfrak{a}},Z] = (0)$.

The above relations give $\hat{\mathfrak{a}}$ a Lie algebra structure, called the *affine Lie algebra* associated with $\mathfrak{a}$.

The element $d_0 \in \mathfrak{g}$ (see §2) acts on $\hat{\mathfrak{a}}$ as the degree derivation so that we can form

$$\hat{\mathfrak{a}}^e = \mathbb{C}d_0 \times \hat{\mathfrak{a}}.$$

Let $\mathfrak{u}$ (resp. $\mathfrak{u}^-$) denote the sum of the positive (resp. negative) rootspaces. Than $\mathfrak{a} = \mathfrak{u}^-\oplus\mathfrak{t}\oplus\mathfrak{u}$. We set $\hat{\mathfrak{t}} = \mathfrak{t}\oplus\mathbb{C}Z$, $\hat{\mathfrak{t}}^e = \hat{\mathfrak{t}}\oplus\mathbb{C}d_0$, $\hat{\mathfrak{u}}^- = \mathfrak{u}^-\oplus\mathfrak{a}\otimes t^{-1}\mathbb{C}[t^{-1}]$, and $\hat{\mathfrak{u}} = \mathfrak{u}\oplus\mathfrak{a}\otimes t\mathbb{C}[t]$. Then $\hat{\mathfrak{a}}^e = \hat{\mathfrak{u}}^-\oplus\hat{\mathfrak{t}}^e\oplus\hat{\mathfrak{u}}$.

A *highest weight* (or vacuum vector) *representation* V of $\hat{a}^e$ is one generated by a vector v_λ, $\lambda \in t^*$, such that $Hv_\lambda = \lambda(H)v_\lambda$ for all $H \in \hat{t}^e$, and $\hat{u} \cdot v_\lambda = 0$; λ is called the *highest weight* of V. It is clear that for any $v \in V$, there exists $n_0 \in \mathbb{N}$ (depending on v) such that $X(n)v = 0$ for all $n > n_0$ and $X \in a$.

The normal-ordered product of $X(p)$ and $Y(q)$, for $X,Y \in a$ and $p,q \in \mathbb{Z}$, is defined by

$$:X(p)Y(q): \begin{cases} X(p)Y(q), & \text{if } p < q \\ \frac{1}{2}(X(p)Y(q) + Y(q)X(p)), & \text{if } p = q \\ Y(q)X(p), & \text{if } p > q \end{cases}$$

Let $\{u_i\}$, $i = 1,\dots,n$, be an orthonormal basis of a relative to (,). Then

$$T_p = \frac{1}{2}\sum_{k\in\mathbb{Z}} \sum_{i=1}^{n} :u_i(k)u_i(p-k):$$

defines an operator on any highest weight representation.

Using Lemmas 1 and 1' we can now state the results of [12] already in normalized form.

<u>Lemma</u> 2 [12]. Let V be a highest weight representation of $\hat{a}^e$ with highest weight λ. Then

$$[X(p), T_q] = (\lambda(Z) + g)pX(p+q) \tag{7}$$

for all $X \in g$, $p,q \in \mathbb{Z}$.

Let V be a highest weight representation of $\hat{a}^e$, λ its highest weight. Assume that $\lambda(Z) \neq -g$. Set

$$D_p = -\frac{1}{\lambda(Z)+g} T_p$$

Lemma 3 [12]. $[D_p,D_q] = (q-p)D_{p+q} + \delta_{p+q,0}\,\frac{p(p^2-1)}{12}\,\tau$,

where $\tau = \frac{n\lambda(Z)}{\lambda(Z)+g}$.

Let $\mathfrak{g}$ denote the Virasoro algebra (§2). It is easy to see (§4) that $\mathfrak{g}$ acts on $\hat{\mathfrak{a}}$. Set $\mathfrak{m} = \mathfrak{g} \times \hat{\mathfrak{a}} \supset \hat{\mathfrak{a}}^e$.

Theorem 4 [12]. Let (π,V) be a highest weight representation of $\hat{\mathfrak{a}}^e$ with highest weight λ such that $\lambda(Z) \neq -g$. Then $\pi|_{\hat{\mathfrak{a}}}$ extends to a representation $\tilde{\pi}$ of $\mathfrak{m}$, with $\tilde{\pi}(d_p) = D_p$ and $\tilde{\pi}(d_0) = \tau$, when $\tau = \frac{n\lambda(Z)}{\lambda(Z)+g}$.

Remarks: 1) If $\mathfrak{a}$ is of type A, D or E, g is the Coxeter number h of $\mathfrak{a}$. Therefore, if in addition $\lambda(Z) = 1$, then $\tau = \frac{n}{1+h}$ = rank of $\mathfrak{g}$ (Compare [5,3]).

2) If V is standard (i.e. V is irreducible and λ is dominant integral) and $\lambda(d_0) \in \mathbb{R}$ then $(\tilde{\pi},v)$ carries a positive definite, Hermitian contravariant form ([12, Corollary 2.4]).

3) In [17] a more general version of Theorem 4 is obtained, which includes the case of twisted affine algebras.

We now return to the study of the representations L((h,c)) or the Virasoro algebra $\mathfrak{g}$. We retain the notation of §2. Setting c = 1 in (6) we obtain:

$$(8)\qquad \det(\ ,\ )_{h,1,m} = \prod_{i=1}^{m} \Big(\prod_{rs=i} (h + \tfrac{1}{4}(r-s)^2)\Big)^{p(m-i)}$$

It is obvious from (8) that M((h,1)) is irreducible if and only if $h \neq -\frac{m^2}{4}$ for all $m \in \mathbb{Z}_+$. We let $\mathfrak{a} = sl(2,\mathbb{C})$ with canonical basis

$$E = \begin{pmatrix} 0 & 1 \\ 0 & 0 \end{pmatrix},\quad H = \begin{pmatrix} 1 & 0 \\ 0 & -1 \end{pmatrix} \quad\text{and}\quad F = \begin{pmatrix} 0 & 0 \\ 1 & 0 \end{pmatrix}$$

Set $h_1 = H\otimes 1$ and $h_0 = H\otimes 1 + Z$ in $\hat{\mathfrak{a}}$. Then h_0, h_1 and d_0 generate $\hat{\mathfrak{t}}^e$. Let $\Lambda_i(h_j) = \delta_{ij}$, $0 \leq i,j \leq 1$. We denote by V_0 (resp. V_1) the irreducible highest weight representation of $\hat{\mathfrak{a}}$ with highest weight Λ_0 (resp. Λ_1). Applying Theorem 4 we obtain

$$(9) \qquad V_0 \oplus V_1 = \bigoplus_{m \in \mathbb{Z}_+} Q(m) \otimes T(m)$$

as a representation of $\mathfrak{g} \times \mathfrak{sl}(2,\mathbb{C})$, where $T(m)$ is the $(m+1)$-dimensional irreducible representation of $\mathfrak{sl}(2,\mathbb{C})$, and $Q(m) = \{v \in V_0 \oplus V_1 \mid Ev = 0, Hv = mv\}$. Using (9), Remark 2 and the representation theory of $\mathfrak{sl}(2,\mathbb{C})$ (see, e.g. [4, Theorem 3.1]) we obtain

Theorem 5. [14,15] For $m \in \mathbb{Z}_+$ one has

$$\mathrm{ch}\, L((-\frac{m^2}{4},1)) = (\prod_{i=1}^{\infty} (1-z^{-i}))^{-1}(z^{-\frac{m^2}{4}} - z^{-\frac{1}{4}(m+2)^2})$$

The above, together with (8) gives all the characters ch $L((h,1))$.

4. VECTOR FIELDS ON THE CIRCLE

Let v denote the Lie algebra of polynomial vector fields on the circle. It was proved by Gelfand and Fuchs [9] that $H^2(v,\mathbb{C}) \cong \mathbb{C}$. A nonzero cocycle on v is given by

$$(10) \qquad C(\frac{1}{i}e^{ki\theta}\frac{d}{d\theta}, \frac{1}{i}e^{\ell i\theta}\frac{d}{d\theta}) = \delta_{k,-\ell}\frac{k(k^2-1)}{12}$$

for all $k,\ell \in \mathbb{Z}$. The virasoro algebra $\mathfrak{g}$ (§2) is clearly isomorphic to the central extension $\hat{v} = v \oplus \mathbb{C}$ of v, corresponding to the cocycle (10). The isomorphism sends d_k to $\left(\frac{1}{i}e^{ki\theta}\frac{d}{d\theta},1\right)$ and d_0' to (0,1). We note that $\frac{1}{i}e^{ki\theta}\frac{d}{d\theta} = \frac{1}{k}\left[\frac{1}{i}\frac{d}{d\theta}, \frac{1}{i}e^{ki\theta}\frac{d}{d\theta}\right]$, if $k \neq 0$, and $\frac{1}{i}\frac{d}{d\theta} = \frac{1}{2}\left[\frac{1}{i}e^{-i\theta}\frac{d}{d\theta}, \frac{1}{i}e^{i\theta}\frac{d}{d\tau}\right]$, i.e., $v = [v,v]$. Therefore, by [8, lemma I.10] $\hat{v}$ is the universal covering of v.[4]

Let $\mathfrak{g}$, $\mathfrak{h}$, λ be as in §2. The highest weight representations of v correspond to the weights λ such that $\lambda(d_0') = 0$, i.e., $c = 0$.

---

[4] We thank H. Garland for pointing this out to us.

For simplicity we drop the variable c in the notation of §2. (6) becomes

$$(11) \qquad \det(\ ,\ )_{h,m} = \prod_{i=1}^{m} \Big(\prod_{rs=i} \{h + \frac{1}{24}[(3r-2s)^2-1]\} \Big)^{p(m-i)}$$

<u>Corollary</u> 1 [14,15]. $M(h) = L(h) \Leftrightarrow h \neq -\frac{1}{24}(m^2-1)$, for all $m \in \mathbb{N}$. By the corollary and formula (11) all the characters ch L(h) will be determined once we compute $\mathrm{ch}\, L\left[-\frac{1}{24}(m^2-1)\right]$ for all $m \in \mathbb{N}$.

The integers of the form $\frac{1}{24}(m^2-1)$, $m \in \mathbb{N}$, are *Euler's pentagonal numbers*, i.e., the numbers of the form $\frac{3k^2-k}{2}$, $k \in \mathbb{Z}$. The integral case, being the most difficult one, well illustrates the methods developed in [22,23]. For this reason, we discuss only this case in this paper, and refer the reader to [23] for the easier case of nonintegral highest weights. We now state the theorem that gives the integral irreducible characters. We set $s_k = -\frac{3k^2+k}{2}$ and $t_k = -\frac{3k^2-k}{2}$, $k \in \mathbb{Z}_+$.

<u>Theorem</u> 2 [22]. Let $\nu_k \in \{s_k, t_k\}$. There exists a resolution of g-modules:

$$(12) \qquad \cdots \xrightarrow{d_{k+2}} M(s_{k+1}) \oplus M(t_{k+1}) \xrightarrow{\eta_{k+1}} M(\nu_k) \xrightarrow{\epsilon_k} L(\nu_k) \longrightarrow 0$$

The following are immediate consequences of theorem 2:

<u>Corollary</u> 3 [22]:

$$H^p(n, L(\nu_k)) = \mathbb{C}(s_{k+p}) \oplus \mathbb{C}(t_{k+p})$$

as h-modules, $p \geqslant 1$.

<u>Corollary</u> 4 [22]:

$$\mathrm{ch}\, L(\nu_k) = \prod_{i=1}^{\infty} (1-z^{-i})^{-1} (z^{\nu_k} + (-1)^k \sum_{i>k} (-1)^i (z^{s_k} + z^{t_k}))$$

For k = 0, (12) is the resolution of the trivial module:

(13) $$\ldots M(s_i)\oplus M(t_i)\xrightarrow{d_i}\ldots\xrightarrow{d_2}M(-1)\oplus M(0)\xrightarrow{d_1}M(0)\xrightarrow{\epsilon_0}L(0)\longrightarrow 0$$

which was obtained in [21] (see also [19]) using Theorem 2.3, Goncharova's result on the cohomology of n^- with trivial coefficients [10,11], and standard arguments of homological algebra. The resolution (13) and the formula of Corollary 4 were conjectured by Kac [16].

We now indicate how to obtain (12) using (13) as a starting point. It was shown in [21] that $M(s_k) + M(t_k)$ is isomorphic to $M(s_{k-1}) \cap M(t_{k-1})$ for $k \geq 2$. Let $k \in \mathbb{N}$ be fixed and let $\nu_k \in \{s_k, t_k\}$. We define $\tilde{L}(\nu_k) = M(\nu_k)/(M(s_{k+1}) + M(t_{k+1}))$. (13) gives rise to a g-module resolution:

(12') $$\ldots \xrightarrow{d_{k+2}} M(s_{k+1})\oplus M(t_{k+1}) \xrightarrow{\eta_{k+1}} M(\nu_k) \xrightarrow{\epsilon_k'} \tilde{L}(\nu_k) \longrightarrow 0$$

In §2 we constructed a filtration of g-modules $M(\nu_k) = M(\nu_k)_{(0)} \supset M(\nu_k)_{(1)} \supset \ldots$. By Corollary 2.2 and (11) we have

(14) $$\sum_{i>0} \mathrm{ch}\ M(\nu_k)_{(i)} = \sum_{\ell\in\mathbb{N},\ \ell\ \mathrm{odd}} (\mathrm{ch}\ M(s_{k+\ell})+\mathrm{ch}\ M(t_{k+\ell}))$$

We set $N(\nu_k) = M(\nu_k)/M(t_{k+1})$. $N(\nu_k)_0$ also posseses a g-module filtation $N(\nu_k) = N(\nu_k)_{(0)} \supset N(\nu_k)_{(1)} \supset \ldots$ such that $N(\nu_k)_{(1)}$ is the largest proper submodule of $N(\nu_k)$, $N(\nu_k)_{(i)}/N(\nu_k)_{(i+1)}$ has a non-degenerate contravariant form, and

(15) $$\sum_{i>0} \mathrm{ch}\ N(\nu_k)_{(i)} = \sum_{i>0} \mathrm{ch}\ M(\nu_k)_{(i)} - \sum_{i>0} \mathrm{ch}\ M(t_{k+1})_{(i)} - \mathrm{ch}\ M(t_{k+1})$$

The construction of this filtation, although it follows the general lines desribed in §2, is of much deeper nature than that of $M(\lambda)$, and it involves the construction of a cross-section on the variety $\psi_{r,s}(h,c) = 0$, for appropriate r, s $\in \mathbb{N}$. Combining (14) and (15) we see that

(16) $$\sum_{i>0} \mathrm{ch}\ N(\nu_k)_{(i)} = \mathrm{ch}\ \tilde{L}(s_{k+1}).$$

Also, $\mathrm{ch}((M(s_{k+1}) + M(t_{k+1}))/M(t_{k+1})) = \mathrm{ch}\ \tilde{L}(s_{k+1})$. Since $(M(s_{k+1}) + M(t_{k+1}))/M(t_{k+1}) \subset N(\nu_k)_{(1)}$ we conclude that this inclusion is in fact an equality. Hence $\mathrm{ch}(\tilde{L}(\nu_k)) = \mathrm{ch}\ (N(\nu_k)/N(\nu_k)_{(1)})$, i.e., $\tilde{L}(\nu_k) = L(\nu_k)$. This concludes the proof of the existence of (12).

5. THE GENERAL CASE

A description of the subrepresentations of the Verma modules $M((h,c))$ over $\mathfrak{g}$, for arbitrary h and c, was recently announced in [2]. This description implies, in particular, the characters derived in §§3 and 4. In this section we state and discuss the results of [2]. We use the notation of §2.

Theorem 1 [2]. Let $\lambda \in \mathfrak{h}^*$ and let M be any submodule of $M(\lambda)$. There exists $\lambda_i \in \mathfrak{h}^*$, $v_i \in M_{\lambda_i}$, $i = 1,\dots,n$ such that $\mathfrak{n} \cdot v_i = 0$, $i = 1,\dots,n$ and $M = \sum_{i=1}^{n} V(\mathfrak{g})v_i$.

Theorem 1 says that any submodule of $M(\lambda)$ is a sum of submodules that are Verma modules. This implies that the knowledge of all the embeddings among Verma modules enables one to describe all the submodules of a given Verma module.

We write $\mu \longleftarrow \lambda$ whenever there is an embedding $M(\mu) \longrightarrow M(\lambda)$, $\mu,\lambda \in \mathfrak{h}^*$. We consider the following types of diagrams of embeddings

1) (a)

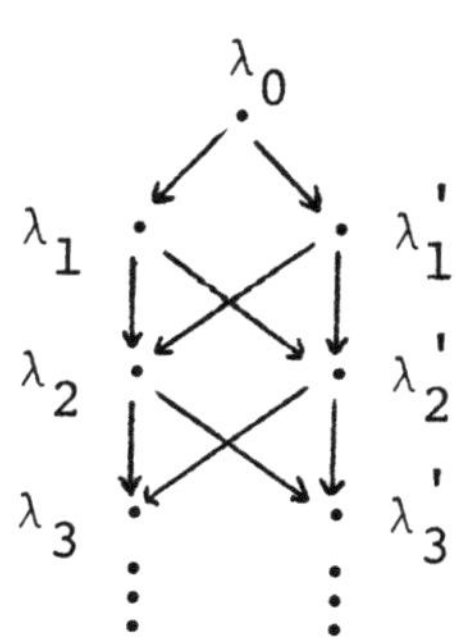

(b)

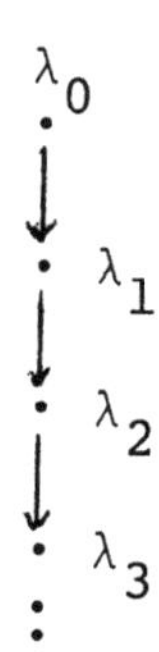

2) (a)

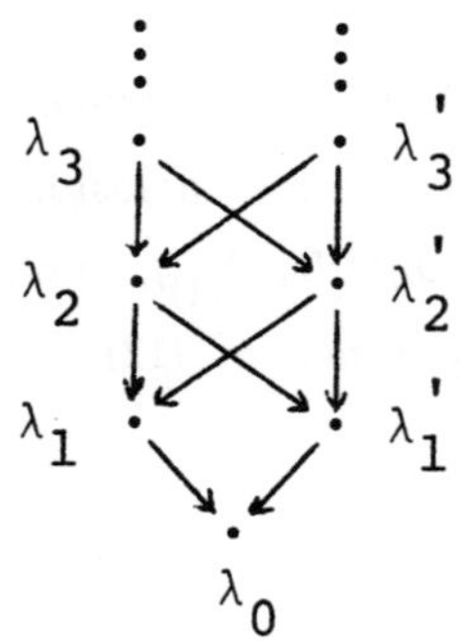

(b)

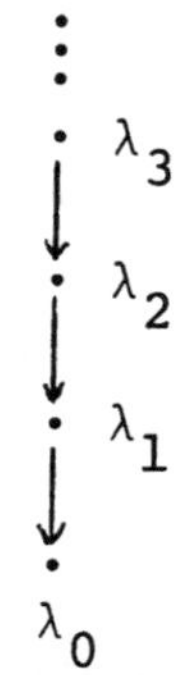

3)

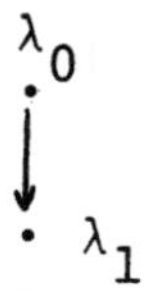

Theorem 3 of [2] states that any embedding among two Verma modules corresponds to an arrow in a diagram of one of the types above.

We now show how to use the above to write down the characters of the $L(\lambda)$. If $L(\lambda) = M(\lambda)$ this is given in §2 (4). So, assume that $L(\lambda) \neq M(\lambda)$. Let $\lambda \in \{\lambda_i, \lambda'_i\}$ in diagram 1) a). Then $M(\lambda_{i+1})$ and $M(\lambda'_{i+1})$ are both contained in $M(\lambda)$ and any other Verma module which is contained in $M(\lambda)$ is also contained in $M(\lambda_{i+1})$ and in $M(\lambda'_{i+1})$. By Theorem 1, $M(\lambda_{i+1}) + M(\lambda'_{i+1})$ is the largest proper submodule of $M(\lambda)$. The same reasoning shows that $M(\lambda_j) \cap M(\lambda'_j) = M(\lambda_{j+1}) + M(\lambda'_{j+1})$, for all j It is now easy to see that ch $L(\lambda)$ = ch $M(\lambda) + (-1)^i \sum_{j>i} (-1)^j$(ch $M(\lambda_j)$ + ch $M(\lambda'_1)$).

Similarly, for $\lambda \in \{\lambda_i, \lambda'_i\}$ in diagram 2) a) we obtain ch $L(\lambda)$ =

ch $M(\lambda) + (-1)^i \left[\sum_{0<j<i} (-1)^i(\text{ch } M(\lambda_j) + \text{ch } M(\lambda'_j)) + \text{ch } M(\lambda_0) \right]$. If $\lambda = \lambda_i$ in diagrams 1) b), 2) b) or 3) it is also clear that ch $L(\lambda)$ = ch $M(\lambda)$ − ch $M(\lambda_{i+1})$. Summing up and using §2(4) we have:

$$\text{(17)} \quad \text{ch } L(\lambda) = \left[\prod_{h=1}^{\infty} (1-z^{-n}) \right]^{-1} \left[z^{\lambda} + (-1)^i \sum_{j>i} (-1)^j (z^{\lambda_j} + z^{\lambda'_j}) \right]$$

$$(18)\quad ch\ L(\lambda) = \left[\prod_{h=1}^{\infty}(1-z^{-n})\right]^{-1}(z^{\lambda}+(-1)^{i}\left[\sum_{0<j<i}(-1)^{j}(z^{\lambda_j}+z^{\lambda'_j})+z^{\lambda_0}\right]$$

$$(19)\quad ch\ L(\lambda) = \left[\prod_{n=1}^{\infty}(1-z^{-n})\right]^{-1}\left[z^{\lambda} - z^{\lambda_{i+1}}\right]$$

for $\lambda \in \{\lambda_i, \lambda'_i\}$ in diagram 1)(a), $\lambda \in \{\lambda_i, \lambda'_i\}$ in diagram 2)(b) and $\lambda = \lambda_i$ in any of the other diagrams, respectively.

In [2, Theorem 3] a procedure is given to find the vertices connected to a given vertex in diagrams 1) - 3). The full statement of Theorem 3 is somewhat long and it involves the separate treatment of various cases. For this reason we chose to state only their result in the situation that is the most interesting to us. This case (subcase $III_{\pm}$ of [2, Theorem 3]) corresponds to the diagrams 1)(a) and 2)(a).

Let h, c $\in \mathbb{C}$. It is easy to see that the set of points (r,s) of any of the four lines representing the equation $\psi_{r,s}(h,c) = 0$ in rs-plane yield the same set of products rs. Let $\ell_{h,c}$ be any one of these four lines and we denote by $S_{h,c}$ the set of points of $\ell_{h,c}$ with integral coordinates. Suppose that

(*) $\quad S_{h,c}$ is infinite and if $(r,s) \in S_{h,c}$ then $rs \neq 0$.

By (*) $ab \neq cd$ if $(a,b) \neq (c,d)$, $(a,b), (c,d) \in S_{h,c}$. We enumerate the elements of $S_{h,c}$: ... $(r_{-1},s_{-1}), (r_0,s_0), (r_1,s_1)$, ... so that ... $< r_{-1}s_{-1} < 0 < r_0s_0 < r_1s_1 < \ldots$. Now, let $\ell'_{h,c}$ be the line passing through $(-r_0,s_0)$ and parallel to $\ell_{h,c}$. Let $S'_{h,c}$ be the set of points of $\ell'_{h,c}$ with integral coordinates. The points of $S'_{h,c}$ can be ordered:

$$\ldots (r'_{-1},s'_{-1}),\ (r'_0,s'_0) = (-r_0,s_0),\ (r'_1,s'_1),\ \ldots ,$$

so that

$$\ldots r'_{-1}s'_{-1} < r'_0s'_0 < 0 < r'_1s'_1 < \ldots$$

By [2, Theorem 3] one has the following diagram of embeddings:

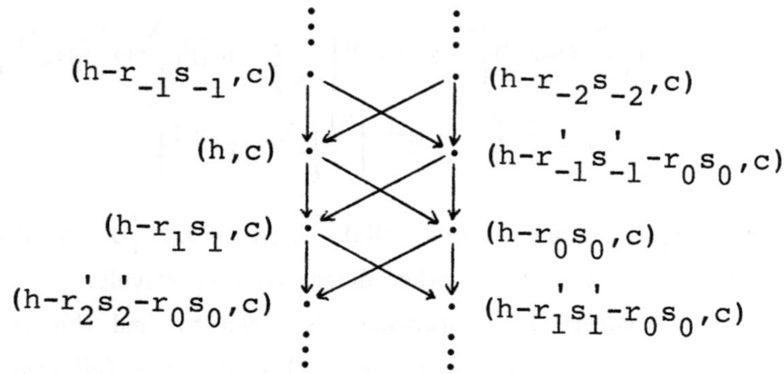

__Remarks__: 1) The assumption that $S_{h,c}$ is infinite implies that $\ell_{h,c}$ is real and so $c \in \mathbb{R}$ with either $c \leq 1$ or $c \geq 25$. The diagram above is of type 1)(a) if $c \leq 1$ and of type 2)(a) if $c \geq 25$.

2) Let $\ell''_{h,c}$ denote the line through $(-r_1,s_1)$ and parallel to $\ell_{h,c}$, and let $S''_{h,c}$ denote the set of points of $\ell''_{h,c}$ with integral coordinates. Then $\{h-r'_is'_i - r_0s_0\} = \{h - r''_is''_i - r_1s_1\}$, where $S''_{h,c} = \{(r''_i,s''_i)\}$.

3) $\{h - r'_is'_i - r_0s_0\} = \{h - r_0s_0 - a_ib_i\}$ where $S_{h-a_0s_0,c} = \{(a_i,b_i)\}$.

4) Using Remark 3) and the above diagram of embeddings we can see that for $\lambda = (h,c)$, $M(\mu) \subset M(\lambda)$ if and only if $\mu \uparrow \lambda$ (see §2).

6. NEW CHARACTER FORMULAS

If $\lambda = (h,c)$, $h,c \in \mathbb{R}$, one can construct a unique, up to scalar, *hermitian contravariant* form on $M(\lambda)$ as follows; let ω be the unique conjugate-linear antiautomorphism of $U(g)$ such that $\omega(d_i) = d_{-i}$ for all $i \in \mathbb{Z}$, and $\omega(d'_0) = d'_0$. Let β be as in §2, and set $\langle Xv_\lambda,Yv_\lambda\rangle_\lambda = (\lambda \circ \beta)(\omega(Y)X)$, for all $X,Y \in U(g)$. As in §2 we can show that distinct weightspaces are orthogonal relative to $\langle\ ,\ \rangle_\lambda$, and that $\langle\ ,\ \rangle_\lambda$ induces a nondegenerate hermitian contravariant form on $L(\lambda)$. $L(\lambda)$ is said to be *unitarizable* if the induced form is positive definite. It is easy to see that if $L(\lambda)$ is unitarizable then

$h \leqslant 0$ and $c \geqslant 0$. Furthermore, if $c = 0$ then $h = 0$. In [6,7], the authors investigate the unitarizable representations $L(\lambda)$, for $0 \leqslant c < 1$. They show that a necessary condition for the unitarizability of $L(\lambda)$ is given by the formulas:

$$(20) \quad \begin{cases} c = 1 - \dfrac{6}{m(m+1)}, & m \in \mathbb{N},\ m \geqslant 2 \\ h = h_{p,q} = -\dfrac{[(m+1)p-mq]^2-1}{4m(m+1)}, & p=1,\dots,m-1,\ q=1,\dots,p. \end{cases}$$

We set $\chi_{p,q}(z) = \mathrm{ch}\ L((h,c))$, where h,c are given in (20). We found that the characters $\chi_{p,q}(z)$ are given by the formulas:

$$(21) \qquad \chi_{p,q}(z) = \left[\prod_{i=1}^{\infty}(1-z^{-i}))^{-1} \sum_{k\in\mathbb{Z}}(-z^{-a(k)} + z^{-b(k)}\right]$$

where

$$(22) \qquad a(k) = \frac{[2m(m+1)k + (m+1)p + mq]^2-1}{4m(m+1)}$$

$$(23) \qquad b(k) = \frac{[2m(m+1)k + (m+1)p - mq]^2-1}{4m(m+1)}$$

In order to convey the combinatorial flavor of (21) we give some examples.

If we set $m = 2$, (20) gives only the trivial module $L((0,0))$. (21) yields

$$(24) \qquad \chi_{1,1}(z) = \mathrm{ch}\ L((0,0)) = \left[\prod_{i=1}^{\infty}(1-z^{-i})\right]^{-1} \sum_{k\in\mathbb{Z}}(-1)^k z^{-\frac{3k^2+k}{2}}$$

Since $\mathrm{ch}\ L((0,0) = 1$, (24) is simply the statement of Euler's pentagonal number theorem (see. e.g. [13]).

For $m = 3$, we obtain from (21):

$$(25) \qquad \chi_{1,1}(z) = \mathrm{ch}\ L((0,\tfrac{1}{2})) =$$

$$\left[\prod_{i=1}^{\infty}(1-z^{-i})\right]^{-1}\sum_{\substack{r\equiv 0,3(\mathrm{mod}\ 4)\\ s\equiv 0,1(\mathrm{mod}\ 4)}}\left[(-1)^r z^{-\frac{3r^2-r}{4}}+(-1)^s z^{-\frac{3s^2+s}{4}}\right],\quad r,s\in\mathbb{N}$$

(26) $$\chi_{2,1}(z) = \mathrm{ch}\ L((-\tfrac{1}{2},\tfrac{1}{2})) = \left[\prod_{i=1}^{\infty}(1-z^{-i})\right)^{-1}$$

$$\sum_{\substack{r\equiv 2,1\ (\mathrm{mod}\ 4)\\ s\equiv 2,3\ (\mathrm{mod}\ 4)}}\left[(-1)^{r+1} z^{-\frac{3r^2-r}{4}}+(-1)^{s+1} z^{-\frac{3s^2+s}{4}}\right],\quad r,s\in\mathbb{N}$$

(27) $$\chi_{2,2}(z) = \mathrm{ch}\ L(-\frac{1}{16},\frac{1}{2})) =$$

$$\left[\prod_{i=1}^{\infty}(1-z^{-i})\right]^{-1}\sum_{k\in\mathbb{Z}}(-1)^k z^{-(3k^2+k)}$$

Using Jacobi's triple product identity (see e.g. [13]) we obtain:

(28) $$\mathrm{ch}\ L((0,\tfrac{1}{2})) \pm \mathrm{ch}\ L(-\tfrac{1}{2},\tfrac{1}{2})) = \prod_{n\in\mathbb{N}-\frac{1}{2}}(1 \pm z^{-n})$$

(29) $$\mathrm{ch}\ L\ ((-\frac{1}{16},\frac{1}{2})) = z^{-\frac{1}{16}}\prod_{n\in\mathbb{N}}(1+z^{-n})$$

Remark: The formulas for ch $L((0,\frac{1}{2}))$ and ch $L((-\frac{1}{2},\frac{1}{2}))$ were known to physicists. The version (28) of them was communicated to the author by D. Friedan before we obtained (21). Inspired by the preceding formulas, I. Frenkel derived the simplified version (29) of (27) by different methods.

For m = 4, (21) gives the following set of formulas.

(30) $$\chi_{1,1}(z) = \mathrm{ch}\ L((0,\frac{7}{10})) = \left[\prod_{i=1}^{\infty}(1-z^{-i})\right]^{-1}$$

$$\sum_{\substack{r\equiv 0,1\ (\mathrm{mod}\ 4)\\ s\equiv 0,3\ (\mathrm{mod}\ 4)}}\left[(-1)^r z^{-\frac{5r^2-r}{4}} + (-1)^s z^{-\frac{5s^2+s}{4}}\right],\quad r,s \in \mathbb{N}$$

$$(31)\quad \chi_{3,1}(z) = \mathrm{ch}\, L((-\frac{3}{2},\frac{7}{10})) = \left[\prod_{i=1}^{\infty}(1-z^{-i})\right)^{-1}$$

$$\sum_{\substack{r\equiv 2,3\ (\mathrm{mod}\ 4)\\ s\equiv 2,1\ (\mathrm{mod}\ 4)}} \left[(-1)^{r+1} z^{-\frac{5r^2-r}{4}} + (-1)^{s+1} z^{-\frac{5s^2+s}{4}}\right],\ r,\ s \in \mathbb{N}$$

$$(32)\quad \chi_{3,3}(z) = \mathrm{ch}\, L((-\frac{1}{10},\frac{7}{10})) = \left[\prod_{i=1}^{\infty}(1-z^{-i})\right)^{-1}$$

$$z^{-\frac{1}{10}} \sum_{\substack{r\equiv 0,3\ (\mathrm{mod}\ 4)\\ s\equiv 0,1\ (\mathrm{mod}\ 4)}} \left[(-1)^{r} z^{-\frac{5r^2-3r}{4}} + (-1)^{s} z^{-\frac{5s^2+3s}{4}}\right],\ r,\ s \in \mathbb{N}$$

$$(33)\quad \chi_{3,2}(z) = \mathrm{ch}\, L((-\frac{3}{5},\frac{7}{10})) = \left[\prod_{i=1}^{\infty}(1-z^{-i})\right)^{-1}$$

$$z^{-\frac{1}{10}} \sum_{\substack{r\equiv 2,1\ (\mathrm{mod}\ 4)\\ s\equiv 2,3\ (\mathrm{mod}\ 4)}} \left[(-1)^{r+1} z^{-\frac{5r^2-3r}{4}} + (-1)^{s+1} z^{-\frac{5s^2+3s}{4}}\right],\ r,\ s \in \mathbb{N}$$

$$(34)\quad \chi_{2,1}(z) = \mathrm{ch}\, L((-\frac{7}{16},\frac{7}{10})) =$$

$$\left[\prod_{i=1}^{\infty}(1-z^{-i})\right)^{-1} z^{-\frac{7}{16}} \sum_{k\in\mathbb{Z}} (-1)^k z^{(5k^2+3k)}$$

$$(35)\quad \chi_{2,2}(z) = \mathrm{ch}\, L((-\frac{3}{80},\frac{7}{10})) =$$

$$\left[\prod_{i=1}^{\infty}(1-z^{-i})\right]^{-1} z^{-\frac{3}{80}} \sum_{k\in\mathbb{Z}} (-1)^k z^{-(5k^2+3k)}$$

Using Jacobi's identity, we obtain

$$(36)\qquad \mathrm{ch}\, L((0,\frac{7}{10})) \pm \mathrm{ch}\, L((-\frac{3}{2},\frac{7}{10})) =$$

$$\prod_{n=0}^{\infty} \frac{(1\pm z^{-(5n+\frac{5}{2})})(1\pm z^{-(5n+\frac{3}{2})})(1\pm z^{-(5n+\frac{7}{2})})}{(1-z^{-(5n+2)})(1-z^{-(5n+3)})}$$

(37) $$\mathrm{ch}\ L\left(\left(-\frac{1}{10},\frac{7}{10}\right)\right) \pm \mathrm{ch}\ L\left(\left(-\frac{3}{5},\frac{7}{10}\right)\right) =$$

$$z^{-\frac{1}{10}} \prod_{n=0}^{\infty} \frac{(1\pm z^{-(5n+\frac{5}{2})})(1\pm z^{-(5n+\frac{1}{2})})(1\pm z^{-(5n+\frac{9}{2})})}{(1-z^{-(5n+1)})(1-z^{-(5n+4)})}$$

(38) $$\mathrm{ch}\ L\left(\left(-\frac{7}{16},\frac{7}{10}\right)\right) =$$

$$z^{-\frac{7}{16}} \prod_{n=0}^{\infty} \frac{(1+z^{-(5n+1)})(1+z^{-(5n+4)})(1+z^{-(5n+5)})}{(1-z^{-(5n+2)})(1-z^{-(5n+3)})}$$

(39) $$\mathrm{ch}\ L\left(\left(-\frac{3}{80},\frac{7}{10}\right)\right) =$$

$$z^{-\frac{3}{80}} \prod_{n=0}^{\infty} \frac{(1+z^{-(5n+2)})(1+z^{-(5n+3)})(1+z^{-(5n+5)})}{(1-z^{-(5n+1)})(1-z^{-(5n+4)})}$$

We note the appearance of the product sides of the Rogers-Ramanujan identities (see e.g. [13]) as factors in (36) - (39).

We now prove (21). We use the notation and the results of §5. By (20) we may take $\ell_{h,c}$ to be:

(40) $$(m+1)r - ms = (m+1)p - mq$$

Let $r(k) = mk+p$, $s(k) = (m+1)k + q$. Then $S_{h,c} = \{(r(k), s(k)) \mid k \in \mathbb{Z}\}$. It is clear from (20) that $S_{h,c}$ satisfies condition (*) of §5. Also, $r(k)s(k) > 0$ for all $k \in \mathbb{Z}$. By remark 2) of §5 we may consider the line

(41) $$(m+1)r - ms = -(m+1)p - mq.$$

The points of (41) with integral coordinates are $(r'(k), s'(k))$, where $r'(k) = mk - p$, $s'(k) = (m+1)k + q$. It follows from §5 that one has the following diagram of embeddings:

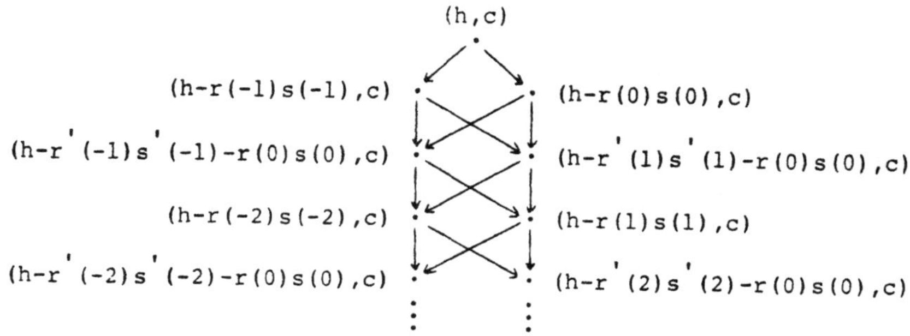

Formula (21) now follows from §5 (17).

REFERENCES

[1] B.L. Feigin and D.B. Fuchs, Functs. Anal. Prilozhen. 16, No. 2 (1982), 47-63.

[2] ————————, Functs. Anal. Prilozhen. 17, No. 3 (1983), 91-92.

[3] I. Frenkel, J. Funct. Anal. 44 (1981), 259-327.

[4] ————————, Representations of Kac-Moody Algebras and Dual Resonance Models, Proc. 1982 Summer Seminar on Applications of Group Theory in Physics and Mathematical Physics, Lectures in Applied Math., A.M.S. 21 (1984).

[5] I. Frenkel and V.G. Kac, Invent. Math. 62 (1980), 23-66.

[6] D. Friedan, Z. Qiu and S.H. Shenker, Conformal Invariance, Unitarity and Two dimensional critical exponents, these Proceedings.

[7] D. Friedan, Z. Qiu and S.H. Shenker, Phys. Rev. Lett. 52 (1984), 1575.

[8] H. Garland, Public. Math. No. 52, I.H.E.S.

[9] I.M. Gelfand and D.B. Fuchs, Functs. Anal. Prilozhen. 2 (1968), 92-93.

[10] L.V. Goncharova, Functs. Anal. Prilozhen. 7, No. 2 (1973), 6-14.

[11] ————————, Functs. Anal. Prilozhen. 7, No. 3 (1973), 33-44.

[12] R. Goodman and N.R. Wallach, Journal für die reine und angew. Math., 347 (1984), 69-133.

[13] G.H. Hardy and E.M. Wright, An Introduction to the theory of numbers, Clarendon Press, Oxford (1965).

[14] V.G. Kac, Proceedings of the International Congress of mathematicians, Helsinki (1978).

[15] ————————, Lecture Notes in Physics 94 (1979), 441.

[16] ————————, Lecture Notes in Math. 933 (1982) 117-126.

[17] V.G. Kac and D.H. Peterson, Infinite-Dimensional Lie Algebras, Theta Functions and Modular Forms, preprint.

[18] E. Looijenga, Invent. Math. 38, (1976), 17-32.

[19] A. Rocha-Caridi, Lecture Notes in Math 933 (1982), 176-190.

[20] A. Rocha-Caridi and N.R. Wallach, Math. Z. 180 (1982), 151-177.

[21] ————————, Transactions of the A.M.S. 277, No. 1 (1983),

133-162.

[22] ——————, Invent. Math. 72 (1983), 57-75.

[23] ——————, Math. Z. 185 (1984), 1-21.

[24] G. Segal, Commun. Math. Phys. 80 (1981), 301-342.

[25] J. Schwarz, Phys. Rep. 8C (1973), 269-335.

CLASSICAL INVARIANT THEORY AND THE VIRASORO ALGEBRA

Nolan R. Wallach

INTRODUCTION.

The Virasoro algebra (a certain central extension of the Lie algebra of vector fields on the circle) has come to play a more important role in such fields as quantum field theory and number theory. In [3], Wakimoto and Yamada gave a new relationship with classical invariant theory. The purpose of this note is to give a more direct relationship between highest weight vectors for the Virasoro algebra and characters of the unitary groups U(n), $n \geq 1$. In particular, our result gives (in principle) a method of calculating all highest weight vectors for all Verma modules for the Virasoro algebra. However, the combinatorial problems in the general case are very difficult. In the special case studied in [3], we give a simple proof of the main result announced in [3].

We thank Alvany Rocha for kindly sending us a copy of the announcement in [3]. The representations in Section 2 appear for the first time in [1].

1. A REALIZATION OF THE WITT ALGEBRA

We set $M_n(\mathbb{C})$ equal to the space of $n \times n$ matrices over $\mathbb{C}$. We look upon $GL(n,\mathbb{C})$ and $U(n)$ as subsets of $M_n(\mathbb{C})$. Let $\underline{u}(n) = \{X \in M_n(\mathbb{C}) \mid X^* = -X\}$. If $X \in M_n(\mathbb{C})$ then we write

$$X = X_1 + iX_2$$

with $X_1, X_2 \in \underline{u}(n)$.

We define for each $m \in \mathbb{Z}$ the vector field d_m on U(n) as follows:

Vertex Operators in Mathematics and Physics - Proceedings of a Conference November 10-17, 1983. Publications of the Mathematical Sciences Research Institute #3, Springer-Verlag, 1984.

$$d_m f(g) = \frac{d}{dt}_{t=0} \{f(ge^{t(g^m)_1}) + if(ge^{t(g^m)_2})\}.$$

If f extends to a holomorphic function on $GL(n,\mathbb{C})$ then

$$d_m f(g) = \frac{d}{dt}_{t=0} f(ge^{tg^m}).$$

One checks that

$$[d_p, d_q] = (q-p)d_{p+q} \ . \tag{1}$$

We set $\underline{d} = \oplus \mathbb{C} d_p$. Then $\underline{d}$ is isomorphic to the Witt algebra. We note that if n=1 then

$$d_m = \frac{1}{i} e^{im\theta} \frac{d}{d\theta},$$

the usual basis of the Witt algebra.

If M is an oriented manifold with volume form ω and if D is a differential operator on M then we define D^T by

$$\int_M (Df)g\omega = \int_M f(D^T g)\omega$$

for, say, $f,g \in C_c^\infty(M)$. If X is a smooth vector field on M then

$$X^T = -X - \phi_{\omega,X}$$

with $\phi_{\omega,X}$ a smooth function on M which is up to sign the divergence of X relative to ω.

On U(n) we take ω to be the normalized invariant measure. We will write

$$\int_{U(n)} f\omega = \int_{U(n)} f(g)dg.$$

We also write ϕ_m for ϕ_{ω,d_m}.

Then we have

$$(2)\quad \begin{aligned} \phi_m(g) &= \sum_{k=0}^{m-1} tr(g^{k+1})tr(g^{m-k-1}) && \text{if } m \geqslant 0 \\ \phi_{-m}(g) &= \sum_{k=0}^{m-1} tr(g^{-m+k+1})tr(g^{-k-1}) && \text{if } m > 0. \end{aligned}$$

The operators d_m are not left invariant on U(n); however, they are central operators. They are intimately connected with the "symbolic operators" of classical invariant theory. In fact, let us set $S_k(g) = tr(g^k)$. It is easy to check that

$$(3)\qquad d_m S_k = kS_{k+m}\,.$$

We define by $\lambda,\mu \in \mathbb{C}$ the operators $\sigma_{\lambda,\mu}(d_m)$ by

$$\sigma_{\lambda,\mu}(d_m)f = d_m f + (\lambda+m\mu)S_m f.$$

Then one checks (using (3))

$$(4)\qquad [\sigma_{\lambda,\mu}(d_p),\ \sigma_{\lambda,\mu}(d_q)] = (q-p)\sigma_{\lambda,\mu}(d_{p+q}).$$

We note that

$$(5)\qquad \sigma_{k,0}(d_m)\cdot \det{}^{-k} = 0 \text{ for all } m \in \mathbb{Z}.$$

Indeed, $(d_m \det^{-k})(g) = \frac{d}{dt}\Big|_{t=0} \det(ge^{tg^m})^{-k}$

$$= \frac{d}{dt}\Big|_{t=0} \det(g)^{-k}e^{-k \, tr\, g^m}$$

$$= -k\ tr(g^m)\det(g)^{-k}.$$

We will see in Section 3 that this observation implies the result of [3].

2. FOCK REALIZATIONS OF HIGHEST WEIGHT MODULES FOR THE VIRASORO ALGEBRA

Let $\hat{\underline{d}}$ be the Lie algebra with basis $\{d_n \mid n \in \mathbb{Z}\}$ and κ a central element with commutation relations

$$(1) \qquad [d_m, d_n] = (n-m)d_{n+m} + \frac{1}{12}n(n^2-1)\delta_{m,-n}\kappa.$$

Let $V = \mathbb{C}[x_1, x_2, \ldots]$. We define operators on V as follows:

$$u(n)f = n\frac{\partial}{\partial x_n}f, \quad n \geq 1$$

$$(2) \qquad u(0)f = 0$$

$$u(-n)f = x_n f, \quad n \geq 1.$$

We set

$$\pi_0(d_{2n}) = \frac{1}{2}u(n)^2 - \sum_{k=1}^{\infty} u(n-k)u(n+k)$$

$$\pi_0(d_{2n+1}) = -\sum_{k=0}^{\infty} u(n-k)u(n+k+1).$$

Then one checks that if we set $\pi_0(\kappa) = I$ then (π_0, V) is a representation of the Virasoro algebra.

We note that

$$(3) \qquad [\pi_0(d_m), u(n)] = nu(n+m).$$

Thus $\pi_0(d_m)$ and $u(q)$ can be considered to be a simultaneous "quantization" of the operators d_m and multiplication by S_q. In light of (3) we can define for $\lambda, \mu \in \mathbb{C}$ the operators

$$(4) \qquad \pi_{\lambda,\mu}(d_m) = \pi_0(d_m) + (\lambda+m\mu)(u(m)) - \left(\frac{\lambda^2}{2} - \frac{\mu^2}{2}\right)\delta_{m,0}I.$$

Then (3) implies

Lemma 2.1. If we set $\pi_{\lambda,\mu}(\kappa) = (1-12\mu^2)I$ then $(\pi_{\lambda,\mu},V)$ is a representation of the Virasoro algebra with

$$\pi_{\lambda,\mu}(d_0)\cdot 1 = -\left(\frac{\lambda^2}{2} - \frac{\mu^2}{2}\right)1.$$

On V we put the pre-Hilbert space structure

(5) $$\langle x^I,x^J\rangle = \delta_{I,J}I!(1^{i_1}2^{i_2}\ldots)$$

Here we have used standard multi-index notation. That is, $x^I = x_1^{i_1}x_2^{i_2}\ldots$, $I! = i_1!i_2!\ldots$. We will also write $|I| = \Sigma i_j$ and $\langle I\rangle = \Sigma j i_j$. One checks that

(6) $$\langle u(n)f,g\rangle = \langle f,u(-n)g\rangle \ .$$

Hence

(7) $$\langle \pi_{\lambda,\mu}(d_n)f,g\rangle = \langle f,\pi_{\bar{\lambda},-\bar{\mu}}(d_{-n})g\rangle \ .$$

Thus if we define $d_n^* = d_{-n}$, $n \in \mathbf{Z}$ and $\kappa^* = \kappa$ and extend this operation to be conjugate linear we find that if $\hat{\underline{d}}_{\mathbb{R}} = \{d \in \hat{\underline{d}} \mid d^* = -d\}$ then if $\lambda \in \mathbb{R}$, $\mu \in i\mathbb{R}$

(8) $$\langle \pi_{\lambda,\mu}(d)f,g\rangle = -\langle f,\pi_{\lambda,\mu}(d)g\rangle, \ d \in \hat{\underline{d}}_{\mathbb{R}}.$$

3. THE INTERTWINING OPERATORS

For $g \in U(n)$, $x_i \in \mathbb{C}$, $\Sigma|x_i| < \infty$ let $\gamma(g,x) = \sqrt{2}\sum_{m=1}^{\infty} x_m \mathrm{tr}(g^m)/m$. Then $\gamma(g,x)$ is a real analytic function on U(n).

If $f \in C^\infty(U(n))$ we define

$$T_n(f)(x) = \int_{U(n)} e^{\gamma(g,x)} f(g)dg.$$

We note that if $f(e^{i\theta}g) = e^{-im\theta}f(g)$, $\theta \in \mathbb{R}$, $g \in U(n)$, then $T_n(f)(x)$ is a polynomial in $x_1, x_2, \ldots, x_m$ of degree at most m.

<u>**Theorem**</u> **3.1.** If $m > 0$ then

$$\pi_{\lambda,\mu}(d_m)T_n(f) = T_n(\sigma_{\sqrt{2}\lambda+n,\sqrt{2}\mu}(d_m)f)$$

for all $\lambda, \mu \in \mathbb{C}$, $f \in C^{\infty}(U(n))$.

<u>**Proof**</u>. Since d_1, d_2 generate $\sum_{m \geqslant 1} \mathbb{C}d_m$ as a Lie algebra it is enough to check the formula for d_1, d_2. In each case this is a straightforward calculation which we now do.

$$\pi_{\lambda,\mu}(d_1) = (\lambda+\mu)\frac{\partial}{\partial x_1} - \sum_{m=1}^{\infty} (m+1)x_m\frac{\partial}{\partial x_{m+1}}.$$

Thus

$$(\pi_{\lambda,\mu}(d_1)T_n(f))(x) = \sqrt{2}(\lambda+\mu)\int_{U(n)} tr(g)e^{\gamma(g,x)}f(g)dg$$

$$- \sqrt{2}\sum_{m=1}^{\infty} x_m\int_{U(n)} tr(g^{m+1})e^{\gamma(g,x)}f(g)dg$$

$$= \sqrt{2}(\lambda+\mu)\int_{U(n)} tr(g)e^{\gamma(g,x)}f(g)dg$$

$$-\int_{U(n)} (d_1 \cdot e^{\gamma(g,x)})f(g)dg$$

$$= \sqrt{2}(\lambda+\mu)\int_{U(n)} tr(g)e^{\gamma(g,x)}f(g)dg$$

$$-\int_{U(n)} e^{\gamma(g,x)}d_1^T f(g)dg.$$

Since $d_1^T = -d_1 - n\,tr(g)$ (formula 1 (2)) the equation is true for d_1. We now look at d_2. $\pi_{\lambda,\mu}(d_2) = (2\pi+4\mu)\frac{\partial}{\partial x_2}$

$$-\frac{1}{2}\frac{\partial^2}{\partial x_1^2} - \sum_{m=1}^{\infty} (m+2)x_m\frac{\partial}{\partial x_{m+2}}.$$

Thus

$$\pi_{\lambda,\mu}(d_2)T_n(f)(x) = \sqrt{2}(\lambda+2\mu)\int_{U(n)} tr(g^2)e^{\gamma(g,x)}f(g)dg$$

$$- \int_{U(n)} (tr(g))^2 e^{\gamma(g,x)}f(g)dg - \sum_{m=1}^{\infty} \sqrt{2}x_m \int_{U(n)} tr(g^{m+2})e^{\gamma(g,x)}f(g)dg$$

$$= \sqrt{2}(\lambda+2\mu)\int_{U(n)} tr(g^2)e^{\gamma(g,x)}f(g)dg$$

$$- \int_{U(n)} (tr\ g)^2 e^{\gamma(g,x)}f(g)dg - \int_{U(n)} (d_2 e^{\gamma(g,x)})f(g)dg.$$

This time $d_2^T = -d_2 - tr(g)^2 - n\,tr(g^2)$. The formula now follows for d_2.

We combine this result with formula 1 (6) to obtain

<u>Corollary</u> 3.2. $\pi_{\frac{p-n}{\sqrt{2}}}(d_m)T_n(\det^{-p}) = 0$ for $p \geq 0$, $n \geq 1$, $m \geq 1$.

We note that the coefficient of x_1^{np} in $T_n(\det^{-p})$ is

$$\frac{2^{\frac{np}{2}}}{(np)!}\int_{U(n)} tr(g)^{np}\det(g)^{-p}dg = 2^{np/2}/(np)!\ .$$

We can give more explicit formulas for the action of T_n using classical results of Frobenius. We parametrize the characters of the finite dimensional irreducible polynomial representations of U(n) by $\Lambda = (m_1,\ldots,m_n)$, $m_1 \geq m_2 \geq \ldots \geq m_n \geq 0$, $m_i \in \mathbb{Z}$. Let χ_Λ be the corresponding character. We also look upon Λ as a partition of $|\Lambda| = m_1+\ldots+m_n$. Let χ^Λ be the corresponding character of $S_{|\Lambda|}$, the symmetric group on $|\Lambda|$ letters. If $I = (i_1,i_2,\ldots)$ and if $\langle I\rangle = \Sigma j i_j = |\Lambda|$ we denote by C_I the conjugacy class in $S_{|\Lambda|}$ which corresponds to i_1 cycles of length 1, i_2 cycles of length 2, ... Let $|C_I|$ be the order of C_I. Then the Weyl character formula implies (see [2], p. 67, 5.2; 8 and its derivation)

$$T_n(\bar{x}_\Lambda) = \frac{2^{|\Lambda|/2}}{|\Lambda|!} \sum_{\langle I\rangle = |\Lambda|} |C_I| \chi^\Lambda(C_I) x^I.$$

We note that if we set $\phi_{n,p} = T_n(\det^{-p})$ then we have

Lemma 3.3. $\displaystyle\int_{SU(n)} e^{\sqrt{2}\Sigma t^m x_m \operatorname{tr}(g^m)/m} dg = \sum_{p=0}^{\infty} t^{np} \phi_{n,p}(x).$

In particular this gives $e^{\sqrt{2}\Sigma t^m x_m/m} = \sum_{p=0}^{\infty} t^p \phi_{1,p}(x).$

Set $V_n = \{f \in V \mid \pi_0(d_0)f = nf\}$. That is $V_n = \sum_{\langle I\rangle = n} \mathbb{C}x^I$.

Set $I_n = \sum_{|\Lambda|=n} \mathbb{C}\bar{x}_\Lambda$. Then the above formula implies that

$$T_n\colon I_n \longrightarrow V_n$$

is bijective. Hence in principle one can compute all of the highest weight vectors for all of the $\pi_{\lambda,\mu}$ using Theorem 3.1.

REFERENCES

[1] R. Goodman and N.R. Wallach, Projective unitary positive energy representations of Diff(S^1), preprint.

[2] D.E. Littlewood, The Theory of Group Characters, Oxford University Press, London, 1940.

[3] M. Wakimoto and H. Yamada, Irreducible decompositions of Fock representations of the Virasoro algebra, preprint.

Rutgers University
New Brunswick, NJ 08903